Coleoptera: Staphylinidae: Scaphidiinae

World Catalogue of Insects

VOLUME 16

The titles published in this series are listed at *brill.com/wci*

Coleoptera: Staphylinidae: Scaphidiinae

By

Ivan Löbl

BRILL

LEIDEN | BOSTON

Cover illustration: *Pseudobironium confusum* Löbl & Tang, 2013. Photo by Liang Tang. Courtesy of *Revue suisse de zoologie*.

Library of Congress Cataloging-in-Publication Data

Names: Löbl, Ivan, author.
Title: Coleoptera: staphylinidae: scaphidiinae / by Ivan Löbl.
Description: Leiden ; Boston : Brill, [2018] | Series: World catalogue of insects ; volume 16 | Includes bibliographical references and index.
Identifiers: LCCN 2018031983 (print) | LCCN 2018032519 (ebook) | ISBN 9789004375956 (E-book) | ISBN 9789004368279 (hardback : alk. paper)
Subjects: LCSH: Beetles--Classification. | Staphylinidae--Classification. | Scarabaeidae--Classification.
Classification: LCC QL577 (ebook) | LCC QL577 .L63 2018 (print) | DDC 595.76--dc23
LC record available at https://lccn.loc.gov/2018031983

Typeface for the Latin, Greek, and Cyrillic scripts: "Brill". See and download: brill.com/brill-typeface.

ISSN 1398-8700
ISBN 978-90-04-36827-9 (hardback)
ISBN 978-90-04-37595-6 (e-book)

This book is printed on acid-free paper and produced in a sustainable manner.

Contents

Acknowledgements

My thanks are due to colleagues who provided information and assisted in various ways: Donald S. Chandler (Durham), Johannes Frisch (Berlin), Ernst Heiss (Innsbruck), Hideto Hoshina (Fukuoka), Pawel Jałoszyński (Wroclaw), Chi-Feng Lee (Taichung), Richard A.B. Leschen (Auckland), Brian Levey (Cardiff), Hans Mejlon (Uppsala), Alfred F. Newton (Chicago), Nikolay B. Nikitsky (Moscow), Ryo Ogawa (Matsuyama), Stewart B. Peck (Ottawa), Rafal Ruta (Wroclaw), Dmitry Schigel (Helsinki), Shigehiko Shiyake (Osaka), Aleš Smetana (Ottawa), Igor A. Solodovnikov (Vitebsk), Azadeh Taghavian (Paris), Piotr Tykarski (Warszawa), Mark G. Volkovitsh (St. Petersburg), Oskar Vorst (Utrecht), and Zi-Wei Yin (Shanghai). I am particularly indepted to Alfred F. Newton (Chicago), Michael A. Ivie (Bozeman) and Michael Schülke (Berlin) who significantly improved the manuscript.

Introduction

The staphylinid subfamily Scaphidiinae, or shining fungus beetles, is a morphologically distinctive group, worldwide in distribution. The conspicuous shape of body of its members led early workers to consider the group to be a separate family, usually related to rove beetles. The more distinctive features of Scaphidiinae are the strongly convex and compact body with tapering abdomen covered in large extent by truncate elytra, the small head, the voluminous pronotum narrowed anteriad and having basal margin sinuate and projecting posterio-medially, the six exposed abdominal sternites and the large first and fifth ventrites. Among other characters considered diagnostic are the 5-segmented tarsi, the 11-segmented antennae with a 5-segmented club, the concealed protrochantins, and the crenulate labral margin in larvae. Most adult scaphidiinae have the 8th antennomere smaller than other club segments (7 and 9–11), a condition similar to that in many other staphylinoids.

Many scaphidiines possess a pair of large spinous prothoracic pockets (corbicula) of as yet unknown function (Leschen & Löbl, 2005). Corbicula that have been examined do not contain spores and are unlikely to be mycangia, as was hypothetized by Newton (1984).

The monophyly of the group is strongly supported (Grebennikov & Newton 2012, McKenna et al. 2015). Scaphidiinae was for the first reduced to subfamily rank by Kasule (1966), based on larval characters. Lawrence & Newton (1982) placed them as a probable member of the oxyteline group of Staphylinidae, and Newton & Thayer (1992) also placed them within that group [see also Leschen and Löbl (2005), Thayer (2005), McKenna et al. 2015]. Naomi (1985) considered them to be the sister taxon to Scydmaenidae (at present Scydmaeninae in Staphylinidae), and Hansen (1997) placed them basally within the Staphylinoidea.

Scaphidiinae feed on fungi and slime moulds, and are encountered in leaf litter, rotten wood and other decaying vegetation, and in the fungal gardens of termites. They may cause damage to oyster mushroom (*Pleurotus*) cultures in some tropical countries. The shift to mycophagy in Scaphidiinae has been accompanied by adaptations of the mouthparts (Leschen 1993). Larvae feeding on harder fungi build fecal canopies (Leschen 1993, 1994). Overviews of mycophagy, fungus and slime mould association and adaptations are included in Lawrence & Newton (1980), Newton (1984, 1991), Lawrence (1989), Handley (1996), Betz et al. (2003), Löbl & Leschen (2003b), Leschen & Löbl (2005), and Thayer (2005). Tang et al. (2014) reported on habits of *Scaphidium* and on night-activity and catalepsy of adults.

Scaphidiines were long considered to be species-poor. Gemminger & Harold (1868) listed 41 valid species, and this number had increased to only 241 and 245 (respectively) in the first catalogues published in the 20th century (Csiki 1908, 1910). Though the number of recognized scaphidiines increased steadily after that, the next attempt to provide a world Catalogue was published almost ninety years later (Löbl 1997).

The unexpectedly high flow of new information since 1997, resulting changes in classification, descriptions of new taxa, redescriptions, new synonymies, and additional distributional records and host data, provided the impetus to publish the present Catalogue. This work provides information on 1797 valid extant species placed in 46 genera, and on five extinct species and two genera. Unlike the previous catalogues, and numerous regional overviews, it gives detailed information on type material of each species-group taxon, records subsequent taxonomic and nomenclatural acts, distributional records, as well as all known fungal and slime mould hosts.

Included Data

The present Catalogue lists all taxa published before January 1, 2018 currently considered as valid, their synonyms, emended names, and misspellings. Names deemed unavailable are in square brackets, extinct taxa are listed separately. The classification of the family-group taxa follows that of Leschen & Löbl (1995, 2005). Taxa ranked below tribal rank are arranged alphabetically but the nominal subspecies in polytypic species are listed first. The names of taxa are given as currently accepted and as originally published; misspellings are in quare brackets. The type genera for all supraspecific groups are given. The gender and the type species of the genus-group names are given, with reference to the respective fixations. The sexes (if known) and depositories of the primary types of the species-group taxa are given. The type localities are given in full, subsequent information about the type localities is in square brackets.

References are provided to all primary descriptions, taxonomic and nomenclatural acts, identification keys, character assessments, records, habitats, immature stages, and fungal and myxomycete hosts. All references cited have been seen by the author and are in full. Records on fungal and myxomycete hosts are provided per scaphidiines species in an Appendix.

The distribution is given by country, arranged alphabetically, with respective geographical subunities included for large and/or insular countries (such as Australia, Canada, China, Greece, India, Indonesia, Japan, Malaysia,

Philippines, Russia, USA, etc.). References to numerous catalogues of individual country or regional faunae are not included in the taxonomic section of the Catalogue, unless made necessary by pertinent information. Such works (e.g., Winkler 1925, Blackwalder 1957, Jelínek 1993, Angelini et al. 1995, Hansen 1996, Silfverberg 2004, Telnov 2004, Löbl 2004b, Vorst 2010, Rassi et al. 2015, Schülke & Smetana 2015, Ahn et al. 2017) are only listed in the reference section of the present Catalogue.

The identity of several European species of *Scaphisoma* were often confused by early workers, records of such species are not referred to if published before to the works of Taminani (1954, 1969a, 1969b, 1970) and Löbl (1963b, 1964b, 1967b, 1967d, 1970c).

Taxa originally assigned to *Scaphidium* and *Scaphisoma* and moved to other families, or to other subfamilies of the Staphylinidae, are listed separately.

Collection Codens

Primary types are deposited in the collections listed below. The term "syntypes" is used in the plural form, even if only a single specimen is known to be found in the listed collections. Multiple collection codens are separated by a slash for syntypes known to be deposited in more than one collection.

AMNZ	Auckland War Memorial Museum, New Zealand
AMSC	Australian Museum, Sydney, Australia
ANIC	Australian National Insect Collection, Canberra, Australia
BPBM	Bernice P. Bishop Museum, Hawaii, USA
CC-UAEH	Colección Coleoptera, Universidad Autónoma del Estado de Hidalgo, Mexico
CMNC	Canadian Museum of Nature, Ottawa, Canada
CNCI	Canadian National Collection of Insects, Ottawa, Canada
CNUIC	Chungnam National University, Insect Collection, Daejeon, Republic of Korea
CUIC	Cornell University Insect Collection, Ithaca, USA
CUMZ	University Museum of Zoology, Cambridge, UK
CZUG	Centro de Estudios Zoología, Universidad de Guadalajara, Jalisco, Mexico
DEIC	Deutsches Entomologisches Institut, Müncheberg, Germany
EMEC	Essig Museum of Entomology, University of California, USA
EUMJ	Ehime University Museum, College of Agriculture, Matsuyama, Japan

FMNH	Field Museum Natural History, Chicago, USA
FSCA	Florida State Collection of Arthropods, Gainsville, USA
FZCH	Fernando de Zayas Collection, Habana, Cuba
HBUM	Museum of Hebei University, China
HMNH	Museum of Nature and Human Activities, Hyogo, Japan
HNHM	Hungarian Natural History Museum, Budapest, Hungary
HUSJ	Hokkaido University, Systematic Entomology, Sapporo, Japan (including collection T. Nakane, Chiba, Japan)
IFAN	Institut Français d'Afrique Noire, Dakar, Senegal
INBIO	Instituto National de Biodiversidad, San Domingo de Heredia, Costa Rica
ISNB	Institut Royal des Sciences Naturelles, Bruxelles, Belgium
KCMI	Kashihara City Museum of Insects, Nara, Japan
KMNH	Kanagawa Pref. Museum of Natural Hististory, Kanagawa, Japan
KUIC	Faculty of Agriculture, Kyushu University, Fukuoka, Japan (including collection M. Chûjô)
LUNZ	Entomology Research Museum, Lincoln University, Canterbury, New Zealand
MACN	Museo Argentino de Ciencias Naturales, Buenos Aires, Argentina
MBRA	La Plata Museum, La Plata, Argentina
MCRC	Museo Civico, Rovereto, Italy
MCSN	Museo Civico di Storia Naturale, Genova, Italy
MCZC	Museum of Comparative Zoology, Harvard, USA
MHNG	Muséum d'histoire naturelle, Geneva, Switzerland
MIMM	Mauritius Institute, Port Louis, Mauritius
MNHA	Museum of Nature and Human Activities, Hyôgo, Japan
MNHN	Muséum National d'Histoire Naturelle, Paris, France
MRAC	Musée Royal de l'Afrique Central, Tervuren, Belgium
MSNM	Museo Civico di Storia Naturale, Milano, Italy
MVMA	Museum of Victoria, Abbotsville, Victoria, Australia
MZBI	Museum Zoologicum Bogoriense, Cibinong, Indonesia
MZLU	Museum of Zoology, Lund University, Lund, Sweden
MZUC	Museo di Zoologia dell'Università, Catania, Italy
MZUN	Museo Zoologico dell'Università, Napoli, Italy
NBCL	Naturalis Biodiversity Center, Leiden, The Netherlands (including former Zoölogisch Museum, Amsterdam and Rijksmuseum van Natuurlijke Historie, Leiden, The Netherlands)
NHMB	Naturhistorisches Museum, Basel, Switzerland
NHML	The Natural History Museum, London, UK (former British Museum (Natural History)

NHMW	Naturhistorisches Museum, Wien, Austria
NHRS	Naturhistoriska Riksmuseet, Stockholm, Sweden
NMEC	Naturkundemuseum, Erfurt, Germany
NMPC	Národní museum, Entomologické oddělení, Praha, Czech Republic
NMNS	National Museum of Nature and Science, Tokyo, Japan
NZAC	New Zealand Arthropod Collection, Auckland, New Zealand
OMNH	Osaka Museum of Natural History, Osaka, Japan
OSUC	Ohio State University, Columbus, USA
PCAP	Private collection A. Pütz, Eisenhütenstadt, Germany
PCMS	Private collection M. Sakai, Matsuyama, Japan
PURC	Purdue University, West Lafayette, Indiana, USA
QMBA	Queensland Museum, Brisbane, Australia
SAMA	South Australian Museum, Adelaide, Australia
SEMC	Snow Entomological Museum, University of Kansas, Lawrence, USA
SIEE	Severtsov's Institute of Ecology and Evolution, Moscow, Russia
SMNS	Staatliches Museum für Naturkunde, Stuttgart, Germany
SMTD	Staatliches Museum für Tierkunde, Dresden, Germany
SNMC	Slovenské narodné muzeum, Bratislava, Slovakia
SNUC	Shanghai Entomology Museum, Chinese Academy of Science, China
TARI	Research Center fir Biodiversity, Academia Sinica, Taipei, Taiwan
UAIC	University of Arizona Insect Collection, Tucson, USA
UNHC	University of New Hampshire, Durham, USA
UQBA	University of Queensland, Queensland, Australia
USNM	United States Nation Museum, Washington, D.C., USA
UZMH	Universitets Helsinki, Helsinki, Finland
ZFMK	Zoologisches Forschunginstitut und Museum "Alexander König", Bonn, Germany
ZIBC	Zoological Institute, Chinese Academy of Sciences, Beijin, China
ZMAS	Zoological Museum, Academy of Sciences, St. Petersburg, Russia
ZMPA	Zoological Museum, Academy of Sciences, Warsawa, Poland
ZMUB	Zoologisches Museum, Museum für Naturkunde, Berlin, Germany
ZMUC	Universitets Museum, University of Copenhavn, Kobenhavn, Denmark
ZMUH	Zoologisches Institut und Zoologisches Museum, Hamburg, Germany
ZMUM	Zoological Museum University of Moscow, Moscow, Russia
ZSMC	Zoologische Staatssammlung, München, Germany

Taxa Removed to Other Families of Coleoptera, or to Other Subfamilies of Staphylinidae

Scaphidium acuminatum Marsham, 1802: 234 (*Cypha*, Aleocharinae, Staphylinidae)
Scaphidium bicolor Fabricius, 1798: 179 (*Scaphidema*, Tenebrionidae)
Scaphidium dubium Marsham, 1802: 234 (*Calyptomerus*, Clambidae)
Scaphidium griseum Marsham, 1802: 233 (unknown)
Scaphidium haemorrhoidale Germar, 1818: 255 (*Eucinetus*, Eucinetidae)
Scaphidium longicorne Paykull, 1800: 340 (*Cypha*, Aleocharinae, Staphylinidae)
Scaphidium punctatum Gyllenhal, 1827: 293 (*Ptenidium*, Ptiliidae)
Scaphidium pusillum Gyllenhal, 1808: 189 (*Ptenidium*, Ptiliidae)
Scaphidium scutellatum Panzer, 1793: 11 (*Brachypterolus*, Kateretidae)
Scaphisoma concinna [sic] Broun, 1880: 158 (*Zearagytodes*, Leiodidae)

Taxa Published as Varieties

The entities below have been described as varieties, and are listed as such in Leschen & Löbl 1995 and/or in Löbl 1997. They are deemed available and have subspecific rank under the provisions of the ICZN (1999), Art. 45.6.4, and are therefore treated as subspecies in the following catalogue. However, revisions of the respective type material and reassessments of the validity of these taxa are wanting.

Cyparium variabile var. *atrocinctum* Pic, 1955: 50
Cyparium variabile var. *diversipenne* Pic, 1955: 50
Diatelum wallacei Pascoe var. *laterale* Achard, 1920d: 123
Scaphidium assamense var. *multimaculatum* Pic, 1915f: 43
Scaphidium assamense var. *semifasciatum* Pic, 1915f: 43
Scaphidium baconi var. *uniplagatum* Achard, 1922d: 263
Scaphidiolum leleupi var. *atropygum* Pic, 1954b: 35
Scaphidiolum longithorax var. *nigriventre* Achard, 1924d: 91
Scaphidium lunatum var. *inconjunctum* Pic, 1921a: 158
Scaphidium lunatum var. *rufithorax* Pic, 1921a: 159
Scaphidium lunatum var. *bioculatum* (Achard, 1924d: 91)
Scaphidium multipunctatum var. *luluanum* (Pic, 1954b: 33)
Scaphidium ocellatum var. *birmanicum* Achard, 1920j: 264
Scaphidium pardale var. *nigripenne* Oberthür, 1883: 11
Scaphidium picconii var. *sexmaculatum* Reitter, 1889: 7

Scaphidium pygidiale var. *bicoloricolle* Pic, 1917b: 3
Scaphidium quadrillum var. *biconjunctum* Pic, 1920b: 4
Scaphidium rufitarse var. *modiglianii* Pic, 1920e: 94.
Scaphidium striatipenne var. *ornatipenne* (Achard, 1922b: 34)
Baeocera argentina var. *tucumana* Pic, 1920i: 50
Baeocera bicolor Achard var. *diluta* Achard, 1920e: 352
Baeocera dufaui var. *tricolor* Pic, 1920b: 3
Baeocera dufaui var. *unicolor* Pic, 1920b: 3
Scaphisoma atrofasciatum var. *motoense* Pic, 1928b: 44
Scaphisoma jeanneli var. *chappuisi* Pic, 1946: 84
Scaphisoma piceicolle var. *boxi* Pic, 1930c: 176
Scaphisoma schoutedeni var. *atropygum* Pic, 1930a: 88
Scaphisoma testaceomaculatum var. *conjunctum* (Pic, 1920e: 97)
Scaphisoma tropicum var. *andreinii* Pic, 1920e: 96
Vituratella cinctipennis var. *boutakoffi* (Pic, 1954a: 38)
Vituratella elongatior var. *obscura* (Pic, 1955: 54)
Vituratella subelongata var. *lata* (Pic, 1928b: 40)

Reversed Priority

Scaphidium nigripes Chevrolat (often cited as Chevrolat, 1844), currently listed as an invalid junior synonym of *S. mexicanum* Laporte, 1840, is illustrated on plate 17 in Guérin-Méneville's Illustrations that was issued in October-December 1834 (Bousquet, 2016) and thus has priority over *Scaphidium mexicanum* Laporte, 1840, and is credited to Guérin-Méneville, 1834.

The Identity of *Scaphisoma agaricinum* (Linnaeus)

Scaphisoma agaricinum (Linnaeus, 1758) is the very first scaphidiine described. It is a common species widely distributed from Great Britain to the Russian Far East. Linnaeus (1758: 360) noted habitat as "in Agaricis" and subsequently (Linnaeus 1760: 148) mentioned this species was from Sweden. Hence, there is little doubt that it is based on Swedish material. Subsequent workers assigned to *S. agaricinum* a common species, also occurring in Sweden, that, unlike other known European species possesses a dark-colored body, small eighth antennomeres and the sutural stria of the elytra not curved laterad along the elytral bases. The concept of this species remained unambiguous in works issued in the 19th and first half of the 20th centuries, as seen in widely used manuals for the identification of beetles (e.g., Stephens 1830: 4, Redtenbacher 1847: 147, Redtenbacher 1874: 335, Fowler 1889: 347, Ganglbauer 1899: 343, Everts 1903: 446, Reitter 1909: 276, Kuhnt 1912: 362, Porta 1926: 365, Joy 1932: 475). An exception from the rule is in Lundblad (1952: 28) who confused *S. agaricinum* with *S. boleti* Panzer, an error repeated by Tamanini (1954: 88) and corrected by Palm (1953: 172). Löbl (1967: 105) described *Scaphisoma inopinatum*, from Transbaikal, a species similar to that commonly identified as *S. agaricinum*. *Scaphisoma agaricinum* and *S. inopinatum* may be readily distinguished by their aedeagal characters, in addition *S. inopinatum* has coarser elytral punctation. *Scaphisoma inopinatum* was subsequently found in a number of European countries, including Sweden. The sympatry of the two species and the discovery that the original material of *S. agaicinum* is untraceable led Löbl (1970c: 738) to fix *S. agaricinum* by a neotype designation. This fixation turned out to be invalid because an original specimen of *S. agaricinum* is in fact housed in the colletion of the Linnean Society, London (see ICZN, 1999, Art. 75.8). Its image is posted on http://linnean-online.org/20926/ but the relevant characters cannot be seen without cleaning and dissecting the specimen. However, the concept of *S. agaricinum* adopted here complies with that in taxonomic works issued subsequently to Lundblad (1952), e.g., Löbl (1965d: 336, 1967d: 34, 1970c: 738, 2012g: 205), Tamanini (1969a: 488, 1969b: 367, 1970: 17), and Freude (1971: 347).

Nomen Dubium

Sphaeridium pulicarium Rossi, 1792: 21, described as "nigrum oblongum, elytris abbreviatis, abdomine acuto … Long ¾ L, Lat. ½ L.", was placed by Gyllenhal 1808: 187 in synonymy of *Scaphidium agaricinum* (Linnaeus, 1758), the present *Scaphisoma agaricinum*. Since, the Rossi's name is listed as a junior synonym of *Scaphisoma agaricinum* (e.g., Gemminger & Harold 1868: 752, Csiki 1910: 10, Löbl 1997: 84, Löbl 2004d: 499). The type material of *S. pulicarium* was not found in ZMUB (J. Frisch, pers. comm.) where it may have been deposited (Horn et al. 1990: 333), and is not in Gyllenhal's collection (H. Mejlon, pers. comm.). As the title of Rossi's work suggests, the type locality of this species is Tuscany, or a nearby area. As *Scaphisoma agaricinum* is replaced in Tuscany by the similar *S. loebli* Tamanini (pers. observation), the synonymy currently accepted is likely erroneous. Gyllenhal (l.c.) noted under *Sphaeridium pulicarium* "Var. b. rufo-piceum totum." and a line below "*Scaphidium boleti* Panz. …". Thus, *Sphaeridium pulicarium* sensu Gyllenhal is possibly *Scaphisoma boleti* Panzer, and the identity of *Sphaeridium pulicarium* Rossi is unknown.

Its size (linea = 4.560 mm) suggest a species larger than *S. agaricinum* and *S. boleti*.

New Records

Scaphium immaculatum (Olivier): Syria bor.occ., Djebel Ansariya Sharkia, env. 700 m, 6.11.88, 1 ex., leg. Jan Macek (NHMB).

Scaphisoma aspectum Löbl, 2015: Indonesia, E Java, Ijen Nat. Park, 1800 m, Sodong, 26-27.II.1994, 8 ex. leg. Bolm (SMNS, MHNG).

Scaphisoma aurorae Löbl, 1992: Nepal, Prov. Janakpur, District Dolakha, upper Simigau village, 2700–2800 m, 01.VI.2000, 2 ex., leg. Joachim Schmidt (NMEC).

Scaphisoma italicum Tamanini, 1955: Montenegro, Sasko Jezero, NE Ulcinj, Hotel Shas 19.-21.05.2014, 41°59′/19°20′ 30 m, Macchie, Felsrasen, 1 ♂, 1 ♀, leg. Erwin Holzer (private collection Erwin Holzer, MHNG).

Scaphisoma perbrincki Löbl, 1971: India, Goa, Bondla Nat. Res., 21–24.2.94, 1 ♂, leg. Ernst Heiss (MHNG).

Catalogue

SCAPHIDIINAE

Scaphidiinae Latreille, 1806: 3 (Scaphidilia). Type genus: *Scaphidium* Olivier, 1790.

Latreille, 1829: 501 (Scaphidites) (characters)
Stephens, 1829: 71 (Scaphididae) (catalogue, Great Britain)
Stephens, 1830: 2 (Scaphididae) (characters)
Kirby, 1837: 108 (Scaphidiadae) (characters)
Laporte, 1840: 18 (Scaphidites) (characters)
Heer, 1841: 371 (Scaphidida) (characters, Switzerland)
Erichson, 1845: 1 (Scaphidilia) (characters)
Redtenbacher, 1847: 147 (Scaphidii) (characters Austria)
Lacordaire, 1854: 236 (Scaphidiles) (characters)
Fairmaire & Laboulbène, 1855: 341 (Scaphididae) (characters, France)
Jacquelin du Val, 1858: 121 (Scaphidiides) (characters, France)
Thomson, 1862: 125 (Scaphidilia) (characters)
Redtenbacher, 1872: 335 (Scaphidiidae) (keys, Austria)
Reitter, 1880a: 35 (Scaphidiidae) (keys)
LeConte & Horn, 1883: 110 (Scaphidiidae) (characters, key to North American genera)
Seidlitz, 1888a: 72, 295; (Scaphidiidae) (characters)
Seidlitz, 1888b: 74, 311 (Scaphidiidae) (characters)
Matthews, 1888: 158 (Scaphidiidae) (characters, Mexico, Central America)
Casey, 1893: 510 (Scaphidiidae) (keys, USA)
Fowler, 1889: 345 (Scaphidiidae) (characters, Great Britain, Ireland)
Ganglbauer, 1899: 335 (Scaphidiidae) (characters, Central Europe)
Stierlin, 1900: 489 (Scaphidiidae) (characters, Switzerland)
Everts, 1903: 444 (Scaphidiidae) (characters, Netherlands)
Csiki, 1908: 151 (Scaphidiidae) (catalogue World)
Csiki, 1910: 3 (Scaphidiidae) (catalogue World)
Blatchley, 1910: 490 (Scaphidiidae) (characters, Indiana)
Reitter, 1909: 275 (Scaphidiidae) (characters, Central Europe)
Jakobson, 1910: 636 (catalogue, Russia, western Europe)
Achard, 1923: 94 (Scaphidiidae) (review, Japan)
Porta, 1923: 364 (Scaphidiidae) (Italy, keys)
Achard, 1924b: 24 (Scaphidiidae) (classification)
Achard, 1924c: 144 (Scaphidiidae) (key to genera, catalogue of Palaearctic taxa)

 | DOI:10.1163/9789004375956_002

Vitale, 1929: 109 (Scaphiidae) (Sicilia)
Hudson, 1934: 48 (Scaphidiidae) (characters)
Miwa & Mitono, 1943: 513 (Scaphidiidae) (review, Taiwan)
Van Emden, 1942: 213, Fig. 39 (larvae)
Vinson, 1943: 177 (Scaphidiidae) (review, Mauritius)
Blackwelder, 1944: 98 (Scaphidiidae) (catalogue, America south of USA)
Scheerpeltz & Höfler, 1948: 294 (Scaphidiidae) (habitat, fungal hosts)
Crowson, 1950: 279, 285 (Scaphidiidae) (characters, relationships)
Nakane, 1955a: 53 (Scaphidiidae) (Japan)
Nakane, 1955b: 50 (Scaphidiidae) (Japan)
Hatch, 1957: 280 (Scaphidiidae) (review, Pacific Northwest)
Kasule, 1966: 22 (Scaphidiinae) (downgraded to subfamily of Staphylinidae, larval characters)
Kasule, 1968: 116 (Scaphidiinae) (larval characters)
Arnett, 1968: 363 (Scaphidiidae) (North America, key to genera)
Tamanini, 1969c: 136 (Scaphidiidae, Scaphisomidae)
Löbl, 1970a: 3 (Scaphidiidae) (review, Poland)
Tamanini, 1970: 5 (Scaphidiidae, Scaphisomidae) (review, Italy)
Britton, 1970: 543 (Scaphidiidae) (characters)
Löbl, 1971c: 937 (Scaphidiidae) (review, Sri Lanka)
Freude, 1971: 343 (Scaphidiidae) (key, Central Europe)
Löbl, 1972b: 108 (Scaphidiidae) (checklist of Philippine species)
Löbl, 1973b: 309 (Scaphidiidae) (review, New Caledonia)
Löbl, 1975a: 369 (Scaphidiidae) (review, New Guinea)
Löbl, 1977e: 4 (Scaphidiidae) (key to Australian genera)
Löbl, 1977f: 39 (Scaphidiidae) (review, La Réunion)
Burakowski et al., 1978: 232 (Scaphidiidae) (catalogue, Poland)
Löbl, 1979a: 77 (Scaphidiidae) (review, South India)
Löbl, 1980c: 379 (Scaphidiidae) (review, Fiji)
Löbl, 1981a: 347 (Scaphidiidae) (New Caledonia, keys)
Löbl, 1981b: 69 (Scaphidiidae) (review, Micronesia)
Lawrence & Newton, 1982: 274 (Scaphidiidae) (relationships)
Löbl, 1982b: 101 (Scaphidiidae) (review, Ryukyus)
Löbl, 1982g: 47 (Scaphidiidae) (review, Israel)
Löbl, 1983b: 161 (Scaphidiidae) (review, Chile)
Löbl, 1984a: 57 (Scaphidiidae) (review, Northeast India, Bhutan)
Newton, 1984: 317 (Scaphidiidae) (fungal hosts, feeding)
Iablokoff-Khnzorian, 1985: 132 (Scaphidiidae) (keys, ex Soviet Union)
Naomi, 1985: 10 (Scaphidiidae) (relationships, characters)

Morimoto, 1985: 254 (Scaphidiidae) (keys, Japan)
Kompantsev & Pototskaya, 1987: 87 (Scaphidiidae) (larvae, biology)
Hammond & Lawrence, 1989: 294 (Scaphidiinae) (mycophagy)
Lafer, 1989: 367 (Scaphidiidae) (keys, Far East Russia)
Löbl, 1989a: 9 (Scaphidiidae) (review, key, Algeria)
Löbl, 1990b: 505 (Scaphidiidae) (review, Thailand)
Newton & Stephenson, 1990: 197 (fungal and slime mould hosts)
Newton, 1991: 337 (Scaphidiidae) (larvae)
Newton & Thayer, 1992: 63 (Scaphidiinae) (type genera)
Löbl, 1992a: 471 (Scaphidiidae) (review, Nepal Himalaya, key to genera of continental Asia)
Leschen, 1993: 73 (Scaphidiinae) (feeding, morphology)
Leschen, 1994: 3 (Scaphidiinae) (larval behavior)
Lawrence & Newton, 1995: 828 (Scaphidiinae)
Leschen & Löbl, 1995: 430 (Scaphidiinae) (character analysis, classification, key to non- Scaphisomatini genera)
Krasutskij, 1996a: 38 (Scaphidiidae, key, Ural & trans-Ural)
Hansen, 1997: 110, 172 (Scaphidiidae) (characters, relationships, classification)
Löbl, 1997: 1 (Scaphidiinae) (catalogue, World)
Newton et al., 2000: 375 (Scaphidiinae) (characters, key to Nearctic genera)
Navarrete-Heredia et al., 2002: 201 (Scaphidiiinae) (characters, key to Mexican genera)
Löbl & Leschen, 2003b: 15 (Scaphidiinae) (review, keys, New Zealand)
Betz et al., 2003: 225 (oxyteline group) (feeding, larvae)
Löbl, 2004b: 495 (Scaphidiinae) (catalogue, Palaearctic)
Thayer, 2005: 328 (Scaphidiinae) (characters)
Fierros-López, 2010: 1 (Scaphidiinae) (characters, immature stages)
Löbl, 2012g: 201 (Scaphidiinae) (characters, keys, Central Europe)
Lawrence & Ślipiński, 2013: 187 (Scaphidiinae) (characters)
Löbl, 2014a: 49 (Scaphidiinae) (review, Maluku Islands)
Schülke & Smetana, 2015: 730 (Scaphidiinae) (catalogue, Palaearctic)
Löbl, 2015b: 165 (Scaphidiinae) (review, key, Maluku Islands)
Ogawa, 2015: 31 (Scaphidiinae) (phylogeny, divergence time, key, feeding habits, Sulawesi)

[References to immature stages based on missidentifications (see Newton, 1984: 330): Böving & Craighead, 1931: pl. 12; Paulian, 1941: 147; Pototskaya, 1964: pl. 175; Dajoz, 1965: 105].

Extinct Taxa

Scaphidium deletum Heer, 1847: 35, Pl. VII, Fig. 20; 1862: 50; type locality: Oeningen [Öhningen, Württemberg, Germany], Tortonian, Miocene.

Scaphidiopsis Handlirsch, 1906: 550. Type species *Scaphidium hageni* Weyenbergh, 1869, by present designation. Gender: feminine.

Scaphidiopsis hageni (Weyenbergh, 1869: 281); type locality: Solenhofen [Bavaria, Germany], Malm, Upper Jurasic.

Scaphidiopsis aegivoca Handlirsch, 1906: 550; type locality: Solenhofen [Bavaria, Germany], Malm, Upper Jurasic [in Handlirsch, l.c., as possible synonym of *S. hageni*].

Scaphisoma gracile Heer, 1862: 49, Pl. III, Fig. 26; type locality: Oeningen [Öhningen, Württemberg, Germany], Tortonian, Miocene.

Seniaulus C. & L. Heyden, 1866: 137. Type species: *Seniaulus scaphioides* C. & L. Heyden, 1866, by monotypy. Gender: masculin.

Seniaulus scaphioides C. & L. Heyden, 1866: 137; type locality: Siebengebirge, Germany, brown coal.

Extant Taxa

CYPARIINI Achard

Cypariini Achard, 1924b: 28. Type genus: *Cyparium* Erichson, 1845
Leschen & Löbl, 1995: 447, 467 (characters, checklist)
Newton et al., 2000: 375 (characters)
Löbl & Leschen, 2003b: 16 (characters)

Cyparium Erichson

Cyparium Erichson, 1845: 3. Type species: *Cyparium palliatum* Erichson, 1845; by monotypy. Gender: neuter.

Yparicum Achard, 1920d: 126. Type species: *Yparicum yunnanum* Achard, 1920; by monotypy. Gender: neuter.
Lacordaire, 1854: 239 (characters)
Reitter, 1880a: 41 (key to species)
Matthews, 1888: 165 (Central America, key to species)
Casey, 1900: 56 (United States, characters, key to North American species)
Achard, 1923: 109 (characters)
Miwa & Mitono, 1943: 534 (Japan, Taiwan, key to species)
Tamanini, 1969c: 130, Figs 5, 6, 22 (characters)
Löbl, 1971c: 941 (record of unidentified species in Sri Lanka)
Löbl, 1979a: 84 (key to South Indian species)
Löbl, 1984a: 59 (key to Indian species)

Newton, 1984: 317 (fungal hosts)
Iablokoff-Khnzorian, 1985: 137 (characters)
Kompantsev & Pototskaya, 1987: 91 (immature stages)
Leschen, 1988a: 225 (Scaphidiidae) (immature stages, natural history, fungal hosts)
Löbl, 1992a: 498 (synonymy of *Yparicum* with *Cyparium*, key to Himalayan species)
Leschen & Löbl, 1995: 448, Figs 13–15, 28, 29, 31, 35 (characters)
Ogawa & Sakai, 2011: 135 (Japan, key to species)
Löbl, 1990c: 128 (Asia, key [p. 128, replace *humerale* Achard by *montanum* Achard])
Newton, 1991: 337, Figs 34, 171–176 (immature stages)
Leschen, 1993: 73 (larval mandible)
Löbl, 1999: 692 (key to Chinese species)
Fierros-López, 2002: 7 (key to Mexican species, fungal hosts)
Löbl & Leschen, 2003b: 16 (characters)
Löbl, 2011b: 201 (key to Asian species)
Ogawa, 2015: 31 (characters, Figs 4–11a-c)
Ogawa et al., 2016b: 200 (key to Sundaland species)

***Cyparium anale* Reitter**

Cyparium anale Reitter, 1880a: 42. Syntypes, MNHN; type locality: St. Domingo.
Cyparium submetallicum Reitter, 1880a: 43. Syntypes, MNHN; type locality: "India or.?".
Achard, 1921b: 86 (synonymy of *C. submetallicum* with *C. anale*)
DISTRIBUTION. Dominican Republic.

***Cyparium atrum* Casey**

Cyparium ater Casey, 1900: 56. Syntypes, USNM; type locality: USA: Texas, Browsville.
DISTRIBUTION. United States: Texas.

***Cyparium basilewskyi* Pic**

Cyparium basilewskyi Pic, 1955: 50. Holotype, MRAC; type locality: Rwanda: Denzezi, 1600 m, terr. Shangugu.
DISTRIBUTION. Rwanda.

***Cyparium bowringi* Achard**

Cyparium bowringi Achard, 1922c: 42. Lectotype female, NHML; type locality: Indonesia: Java.

Löbl, 1979a: 84 (lectotype designation)
Löbl, 1984a: 59 (records)
Löbl, 1990c: 126, Fig. 1 (characters)
Löbl, 1992a: 499, Fig. 86 (characters)
Ogawa et al., 2016: 196, Fig. 3A

DISTRIBUTION. India: Himachal Pradesh, Meghalaya, Tamil Nadu; Indonesia: Java.

***Cyparium celebense* Ogawa & Löbl**

Cyparium celebense Ogawa & Löbl, 2016b: 196, Figs 1A-C, 2A-C. Holotype, NHML; type locality: Indonesia: Sulawesi Utara, Dumoga Bone National Park, ca 200 m.

DISTRIBUTION. Indonesia: Sulawesi.

***Cyparium championi* Matthews**

Cyparium championi Matthews, 1888: 167, Pl. 4, Fig. 9. Holotype, NHML; type locality: Panama: Bugaba.
Fierros-López, 2006b: 39, Figs 1a, 2a, 3a (characters, records)

DISTRIBUTION. Bolivia; Costa Rica; Ecuador; Nicaragua; Panama; Suriname.

***Cyparium collare* Pic**

Cyparium collare Pic, 1920b: 4. Syntypes, MNHN; type localities: Brésil [Matu Sinhos, Minas; Matto Grosso; Pery-Pery, Pernambuco; right bank of Parahyba].
Note. Pic (l.c.) credited the species epithet to Portevin.

DISTRIBUTION. Brazil.

***Cyparium concolor* (Fabricius)**

Scaphidium concolor Fabricius, 1801: 576. Neotype female, MCZC; type locality: USA: "Southern States".

Cyparium flavipes LeConte, 1860: 322. Lectotype female, MCZC; type locality: USA: "Southern States".

Cyparium substriatum Reitter, 1880a: 42. Syntypes, MNHN; type locality: USA: Alabama.
Casey, 1900: 56 (synonymy of *C. substriatum*)
LeConte, 1860: 322 (characters)
Casey, 1893: 512, 513 (characters, records)
Achard, 1920k: 307 (synonymy of *C. flavipes*)
Brimley, 1938: 149 (records)
Cornell, 1967: 2 (lectotype designation for *C. flavipes*, neotype designation for *S. concolor*)

Kirk, 1969: 30 (records)
Kirk, 1970: 26 (records)
Leschen, 1988a: 231 (fungal hosts)
Downie & Arnett, 1996: 365 (characters, records)

DISTRIBUTION. United States: Alabama, Florida, Illinois, Indiana, Maryland, North Carolina, Ohio, Pennsylvania, South Carolina.

***Cyparium earlyi* Löbl & Leschen**

Cyparium earlyi Löbl & Leschen, 2003b: 16, Fig. 2, Map 17. Holotype male, LUNZ; type locality: New Zealand: FD, Fiordland Nat. Park, south Borland V, 750 m.

DISTRIBUTION. New Zealand: South Island.

***Cyparium ferrugineum* Pic**

Cyparium ferrugineum Pic, 1920b: 5. Syntypes, MNHN; type locality: Brazil [Pery-Pery, Pernambuco, Matto Grosso].
Note. Pic (l.c.) credited the species epithet to Portevin.

DISTRIBUTION. Brazil.

***Cyparium flavosignatum flavosignatum* Zayas**

Cyparium flavosignata Zayas, 1988: 23, Fig. 13. Syntypes, FZCH; type locality: Cuba.

DISTRIBUTION. Cuba.

***Cyparium flavosignatum bicolor* Zayas**

Cyparium flavosignata bicolor Zayas, 1988: 24, Fig. 15. Syntypes, FZCH; type locality: Cuba.

DISTRIBUTION. Cuba.

***Cyparium flavosignatum funebre* Zayas**

Cyparium flavosignata funebris Zayas, 1988: 24, Fig. 14. Syntypes, FZCH; type locality: Cuba.

DISTRIBUTION. Cuba.

***Cyparium flavosignatum splendidum* Zayas**

Cyparium flavosignata splendidum Zayas, 1988: 25, Fig. 16. Syntypes, FZCH; type locality: Cuba.

DISTRIBUTION. Cuba.

***Cyparium formosanum* Miwa & Mitono**

Cyparium formosanum Miwa & Mitono, 1943: 534. Lectotype male, TARI; type locality: Taiwan: Wushe, Nantou.

Löbl, 2011b: 200 (lectotype designation, characters)
Note. Miwa & Mitono, 1943: 555, Fig. G suggests *Pseudobironium*.
DISTRIBUTION. Taiwan.

Cyparium grilloi **Pic**
Cyparium grilloi Pic, 1920e: 94. Syntypes, MCSN; type locality: Brazil: Paranà, Palmeira.
DISTRIBUTION. Brazil.

Cyparium grouvellei **Pic**
Cyparium grouvellei Pic, 1920b: 5. Syntypes, MNHN; type locality: Brazil.
Note. Pic (l.c.) credited the species epithet to Portevin.
DISTRIBUTION. Brazil.

Cyparium humerale **Achard**
Cyparium humerale Achard, 1922c: 40. Syntypes, NMPC; type locality: Bolivia: Cochabamba.
DISTRIBUTION. Bolivia.

Cyparium inclinans **Kirsch**
Cyparium inclinans Kirsch, 1873: 135. Syntypes, SMTD; type locality: Peru.
DISTRIBUTION. Peru.

Cyparium javanum **Löbl**
Cyparium javanum Löbl, 1990c: 126, Figs 2–6. Holotype male, MZBI; type locality: Indonesia: Java, Preanger, Gn. Tangkoeban Prahoe, 4000–5000 ft.
Ogawa et al., 2016b: 196, Fig. 3B (characters)
DISTRIBUTION. Indonesia: Java.

Cyparium jiroi **Ogawa & Sakai**
Cyparium jiroi Ogawa & Sakai, 2011: 130, Figs 1a, 2a, 3e, 4a, d. Holotype male, EUMJ; type locality: Japan: Kyushu Mt. Sobo-san, Takachino Town, Miyazaki Pref.
DISTRIBUTION. Japan: Kyushu.

Cyparium khasianum **Löbl**
Cyparium khasianum Löbl, 1984a: 60, Fig. 2. Holotype male, MHNG; type locality: India: Meghalaya, above Nongpoh, 700 m, Khasi Hills.
DISTRIBUTION. India: Meghalaya.

***Cyparium laevisternale* Nakane**

Cyparium laevisternale Nakane, 1956: A 161. Holotype male, HUSJ; type locality: Japan: Honshu, Noziri, N. Shinano.
Nakane, 1963b: 79 (characters)
Morimoto, 1985: Pl. 45, Fig. 26 (characters)
Löbl, 1990c: 128 (as *laevistriatum*, misspelled)
Ogawa & Sakai, 2011: 134, Figs 1b, 2b 3b, d, 4b, e (characters, records)

DISTRIBUTION. Japan: Honshu, Kyushu, Shikoku.

***Cyparium mathani* Oberthür**

Cyparium mathani Oberthür, 1883: 12. Syntypes, MNHN; type localities: Peru: Iquitos and Amazones.

DISTRIBUTION. Peru.

***Cyparium mikado* Achard**

Cyparium mikado Achard, 1923: 109. Syntypes, MNHN; type locality: Japan: Alps of Nikko.
Lewis, 1893: 294 (as *C. sibiricum*, records)
Miwa & Mitono, 1943: 535 (records)
Nakane, 1955b: 53 (characters)
Chûjô, 1961: 5 (records)
Nakane, 1963b: 79 (characters)
Morimoto, 1985: Pl. 45, Fig. 27 (characters)
Hayashi, 1986: Pl. 18 (larva)
Löbl, 1999: 693 (records)
Hwang & Ahn, 2001: 370, Figs 1, 2 (misspelled *mikardo*, characters, records)
Ogawa & Sakai, 2011: 134, 135, Figs 1c, 2c, 3a, c, 4c, f (characters, records)

DISTRIBUTION. China: Beijing, Shaanxi; Japan: Hokkaido, Honshu, Kyushu, Shikoku, Tsushima; South Korea.

***Cyparium minutum* Pic**

Cyparium minutum Pic, 1931a: 2, Syntypes, MNHN; type locality: Jamaica: Jackson Town.

DISTRIBUTION. Jamaica.

***Cyparium montanum* Achard**

Cyparium montanum Achard, 1922c: 41. Lectotype male, NMPC; type locality: India: Himachal Pradesh, Simla.
Champion, 1927: 272 (lectotype fixed by inference, characters, records)

Löbl, 1984a: 60 (records)
Löbl, 1986c: 343 (records)
Löbl, 1992a: 499, Fig. 85 (characters, records)
Löbl, 1999: 693 (records)
Löbl, 2011c: 201 (records)

Distribution. Bhutan; China: Yunnan; India: Himachal Pradesh, Uttarakhand (Kumaon), West Bengal (Darjeeling District); Taiwan.

Cyparium multistriatum **Pic**

Cyparium multistriatum Pic, 1954b: 35. Holotype, MRAC; type locality: Democratic Republic of the Congo: Kapanga, Lulua.

Distribution. Democratic Republic of the Congo.

Cyparium navarretei **Fierros-López**

Cyparium navarretei Fierros-López, 2002: 8, Figs 1, 3–5. Holotype male, CZUG; type locality: Mexico: Veracruz, Cerro Acatlán, Naolinco.

Distribution. Mexico.

Cyparium nigronotatum **Pic**

Cyparium nigronotatum Pic, 1931a: 2. Syntypes, MNHN; type locality: Peru: [Puno].

Distribution. Peru.

Cyparium oberthueri **Pic**

Cyparium oberthüri Pic, 1956a: 175. Syntypes, MNHN; type locality: Brazil: Matto Grosso.

Distribution. Brazil.

Cyparium palliatum **Erichson**

Cyparium palliatum Erichson, 1845: 4. Syntypes ZMUB [single female, labelled as holotype]; type locality: Mexico.
Matthews, 1888: 166 (characters)
Márquez, 2007: 8 (characters)

Distribution. Mexico.

Cyparium pallidum **Pic**

Cyparium pallidum Pic, 1955: 50. Holotype, MRAC; type locality: Rwanda: Denzezi, 1600 m, terr. Shangugu.

Distribution. Rwanda.

***Cyparium peruvianum* Pic**

Cyparium peruvianum Pic, 1947b: 7. Syntypes, MNHN; type locality: Peru.

DISTRIBUTION. Peru.

***Cyparium piceum* Reitter**

Cyparium piceum Reitter, 1880a: 41. Syntypes, MNHN; type locality: Republic of South Africa: Cape of Good Hope.

DISTRIBUTION. Republic of South Africa.

***Cyparium plagipenne* Achard**

Cyparium plagipenne Achard, 1922c: 41. Lectotype male, NHML; type locality: India.
Achard, 1920d: 127 (characters)
Champion, 1927: 271 (lectotype fixed by inference, characters, records)
Löbl, 1984a: 60 (doubtful identification)
Löbl, 1992a: 500, Fig. 87 (characters)

DISTRIBUTION. India: Uttarakhand (Kumaon), West Bengal (Darjeeling District).

***Cyparium punctatum* Pic**

Cyparium punctatum Pic, 1916c: 18. Syntype female, MNHN; type locality: East Malaysia: Banguey [Banggi Is.].
Achard, 1920d: 128 (characters)
Löbl, 2006: 24 (characters, records)
Ogawa et al., 2016b: 196, Fig. 3C (characters)

DISTRIBUTION. East Malaysia: Banggi Island, Sabah; Philippines.

***Cyparium pygidiale* Achard**

Cyparium pygidiale Achard, 1922c: 40. Syntypes, NMPC; type localities: Brazil: Goyaz, Jatahy.

DISTRIBUTION. Brazil.

***Cyparium ruficolle* Achard**

Cyparium ruficolle Achard, 1922c: 39. Syntypes, NMPC; type locality: Brazil [Corumba, Matto Grosso].

DISTRIBUTION. Brazil.

***Cyparium rufohumerale* Pic**

Cyparium rufohumerale Pic, 1931a: 2. Syntypes, MNHN; type locality: Brazil.

DISTRIBUTION. Brazil.

***Cyparium rufonotatum* Pic**

Cyparium rufonotatum Pic, 1916c: 18. Holotype, MNHN; type locality: Colombia.

DISTRIBUTION. Colombia.

***Cyparium sallaei* Matthews**

Cyparium sallaei Matthews, 1888: 166, Pl. 4, Fig. 10. Syntypes, NHML; type locality: Mexico: Cordoba.

Navarrete-Heredia et al., 2002: 204 (records)

DISTRIBUTION. Mexico.

***Cyparium semirufum* Pic**

Cyparium semirufum Pic, 1917a: 3. Syntypes, MNHN; type locality: Vietnam: Tonkin [Hoa Binh].

Löbl, 1990b: 513 (characters, records)

DISTRIBUTION. Thailand; Vietnam.

***Cyparium siamense* Löbl**

Cyparium siamense Löbl, 1990b: 511, Fig, 2. Holotype male, MHNG; type locality: Thailand: Khao Yai Nat. Park, near Headquarters, 750–850 m.

Löbl, 1999: 693, Fig. 7 (characters, records)

DISTRIBUTION. China: Yunnan, Thailand.

***Cyparium sibiricum* Solsky**

Cyparium sibiricum Solsky, 1871: 350. Holotype, ZMAS; type locality: Russia: Irkutsk.

Heyden, 1880: 88 (records)

Heyden, 1893: 54 (records)

Jakobson, 1910: 636 (records)

Iablokoff-Khnzorian, 1985: 134, 138 (characters, records)

Kompantsev & Pototskaya, 1987: 91, 94, Figs 2, 3 (larva, fungal hosts, records)

Löbl, 1999: 693, Figs 1–3 (characters, records)

DISTRIBUTION. China: Shaanxi, Sichuan, Yunnan; Russia: Far East, Siberia, Transbaikal.

***Cyparium sichuanum* Löbl**

Cyparium sichuanum Löbl, 1999: 694, Figs 4–6. Holotype male, MHNG; type locality: China: Sichuan, Wolong Nat. Res., 100 m

DISTRIBUTION. China: Sichuan.

***Cyparium tamil* Löbl**

Cyparium tamil Löbl, 1979a: 85. Holotype female, MHNG; type locality: India: Tamil Nadu, 15 km east Coonoor, 900 m, Nilgiri Hills.

DISTRIBUTION. India: Tamil Nadu.

***Cyparium tenenbaumi* Pic**

Cyparium tenenbaumi Pic, 1926c: 5. Syntypes, MNHN/ZMPA; type locality: Russia: Far East, Ussuri.
Iablokoff-Khnzorian, 1985: 138 (characters, records)

DISTRIBUTION. Russia: Far East, Siberia, Transbaikal.

***Cyparium terminale* Matthews**

Cyparium terminale Matthews, 1888: 167, Pl. 4, Fig. 9. Syntypes, NHML; type localities: Mexico: Jalapa, Cordoba; Guatemala, Zapote, Capetillo, San Juan in Vera Paz; Panama, Bugaba.
Lawrence & Newton, 1980: 137 (fungal hosts)
Newton, 1984: 317, 338 (fungal and slime mould hosts)
Leschen, 1988a: 231 (fungal hosts)
Navarrete-Heredia, 1991: 126 (habitat)
Navarrete-Heredia et al., 2002: 204, Fig. 10.7 (records)
Fierros-López, 2002: 14, Fig. 8, 10 (characters)
Márquez, 2006: 182 (records)
Márquez, 2007: 2 (characters, records)

DISTRIBUTION. Guatemala; Mexico; Panama.

***Cyparium testaceicorne* Pic**

Cyparium testaceicorne Pic, 1931a: 2. Syntypes, MNHN; type locality: Bolivia: [Coroico].

DISTRIBUTION. Bolivia.

***Cyparium testaceum* Pic**

Cyparium testaceum Pic, 1920f: 23. Syntypes, MCSN/MNHN; type locality: Myanmar: [Palon Pégou].
Pic, 1921a: 162 (characters)

DISTRIBUTION. Myanmar.

***Cyparium thorpei* Löbl & Leschen**

Cyparium thorpei Löbl & Leschen, 2003b: 17, Fig. 3, Map 18. Holotype female, AMNZ; type locality: New Zealand: TO, Mt. Ruapehu, 1160 m, Whakapapa Village.

DISTRIBUTION. New Zealand: North Island.

Cyparium variabile variabile **Pic**

Cyparium variabile Pic, 1955: 50. Syntypes, MRAC; type localities: Rwanda: Tshurugaya, 2400 m, terr. Astrida; east slope of Muhavura, 2100 m, terr. Ruhengeri; Rubengera, 1900 m, terr. Kibuye.

DISTRIBUTION. Rwanda.

Cyparium variabile atrocinctum **Pic**

Cyparium variabile var. *atrocinctum* Pic, 1955: 50. Syntypes, MRAC; type locality: Rwanda.

DISTRIBUTION. Rwanda.

Cyparium variabile diversipenne **Pic**

Cyparium variabile var. *diversipenne* Pic, 1955: 50. Syntypes, MRAC; type locality: Rwanda.

DISTRIBUTION. Rwanda.

Cyparium variegatum **Achard**

Cyparium variegatum Achard, 1920d: 127. Syntype, MNHN; type locality: Indonesia: Java, Meuwen Bay.

Ogawa et al., 2016b: 196, Fig. 3D (characters)

DISTRIBUTION. Indonesia: Java.

Cyparium yapalli **Fierroz-López**

Cyparium yapalli Fierroz-López, 2002: 9, Figs 2, 6, 7, 9, 11. Holotype male, CZUG; type locality: Mexico: Oaxaca, km 164 carr. Sola de Vega-Puerto Escondido.

Márquez, 2007: 9 (records)

DISTRIBUTION. Mexico.

Cyparium yunnanum **(Achard)**

Yparicum yunnanum Achard, 1920d: 126. Lectotype male, NMPC type locality: China: Yunnan, Tali.

Löbl, 1992a: 498 (transfer to *Cyparium*)

Löbl, 1999: 694 (lectotype designation, characters, records)

DISTRIBUTION. China: Yunnan.

SCAPHIINI Achard

Scaphiini Achard, 1924b: 27 (as Scaphiitae). Type genus: *Scaphium* Kirky, 1837.

Leschen & Löbl, 1995: 449 (characters)

Newton et al., 2000: 375 (characters)

Löbl & Ogawa, 2016c: 38 (key to genera)

Ascaphium Lewis

Ascaphium Lewis, 1893: 288. Type species: *Ascaphium sulcipenne* Lewis, 1893, by subsequent designation. Gender: neuter.
Achard, 1923: 98 (key to Japanese species)
Löbl, 1992a: 482 (type species fixation, key to species)
Leschen & Löbl, 1995: 450, 468 (characters, checklist)
Löbl, 1999: 696 (key Chinese to species)
He et al., 2008d: 63 (key to Chinese species)
Tang & Li, 2009: 92 (key to Chinese species)

Ascaphium alienum **Tang & Li**

Ascaphium alienum Tang & Li, 2009: 96, Figs 1c, 4a-d. Holotype male, SNUC; type locality: China: Taibaishan, Shaanxi, 2350–2750 m.
DISTRIBUTION. China: Hubei, Shaanxi.

Ascaphium alticola **Löbl**

Ascaphium alticola Löbl, 1999: 699. Holotype female, MHNG; type locality: China: Gonggashan-Hailuogou, 2900–3200 m, Sichuan.
He et al., 2008d: 67, Figs 3, 7–9 (characters, records)
DISTRIBUTION. China: Sichuan.

Ascaphium apicale **Lewis**

Ascaphium apicale Lewis, 1893: 290. Syntypes, NHML; type localities: Japan: Miganoshita, Subashiri, Nikko, Oyayama.
Achard, 1923: 100 (characters, records)
Miwa & Mitono, 1943: 517 (characters, records)
Nakane, 1955a: 55 (characters)
Nakane, 1963b: 78 (characters)
Morimoto, 1985: Pl. 45, Fig. 12 (characters)
DISTRIBUTION. Japan: Honshu, Kyushu, Shikoku.

Ascaphium hisamatsui **Kimura**

Ascaphium hisamatsui Kimura, 2009: 227, Figs 1–6. Holotype male, KCMI; type locality: Taiwan: Meifeng, Nantou Hsien.
DISTRIBUTION. Taiwan.

Ascaphium huanghaoi **Tang & Li**

Ascaphium huanghaoi Tang & Li, 2009: 95, Figs 1b, 3a-e. Holotype male, SNUC; type locality: China: Qinling Daoban, Zhouzhi County, Shaanxi, 1900 m.
DISTRIBUTION. China: Shaanxi.

Ascaphium ingentis **He, Tang & Li**

Ascaphium ingentis He, Tang & Li, 2008d: 65, Figs 2, 11. Holotype female, ZIBC; type locality: China: Longjiang, 20067 m, Xiaoheishan, Longling Co., Yunnan.

DISTRIBUTION. China: Yunnan.

Ascaphium irregulare **Löbl**

Ascaphium irregulare Löbl, 1999: 697, Figs 8–10. Holotype male, MHNG; type locality: China: Jizu Shan, 2500–2700 m, Yunnan.

DISTRIBUTION. China: Yunnan.

Ascaphium longlingense **He, Tang & Li**

Ascaphium longlingense He, Tang & Li, 2008d: 64, Figs 1, 4–6, 10. Holotype male, ZIBC; type locality: China: Longjiang, Xiaoheishan, Yunnan, 2067 m.

DISTRIBUTION. China: Yunnan.

Ascaphium minus **Pic**

Scaphium sinense Pic var. *minor* Pic, 1954c: 57. Lectotype male, MHNG; type locality: China: Fujian, Kuatun.

Löbl, 1992a: 483 (as *A. minor*, raised to species)

Löbl, 1999: 697 (lectotype designation)

DISTRIBUTION. China: Fujian.

Ascaphium ochripes **Löbl**

Ascaphium ochripes Löbl, 1992a: 483, Fig. 9. Holotype female, MHNG; type locality: Nepal: Chichila, 2200 m, Sankhuwasabha District.

DISTRIBUTION. Nepal.

Ascaphium parvulum **Tang & Li**

Ascaphium parvulum Tang & Li, 2009: 93, Figs 1a, 2a-c. Holotype male, SNUC; type locality: China: Qinling west Sangongligou, Zhouzhi County, Shaanxi, 1336 m.

DISTRIBUTION. China: Shaanxi.

Ascaphium sinense **Pic**

Ascaphium sinense Pic, 1954c: 56. Lectotype male, NHRS; type locality: China: Kuatun.

Löbl, 1999: 697 (lectotype designation)

DISTRIBUTION. China: Fujian.

***Ascaphium sulcipenne* Lewis**

Ascaphium sulcipenne Lewis, 1893: 289. Syntypes, NHML; type localities: Japan: Miyanoshita, Nikko.
Achard, 1923: 99 (characters, records)
Miwa & Mitono, 1943: 517 (characters, records)
Nakane, 1955a: 55, Fig. 9 (characters, records)
Nakane, 1963b: 78 (characters)
Morimoto, 1985: Pl. 45, Fig. 10 (characters)
DISTRIBUTION. Japan: Honshu.

***Ascaphium tibiale* Lewis**

Ascaphium tibiale Lewis, 1893: 289. Syntypes, NHML; type localities: Japan: Nikko, Miyanoshita, Oyayama, Subashiri.
Achard, 1923: 99, 100 (characters, records)
Miwa & Mitono, 1943: 517, 518 (characters, records)
Nakane, 1955a: 55 (characters, records)
Nakane, 1963b: 78 (characters)
Morimoto, 1985: Pl. 45, Fig. 11 (characters)
DISTRIBUTION. Japan: Honshu, Kyushu, Shikoku.

***Ascaphium tonkinense* Achard**

Ascaphium tonkinense Achard, 1921a: 93. Syntypes, NMPC; type locality: Vietnam: Tonkin.
Miwa & Mitono, 1943: 518 (characters)
DISTRIBUTION. Vietnam.

Episcaphium Lewis

Episcaphium Lewis, 1893: 290. Type species: *Episcaphium semirufum* Lewis, 1893; by monotypy. Gender: neuter.
Phenoscaphium Achard, 1922c: 35. Type species: *Phenoscaphium callosipenne* Achard, 1922; by monotypy. Gender: neuter.
Miwa & Mitono, 1943: 519 (characters)
Leschen & Löbl, 1995: 450, Fig. 38, 468 (characters, synonymy of *Phenoscaphium* with *Episcaphium*, checklist)
Löbl, 1992a: 484 (key to species)
Löbl, 1999: 700 (key to species)
Löbl, 2002b: 284 (key to species)
Sheng & Gu, 2009: 35 (key to Chinese species)
Tang, Tu & Li, 2016a: 54 (key to Chinese species)

Episcaphium callosipenne (Achard)

Phenoscaphium callosipenne Achard, 1922c: 36. Holotype, NMPC; type locality: Borneo.
Lechen & Löbl, 1995: 468 (transfer to *Episcaphium*)

DISTRIBUTION. Indonesia: Java, Sumatra; East Malaysia: Sarawak; West Malayasia.

Episcaphium catenatum Löbl

Episcaphium catenatum Löbl, 1999: 700, Figs 11–13. Holotype male, MHNG; type locality: China: Wolong Nat. Res., 1500 m, Sichuan.

DISTRIBUTION. China: Sichuan.

Episcaphium changchini Sheng & Gu

Episcaphium changchini Sheng & Gu, 2009: 36, Figs 1–7. Holotype male, SNUC; type locality: China: Qingling, Houzhi County, Shaanxi, 1260–1336 m.

DISTRIBUTION. China: Shaanxi.

Episcaphium dabashanum Sheng & Gu

Episcaphium dabashanum Sheng & Gu, 2009: 37, Figs 8–14. Holotype male, SNUC; type locality: China: Chongqing, lower Huang'an-Gou, east Daba-shan, Chengkou County, 2039 m.

DISTRIBUTION. China: Chongqing.

Episcaphium grande Löbl

Episcaphium grande Löbl, 2002b: 292, Figs 4–6. Holotype male, MHNG; type locality: Vietnam: Hoa Binh.

DISTRIBUTION. Vietnam.

Episcaphium haematoides Löbl

Episcaphium haematoides Löbl, 1999: 703, Figs 18–20. Holotype male, MHNG; type locality: China: Heishui, 35 km west Lijiang, Yunnan.
Tang et al., 2016: 54 (records)

DISTRIBUTION. China: Gansu, Sichuan, Yunnan.

Episcaphium saucineum (Motschulsky)

Scaphidium saucineum Motschulsky, 1860: 94. Lectotype female, ZMUM; type locality: Sri Lanka: Noura-Ellia [Nuwara Eliya].

Scaphidium saucineum var. *reductum* Pic, 1920f: 23. Lectotype female, MNHN; type locality: Sri Lanka.

Löbl, 1971c: 940 (as *Scaphidium*, lectotype designations, synonymy of var. *reductum* with *S. saucineum*, records)
Löbl, 1992a: 484 (transfer to *Episcaphium*)
DISTRIBUTION. Sri Lanka.

Episcaphium semirufum **Lewis**

Episcaphium semirufum Lewis, 1893: 291. Syntypes, NHML; type localities: Japan: Nikko, Kiga, Miyanoshita.
Episcaphium semirufum var. *ruficolle* Lewis, 1893: 291. Syntypes, NHML; type localities: Japan: Kyushu: Yuyama, Ichiuchi, Konore.
Achard, 1923: 101 (*ruficolle* as var., characters, records)
Miwa & Mitono, 1943: 519 (*ruficolle* as var., characters)
Nakane, 1955a: 55, Fig. 7 (*ruficolle* as ssp., characters, records)
Nakane, 1963b: 78 (characters)
Morimoto, 1985: 254, Pl. 45, Fig. 13a (*ruficolle* as ssp., characters)
Hoshina, 2013: 124 (*ruficolle* as synonym of *E. semirufum*)
Tang, Tu & Li, 2016a: 51, Figs 3, 4, 8–10 (characters)
DISTRIBUTION. Japan: Honshu, Kyushu, Shikoku.

Episcaphium strenuum **Löbl**

Episcaphium strenuum Löbl, 1999: 701, Figs 14–17. Holotype male, MHNG; type locality: China: Yulong Shan, 3000–3500 m, Camaizi Pass, Yunnan.
DISTRIBUTION. China: Sichuan, Yunnan.

Episcaphium unicolor **Löbl**

Episcaphium unicolor Löbl, 1992a: 484. Holotype male, MHNG; type locality: Nepal: Arun Valley, Lamobagar Gola, 1000–1400 m, Sankhuwasabha District.
DISTRIBUTION. Nepal.

Episcaphium watanabei **Löbl**

Episcaphium watanabei Löbl, 2002b: 289, Figs 1–3. Holotype male, ZMUB; type locality: China: Xiaoxiang Ling, side valley above Nanya Cun, 11 km south Shimian, 1250 m, Ya'an Pref. Sichuan, China.
DISTRIBUTION. China: Sichuan; Taiwan.

Episcaphium zhuxiaoyui **Tang, Tu & Li**

Episcaphium zhuxiaoyui Tang, Tu & Li, 2016a: 50, Figs 1, 2, 5–7, 11. Holotype male, SNUC; type locality: China: Yunnan, Gongshan county, Heiwadi, 2000 m.
DISTRIBUTION. China: Yunnan.

Persescaphium Löbl & Ogawa

Persescaphium Löbl & Ogawa, 2016c: 35. Type species *Persescaphium pari* Löbl & Ogawa, 2016; by original designation. Gender: neuter.

Persescaphium pari **Löbl & Ogawa**

Persescaphium pari Löbl & Ogawa, 2016c: 37, 1–5. Holotype female, MHNG; type locality: Iran: Mazāndarān Prov., Najjardeh, ca 36.527°N, 51.692°E.

Distribution. Iran.

Scaphium Kirby

Scaphium Kirby, 1837: 108, Pl. 5, Fig. 1. Type species: *Scaphium castanipes* Kirby, 1837; by monotypy. Gender: neuter.
Erichson, 1845: 6 (characters)
Redtenbacher, 1847: 147 (characters)
Lacordaire, 1854: 239 (characters)
Jacquelin du Val, 1858: 122 (characters)
Gutfleisch, 1859: 222 (characters)
Redtenbacher, 1872: 335 (characters)
LeConte & Horn, 1883: 111 (characters)
Casey, 1893: 511 (characters)
Ganglbauer, 1899: 339 (characters)
Everts, 1903: 445 (characters)
Freude, 1971: 344 (characters)
Iablokoff-Khnzorian, 1985: 136 (characters)
Tamanini, 1969c: Figs 3, 7, 13, 19, 20
Löbl, 1970a: 8 (characters)
Leschen & Löbl, 1995: 451, 468, Figs 16, 21, 23, 24, 26, 34 (characters, checklist)
Löbl, 2012g: 203 (characters)

Scaphium castanipes **Kirby**

Scaphium castanipes Kirby, 1837: 109. Holotype, NHML; type locality: United States: "journey from New York to Cumberland-house".
Casey, 1893: 512 (characters)
Bowditch, 1896: 5 (records)
Hatch, 1957: 281 (characters, records)
Ashe, 1984: 366, Figs 18–34 (larva, natural history, fungal hosts)
Campbell, 1991: 125 (records)
Leschen, 1993: 73 (larval mandible)
Downie & Arnett, 1996: 364 (characters, records)
Newton et al., 2000: 375 (records)

Betz et al., 2003: 203, Fig. 13 (characters: mouth-parts)
Löbl, 2011: 203 (characters)
Webster et al., 2012: 243 (habitat, records)

DISTRIBUTION. Throughout Canada; USA: Alaska, Northern United States.

Scaphium ferrugineum Reitter

Scaphium ferrugineum Reitter, 1880a: 41. Syntypes, MNHN; type locality: "?Cap bon. spei" [= ?South Africa, Cape of Good Hope].
Note. The origin of the species is incertain.

DISTRIBUTION. ?Republic of South Africa.

Scaphium immaculatum (Olivier)

Scaphidium immaculatum Olivier, 1790: 20: 5. Syntypes, MNHN (not traced); type locality: France: near Paris.

Scaphium rufipes Reitter, 1883: 41. Holotype female, MNHN; type locality: Turkey: Kars.
Fabricius, 1801: 576 (*Scaphidium*, characters)
Latreille, 1804: 247 (*Scaphidium*, characters)
Dufschmid, 1825: 71 (*Scaphidium*, characters)
Laporte, 1840: 18 (*Scaphidium*, characters)
Heer, 1841: 372 (*Scaphidium*, characters, records)
Erichson, 1845: 7 (transfer to *Scaphidium*, characters)
Redtenbacher, 1847: 147 (characters)
Fairmaire & Laboulbène, 1855: 343 (*Scaphidium*, characters, records)
Jacquelin du Val, 1858: 122, Pl. 34, Fig. 166 (characters)
Gutfleisch, 1859: 222 (characters)
Redtenbacher, 1872: 335 (characters, records)
Kuthy, 1897: 86 (records)
Ganglbauer, 1899: 340 (characters)
Stierlin, 1900: 490 (records)
Everts, 1903: 445 (characters)
Reitter, 1909: 276, pl. 65 (characters)
Harwood, 1918: 131 (characters, records)
Vitale, 1929: 110 (records)
Roubal, 1930: 297 (records)
Normand, 1934: 45 (records)
Horion, 1949: 253 (records, habitat, fungal hosts)
Benick, 1952: 46 (fungal hosts)
Koch, 1968: 84 (records)
Tamanini, 1969a: 485 (records)

Tamanini, 1969b: 352, Figs 1A-E (characters, records)
Löbl, 1970a: 8, Figs 1–3 (characters, records)
Tamanini, 1970: 7, Figs 1A, 2A-E (characters, records, habitat)
Franz, 1970: 266 (records)
Freude, 1971: 344, Fig. 1: 1 (characters)
Schawaller, 1974; 60 (records)
Löbl, 1974a: 61 (records)
Holzschuh, 1977: 30 (records)
Burakowski et al., 1978: 232 (records, habitat, fungal hosts)
Löbl, 1982g: 47 (records)
Iablokoff-Khnzorian, 1985: 136, Figs 1, 2 (synonymy of *S. rufipes* with *S. immaculatum*, characters, records)
Plaisier, 1986: 75 (records, habitat)
Angelini, 1986: 55 (records)
Merkl, 1987: 113 (records)
Löbl, 1989a: 9 (records)
Finkel et al., 2002: 73 (records)
Bellman et al., 2003: 446 (records)
Nikitsky et al., 2008: 120 (records, habitat)
Lott, 2009: 20 (records)
Delwaide & Thieren, 2010: 5 (records)
Zamotajlov & Nikitsky, 2010: 105 (records)
Samin et al., 2011: 281 (records)
Löbl, 2012g: 203 (characters)
Majzlan, 2016: 108 (records)

Distribution. Albania; Algeria; Austria; Belarus; Belgium; Bosnia-Herzegovina; Cyprus; Czech Republic; France; Hungary; Georgia: Abkhazia; Great Britain; Germany; Greece; Iran; Israel, Italy; Lebanon; Netherlands; Poland; Portugal; Romania; Russia; Slovakia; Spain; Switzerland; Syria; Tunisia; Turkey; Ukraine.

***Scaphium quadraticolle* Solsky**

Scaphium quadraticolle Solsky, 1874: 221. Holotype male, ZMAS; type locality: Uzbekistan, Sarafschan Valley, near Aksay.
Heyden, 1880: 88 (records)
Heyden, 1893: 54 (records)
Heyden, 1896: 39 (records)
Jakobson, 1910: 636 (records)
Iablokoff-Khnzorian, 1985: 137, Fig. 2b (characters, records)
Löbl, 1986c: 343 (records)

Kompantsev & Pototskaya, 1987: 95, 97, Fig. 4 (larva, fungal hosts, natural history, records)
Löbl, 1992a: 482 (records)

DISTRIBUTION. Afghanistan; India: Kashmir; Khazakhstan; Kyrgyzstan; Pakistan; Tajikistan; Turkmenistan; Uzbekistan.

SCAPHIDIINI Latreille

Scaphidiini Latreille, 1806: 3. Type genus: *Scaphidium* Olivier, 1790.
Cerambyciscaphina Pic, 1915b: 30. Type genus: *Cerambyciscapha* Pic, 1915
[Cerambyscaphini Achard, 1915c: 291, misspelling, note]
[Cerambyscaphini Achard, 1924b: 28, misspelling]
Diateliitae Achard, 1924b: 28. Type genus: *Diatelium* Pascoe, 1863
Casey, 1893: 510 (characters)
Achard, 1915c: 291 (synonymy of Cerambyciscaphina with Scaphidiini)
Achard, 1922a: 10 (key to genera)
Ganglbauer, 1899: 339 (characters)
Achard, 1924b: 27 (characters)
Tamanini, 1969c: 129 (characters)
Leschen & Löbl, 1995: 447, 452, 469 (synonymy of Cerambyciscaphini and Diateliitae with Scaphidiini, characters, checklist)
Newton et al., 2000: 375 (characters)
Fierros-López, 2010: 5 (characters)

Cerambyciscapha Pic

Cerambyciscapha Pic, 1915b: 30. Type species: *Cerambyciscapha dohertyi* Pic, 1915; by monotypy. Gender: feminine.
Achard, 1915c: 291 (misspelled *Cerambyscapha*, as synonym with *Heteroscapha* Achard)
Pic, 1925a: 193 (resurrected from synonymy)
Leschen & Löbl, 1995: 455, Figs 20, 25, 39 (characters, relationships)

Cerambyciscapha dohertyi **Pic**

Cerambyciscapha dohertyi Pic, 1915b: 30. Syntypes, MNHN; type locality: Indonesia: Kalimantan, Martapura [Riam Kanan].
Leschen & Löbl, 1995: 455, Figs 20, 25, 39 (characters)

DISTRIBUTION. Indonesia: Kalimantan.

Diatelium Pascoe

Diatelum Pascoe, 1863: 27. Type species: *Diatelium wallacei* Pascoe, 1863; by monotypy. Gender: neuter.
Leschen & Löbl, 1995: 455, Fig. 1 (characters)

Diatelium wallacei wallacei **Pascoe**

Diatelium wallacei Pascoe, 1863: 27, Pl. II, Fig. 2. Syntypes, NHML; type localities: Indonesia: Sumatra, Malaysia: Sarawak.

Apoderus spectrum Vollenhoven, 1865: 159. Lectotype female, NBCL; type locality: Indonesia: Sumatra.

Gestro, 1879b: 59 (synonymy of *Apoderus spectrum* with *Diatelium wallacei*, characters, records)

Achard, 1920d: 123 (characters)

Achard, 1921b: 84 (lectotype fixed by inference for *A. spectrum*, records, habitat)

Leschen & Löbl, 1995: 427, Fig. 1 (characters)

DISTRIBUTION. East Malaysia: Sarawak; Indonesia: Kalimantan, Sumatra.

Diatelium wallacei laterale **Achard**

Diatelum wallacei Pascoe var. *laterale* Achard, 1920d: 123. Syntypes males, NMPC; type locality: Indonesia: Kalimantan, Pontianak.

DISTRIBUTION. Indonesia: Kalimantan.

Euscaphidium Achard

Euscaphidium Achard, 1922a: 12. Type species: *Euscaphidium tuberosum* Achard, 1922; by original designation. Gender: neuter.

Leschen & Löbl, 1995: 453, Fig. 1a (characters)

Euscaphidium tuberosum **Achard**

Euscaphidium tuberosum Achard, 1922a: 13. Syntypes, MCSN; type locality: Borneo.

DISTRIBUTION. "Borneo"; Indonesia: Sumatra.

Scaphidium Olivier

Scaphidium Olivier, 1790: 20: 1. Type species: *Scaphidium quadrimaculatum* Olivier, 1790; fixation by Latreille, 1810. Gender: neuter.

Ascaphidium Pic, 1915a: 24. Type species: *Ascaphidium sikorai* Pic, 1915; by monotypy. Gender: neuter.

Cribroscaphium Pic, 1920e: 93 (subgenus of *Scaphidium*). Type species: *Scaphidium irregulare* Pic, 1920; by monotypy. Gender: neuter.

Falsoascaphidium Pic, 1923b: 16. Type species: *Scaphidium subdepressum* Pic, 1921; by original designation. Gender: neuter.

Hemiscaphium Achard, 1922a: 12. Type species: *Scaphidium striatipenne* Gestro, 1879; by original designation. Gender: neuter.

Hyposcaphidium Achard, 1922a: 12 (subgenus of *Scaphidium*). Type species: *Scaphidium rufopygum* Lewis, 1893, designated by Löbl, 2015e: 21. Gender: neuter.
Isoscaphium Achard, 1922a: 12 (subgenus of *Scaphidium*). Type species: *Scaphidium quadriguttatum* Say, 1823, designated by Löbl, 2015e: 21. Gender: neuter.
Pachyscaphidium Achard, 1922a: 12 (subgenus of *Scaphidium*). Type species: *Scaphidium arrowi* Achard, 1920; by monotypy. Gender: neuter.
Parascaphium Achard, 1923: 97. Type species: *Scaphium optabile* Lewis, 1893; by monotypy. Gender: neuter.
Scaphidiolum Achard, 1922a: 12. Type species: *Scaphidium basale* Laporte, 1840; by original designation. Gender: neuter.
Scaphidopsis Achard, 1922a: 12. Type species: *Scaphidium pardale* Laporte, 1840; by original designation. Gender: feminine.

Fabricius, 1792: 509 (characters)
Herbst, 1793: 131 (characters)
Paykull, 1800: 338 (characters)
Fabricius, 1801: 575 (characters)
Latreille, 1804: 246 (characters)
Latreille, 1810: 427 (fixation of type species)
Leach, 1815: 89 (characters)
Duftschmid, 1825: 71 (characters)
Peletier & Serville, 1825: 344 (characters)
Latreille, 1829: 501 (characters)
Stephens, 1830: 2 (characters)
Westwood, 1838: 11 (characters)
Laporte, 1840: 18 (characters)
Heer, 1841: 372 (characters)
Erichson, 1845: 4 (characters)
Redtenbacher, 1847: 147 (characters)
Lacordaire, 1854: 238 (characters)
Jacquelin du Val, 1858: 121, Pl. 34, Fig. 166 (characters)
Gutfleisch, 1859: 222 (characters); Gyllenhal, 1808: 186 (characters)
Thomson, 1862: 126 (characters)
Redtenbacher, 1872: 335 (characters)
Reitter, 1880a: 36 (key to species)
Matthews, 1888: 159, 189 (characters, key to Central American and Mexican species)
Casey, 1893: 513 (characters, key to North American species)
Casey, 1900: 55 (characters, key to North American species)

Everts, 1903: 445 (characters)
Achard, 1920c: 48 (key to West Malaysian species)
Achard, 1920f: 209 (key to "Indochina" and Yunnan species)
Achard, 1922a: 12 (characters)
Achard, 1922b: 30 (characters)
Miwa & Mitono, 1943: 520 (characters, key to Taiwanese species)
Paulian, 1951: 196 (characters, key to African *Scaphidiolum*))
Shirôzu & Nomura, 1963: 56 (key to Japanese species)
Kasule, 1966: 267 (larvae)
Löbl, 1968c: 386 (synonymy of *Parascaphium* with *Scaphidium*, key to Palaearctic species with entirely black pronotum and elytra)
Kasule, 1968: 117 (larvae)
Tamanini, 1969c: Figs 1, 4, 8–12, 14, 18, 21 (characters)
Löbl, 1970a: 9 (characters)
Freude, 1971: 344, Figs 2: 1 (characters)
Löbl, 1971c: 940 (key to Sri Lankan species)
Ganglbauer, 1899: 340 (characters)
Löbl, 1975a: 370 (key to New Guinean species)
Löbl, 1977e: 287 (key to Australian species)
Löbl, 1978c: 113 (key to New Guinean species)
Löbl, 1979a: 80 (key to South Indian species)
Iablokoff-Khnzorian, 1985: 137 (characters)
Hayashi, 1986: pl. 17 (larva)
Löbl, 1992a: 485 (key to Himalayan species)
Leschen, 1993: 73 (larval mandible)
Leschen, 1994: 3 (larvae, behavior)
Newton, 1984: 319 (fungal hosts)
Leschen & Löbl, 1995: 453, Figs 3, 5, 8, 10–12, 17–19, 27, 30, 32 (synonymy of *Ascaphidium*, *Cribroscaphium*, *Falsoascaphidium*, *Hemiscaphium*, *Hyposcaphidium*, *Isoscaphium*, *Pachyscaphidium*, *Scaphidiolum*, and *Scaphidopsis* with *Scaphidium*, characters, checklist)
Navarrete-Heredia, 1991: 126 (fungal hosts)
Löbl, 1999: 705 (key to Chinese species)
Hoshina & Maruyama, 1999: 480 (key to Ryukyu species)
Hoshina, 2001: 105 (key to Ryukyu species)
Fierros-López, 2005: 12, 14, Figs 1, 5, 8–12 (characters, key to Central American and Mexican species)
Löbl, 2006: 25 (key to Philippine species)
He et al., 2008: 56 (key to Tienmushan species)
Tang & Li, 2010a: 67 (key to *S. grande* group of China)

Löbl, 2012g: 202 (characters)
Tang et al., 2014: 82 (key to East Chinese species)
Ogawa, 2015: 49 (characters, key, Sulawesi)
Tang et al., 2016b: 281 (modified key Chinese *S. grande* group)
Tu & Tang, 2017: 596 (key to East Chinese species)

Scaphidium abdominale **Achard**

Scaphidium abdominale Achard, 1920i: 239. Syntypes, NMPC; type locality: Myanmar: Pegu, Palon.
DISTRIBUTION. Myanmar.

Scaphidium abyssinicum **(Pic)**

Scaphidiolum abyssinicum Pic, 1954a: 11. Syntypes, MNHN; type locality: Ethiopia.
Leschen & Löbl, 1995: 469 (transfer to *Scaphidium*)
DISTRIBUTION. Ethiopia.

Scaphidium ahrensi **Tu & Tang**

Scaphidium ahrensi Tu & Tang, 2017: 593, Figs 1–6. Holotype male, ZFMK; type locality: China: Fujian, Kuatun, 2300 m.
DISTRIBUTION. China: Fujian.

Scaphidium alpicola **Blackburn**

Scaphidium alpicola Blackburn, 1891: 90. Lectotype female, NHML; type locality: Australia: Alpin District, Victoria.
Löbl, 1976b: 291, Fig. 9 (as *S. alpicolum*, lectotype fixed by inference, characters)
Löbl, 1976e (records)
DISTRIBUTION. Australia: New South Wales, Tasmania, Victoria.

Scaphidium alternans **Löbl**

Scaphidium alternans Löbl, 1978c: 115, Figs 2, 3. Holotype male, BPBM; type locality: Indonesia: Western New Guinea, Bodem, 11 km SE of Oerberfaren.
DISTRIBUTION. Indonesia: Western New Guinea; Papua New Guinea.

Scaphidium amurense **Solsky**

Scaphidium amurense Solsky, 1871: 350. Syntypes, ZMAS; type locality: Russia: "Sibérie orientale".
[*Scaphidium amurense* ab. *bodemeyeri* Reitter, 1913: 122, unavailable]
[*Scaphidium amurense* var. *bodemeyeri*; Achard, 1924c: 148, unavailable]

Scaphidium bodemeyeri Löbl, 1965c: 1, Fig. 2. Holotype male, HNHM; type locality: Russia: Amur Region, Sotka Gora.
Scaphidium tsushimense Shirôzu & Morimoto, 1963: 74. Syntypes, unknown; type locality: Japan: Sasuna-Shushi, Tsushima.
Heyden, 1893: 54 (records)
Jakobson, 1910: 636 (records)
Shirôzu & Morimoto, 1963: 75 (records)
Löbl, 1965c: Fig. 1 (characters)
Löbl, 1968b: 419 (records)
Löbl, 1968c: 389 (synonymy *S. bodemeyeri* with *S. amurense*)
Tamanini, 1969b: 355, Figs 2A-C (characters, records)
Morimoto, 1985: Pl. 45, Fig. 22 (as *S. tsushimense*, characters)
Iablokoff-Khnzorian, 1985: 137, Fig. 2a (characters, records)
Kompantsev & Pototskaya, 1987: 88, 91 (larva, natural history, fungal hosts)
Löbl, 1999: 708 (synonymy of *S. tsushimense* with *S. amurense*)
Hwang & Ahn, 2001: 371, Figs 1, 4, 5 (characters, records)
Li, 2015: 38, Fig. 1 (records)
DISTRIBUTION. China: "Manchuria"; Heilongjiang; Japan: Tsushima; North and South Korea; Russia: Amur Region.

Scaphidium andrewesi (Achard)
Scaphidiolum andrewesi Achard, 1922c: 38. Lectotype male, NHML; type localities: India: Mysore, Kanara.
Scaphidiolum andrewesi var. *femoratum* Achard, 1922c: 38. Lectotype female, NHML; type locality: India: Tamil Nadu, Nilgiri Hills.
Löbl, 1979a: 81 (lectotype designation, transfer to *Scaphidium*, synonymy of *S. femoratum* with *S. andrewesi*, records)
DISTRIBUTION. India: Karnataka (Mysore), Tamil Nadu.

Scaphidium angolense Pic
Scaphidium angolense Pic, 1940b: 359. Syntypes, MNHN; type locality: Angola.
DISTRIBUTION. Angola.

Scaphidium angustatum Pic
Scaphidium angustatum Pic, 1920b: 4. Syntypes, MNHN/NMPC; type locality: Madagascar [Antongil Bay].
Note. Pic (l.c.) credited the species epithet to Portevin.
DISTRIBUTION. Madagascar.

Scaphidium antennatum Reitter

Scaphidium antennatum Reitter, 1880a: 37. Syntypes, MNHN; type locality: USA: Texas.

DISTRIBUTION. United States: Texas.

Scaphidium anthrax Achard

Scaphidium anthrax Achard, 1920b: 127. Syntypes, MNHN; type localities: Indonesia: Java, Pasoeroean: Boelve-Lavang [Pasuruan].

DISTRIBUTION. Indonesia: Java.

Scaphidium apicicorne (Pic)

Scaphidiolum apicicorne Pic, 1953: 271. Holotype, ?MNHN, not traced; type locality: Madagascar: Mont Tsaratanana.

Leschen & Löbl, 1995: 469 (transfer to *Scaphidium*)

DISTRIBUTION. Madagascar.

Scaphidium arrowi Achard

Scaphidium arrowi Achard, 1920j: 264. Lectotype male, NMPC; type locality: India: Darjeeling District, Pedong [Bhotan/British Bootan, Padong].

Löbl, 1992a: 492 (lectotype designation, characters, records)

Löbl, 2005: 178 (records)

DISTRIBUTION. India: "Assam", West Bengal: Darjeeling District; Nepal.

Scaphidium ashei Fierroz-López

Scaphidium ashei Fierroz-López, 2005: 21, Figs 16, 17, 109a-c. Holotype male, SEMC; type locality: Nicaragua: 6 km north Matagalpa, Hotel Selva Negra, 1240 m.

DISTRIBUTION. Costa Rica; Nicaragua.

Scaphidium aterrimum Reitter

Scaphidium aterrimum Reitter, 1879: 41. Lectotype female, in NBCL or NMPC; type locality: Indonesia: Sumatra, Rawas District.

Scaphidium gracile Achard, 1920b: 127. Syntypes, MNHN/NMPC; type locality: Indonesia: Sumatra, Palembang. Synonymy: Achard, 1921b: 85.

Achard, 1921b: 85 (synonymy of *S. gracile* with *S. aterrimum*, records)

Achard, 1922d: 262, 263 (lectotype designation for *S. aterrimum*, characters, records)

DISTRIBUTION. India: Andaman, Assam; Indonesia: "Borneo occidental", Sumatra.

***Scaphidium atricolor* Pic**

Scaphidium atricolor Pic, 1915e: 2. Syntypes, MNHN; type locality: Malaysia: Sabah, Kinabalu.
Achard, 1920d: 126 (characters, records)
Achard, 1922d: 262 (characters, records)
DISTRIBUTION. Brunei; East Malaysia: Sabah, Sarawak; Indonesia: Kalimantan.

***Scaphidium* atripenne Gestro**

Scaphidium atripenne Gestro, 1879a: 561. Syntypes, MCSN/MHNG; type locality: Australia: Somerset, Cap York.
Löbl, 1976e: 288 (characters, records)
Löbl, 1978c: 114, Fig. 1 (characters, records)
DISTRIBUTION. Australia: Queensland; Papua New Guinea.

***Scaphidium atripes* Pic**

Scaphidium atripes Pic, 1946: 82. Syntypes, MNHN; type locality: Kenya: Kapenguria [Turkana, 2300 m].
DISTRIBUTION. Kenya.

***Scaphidium atrosuturale* Pic**

Scaphidium atrosuturale Pic, 1915d: 40. Syntypes, MNHN; type locality: Indonesia: Kalimantan [Riam Kanan].
Pic, 1920g: 189 (characters)
DISTRIBUTION. Indonesia: Kalimantan.

***Scaphidium atrum* Matthews**

Scaphidium atrum Matthews, 1888: 164, Pl. 4, Figs 7, 8. Lectotype male, NHML; type locality: Mexico: Córdoba.
Navarrete-Heredia, 1991: 126 (fungal hosts)
Fierroz-López, 2005: 24, Figs 18, 19, 109c, d (lectotype designation, characters)
Márquez, 2006: 182 (records)
DISTRIBUTION. Guatemala; Honduras; Mexico.

***Scaphidium baconi baconi* Pic**

Scaphidium baconi Pic, 1915f: 43. Lectotype female, MNHN; type locality: "Indes boréales".
Scaphidium assamense Pic, 1915f: 43. Lectotype male, MNHN; type locality: India: Assam [Margherita].
Achard, 1922d: 263 (characters, records)

Löbl, 1990b: 511 (*S. assamense* in synonymy with *S. baconi*, records)
Löbl, 1992a: 496, Figs 23, 47, 81 (lectotype designation for *S. baconi* and *S. assamense*, characters, records)
DISTRIBUTION. India: Assam, Meghalaya, "North India"; Nepal; Thailand, Vietnam.

Scaphidium baconi multimaculatum Pic
Scaphidium assamense var. *multimaculatum* Pic, 1915f: 43. Syntypes, MNHN; type locality: India: Assam.
DISTRIBUTION. India: Assam.

Scaphidium baconi semifasciatum Pic
Scaphidium assamense var. *semifasciatum* Pic, 1915f: 43. Syntypes, MNHN; type locality: India: Assam.
DISTRIBUTION. India: Assam.

Scaphidium baconi uniplagatum Achard
Scaphidium baconi var. *uniplagatum* Achard, 1922d: 263. Syntypes, NMPC; type locality: India: Assam, Patkai Mts.
DISTRIBUTION. India: Assam.

Scaphidium badium Heller
Scaphidium badium Heller, 1917: 43. Lectotype female, SMTD; type locality: Philippines: Luzon, Mt. Makiling.
Löbl, 2006: 27, Figs 1, 2 (lectotype designation, characters)
DISTRIBUTION. Philippines: Luzon.

Scaphidium baezi Fierroz-López
Scaphidium baezi Fierroz-López, 2005: 28, Figs 20, 21, 109e-f. Holotype male, FMNH; type locality: Mexico: La Huerta, km 5.7 from Rte. 200, 40 m, El Tecuán, Jalisco.
DISTRIBUTION. Mexico.

Scaphidium basale Laporte
Scaphidium basale Laporte, 1840: 19. Syntypes, MNHN; type locality: Madagascar.
Achard, 1922a: 12 (transfer to *Scaphidiolum*)
Leschen & Löbl, 1995: 469 (transfer to *Scaphidium*)
DISTRIBUTION. Madagascar.

Scaphidium basilewskyi **(Pic)**
Scaphidiolum basilewskyi Pic, 1955: 49. Holotype, MRAC; type locality: Rwanda: Rubengera, 1900 m, terr. Kibuye.
Leschen & Löbl, 1995: 469 (transfer to *Scaphidium*)
DISTRIBUTION. Rwanda.

Scaphidium bayibini **Tang, Li & He**
Scaphidium bayibini Tang, Li & He, 2014: 72, Figs 42, 43, 114–117. Holotype male, SNUC; type locality: China: Yuexi County, Yaoluoping N.R., Ximianzi Vil., Anhui, 1050 m.
DISTRIBUTION. China: Anhui.

Scaphidium becvari **Löbl**
Scaphidium becvari Löbl, 1999: 715, Figs 29, 41, 42. Holotype male, MHNG; type locality: China: Habashan, southeastern slope, 2500–3800 m, Yunnan.
DISTRIBUTION. China: Sichuan, Yunnan.

Scaphidium benitense **(Achard)**
Scaphidiolum benitense Achard, 1922e: 489. Holotype female, MNHN; type locality: Democratic Republic of the Congo: Benito.
Paulian, 1951: 199 (*Scaphidiolum*, characters, records)
Leschen & Löbl, 1995: 469 (transfer to *Scaphidium*)
DISTRIBUTION. Democratic Republic of the Congo; Gabon.

Scaphidium bicinctum **Achard**
Scaphidium bicinctum Achard, 1920j: 264. Lectotype female, NHML; type locality: India: Nilgiri Hills, Tamil Nadu.
Scaphidium anamalaiense Löbl, 1971b: 1, Figs 1, 1a. Holotype male, ZSMC; type locality: India: Cinchona, 3500 ft. Anamalai Hills
Scaphidium nathani Löbl, 1971b: 2, Figs 2, 2a. Holotype male, ZSMC; type locality: India: Cinchona, 3500 ft, Anamalai Hills
Löbl, 1979a: 82 (lectotype designation for *S. bicinctum*, synonymy of *S. anamalaiense* and *S. nathani* with S. *bicinctum*, characters, records)
DISTRIBUTION. India: Kerala, Tamil Nadu.

Scaphidium bicolor **Laporte**
Scaphidium bicolor Laporte, 1840: 19. Syntypes, MNHN; type locality: Madagascar.
Scaphidium unicolor Laporte, 1840: 19. Syntypes, MNHN; type locality: Madagascar.
Reitter, 1880a: 38 (synonymy of *S. unicolor* with *S. bicolor*)
DISTRIBUTION. Madagascar.

Scaphidium bifasciatum Pic

Scaphidium bifasciatum Pic, 1915d: 40. Syntypes, MNHN; type locality: Malaysia: Perak, Malacca.

Achard, 1920c: 52 (characters)

DISTRIBUTION. West Malaysia.

Scaphidium bilineatithorax (Pic)

Scaphidiolum bilineatithorax Pic, 1931a: 1. Syntypes, MNHN; type locality: Vietnam: Tonkin, Hoa Binh.

Leschen & Löbl, 1995: 460 (transfer to *Scaphidium*)

DISTRIBUTION. Vietnam.

Scaphidium binhanum (Pic)

Scaphidiolum binhanum Pic, 1926a: 322. Syntypes, MNHN; type locality: Vietnam: Tonkin, Hoa Binh.

Leschen & Löbl, 1995: 469 (transfer to *Scaphidium*)

DISTRIBUTION. Vietnam.

Scaphidium binigronotatum (Pic)

Scaphidiolum binigronotatum Pic, 1931a: 1. Syntypes, MNHN; type locality: Congo.

Paulian, 1951: 200 (*Scaphidiolum*, characters)

Pic, 1955: 49 (records)

Leschen & Löbl, 1995: 469 (transfer to *Scaphidium*)

DISTRIBUTION. Burundi; "Congo".

Scaphidium binominatum Achard

Scaphidium notaticolle Pic, 1915d: 40. Syntypes, MNHN; type locality: Borneo.

Scaphidium binominatum Achard, 1915c: 292; replacement name for *Scaphidium notaticolle* Pic, 1915d (nec *Scaphidium notaticolle* Pic, 1915b).

Achard, 1922d: 262 (characters)

DISTRIBUTION. "Borneo".

Scaphidium bipunctatum Redtenbacher

Scaphidium bipunctatum Redtenbacher, 1868: 31. Syntypes, NHMW; type locality: Brazil: Rio de Janeiro.

DISTRIBUTION. Brazil.

Scaphidium bisbimaculatum Pic

Scaphidium bisbimaculatum Pic, 1917c: 3. Syntypes, MNHN; type locality: Brazil [Esperito Santo].

DISTRIBUTION. Brazil.

***Scaphidium biseriatum* Champion**

Scaphidium biseriatum Champion, 1927: 268. Holotype male, NHML; type locality: India: Sunderdhung Valley, 8.000–12.000 ft, W. Almora.
Löbl, 1992a: 489, Figs 51, 58, 76 (characters, records)
DISTRIBUTION. India: Uttarakhand, West Bengal (Darjeeling District); Nepal.

***Scaphidium bituberculatum* Tang & Li**

Scaphidium bituberculatum Tang & Li, 2012: 190, Figs 5, 6, 13–16. Holotype male, HBUM; type locality: China: Hainan Prov., Lingshui county, Diaolushan National Reserve.
DISTRIBUTION. China: Hainan.

***Scaphidium biundulatum* Champion**

Scaphidium biundulatum Champion, 1927: 269. Holotype male, NHML; type locality: India: Darjeeling District, Tonglu, 10.074 feet.
Löbl, 1992a: 492, Figs 10, 48, 68 (characters, records)
DISTRIBUTION. India: Sikkim, West Bengal (Darjeeling District); Nepal.

***Scaphidium biwenxuani* He, Tang & Li**

Scaphidium biwenxuani He, Tang & Li, 2008b: 178, Figs 1, 4–6. Holotype male, SNUC; type locality: China: Longwangshan, Anji County, Zhejiang, 950–1200 m.
Tang & Li, 2013: 180 (records)
Tang et al., 2014: 68, Figs 37–39, 106–109, 156–162 (characters, records, habitat, immature stage)
DISTRIBUTION. China: Anhui, Guangxi, Guizhou, Hubei, Hunan, Jiangxi, Sichuan, Yunnan, Zhejiang.

***Scaphidium bolivianum* (Pic)**

Scaphidiolum bolivianum Pic, 1931a: 1. Syntypes, MNHN; type locality: Bolivia [Coroico].
Leschen & Löbl, 1995: 470 (transfer to *Scaphidium*)
DISTRIBUTION. Bolivia.

***Scaphidium borneense* Pic**

Scaphidium borneense Pic, 1915d: 40. Syntypes, MNHN; type locality: Indonesia: Kalimantan [Riam Kanan, Martapura].
Achard, 1922b: 35 (transfer to *Hemiscaphium*, note)
Leschen & Löbl, 1995: 470 (transfer to *Scaphidium*)
DISTRIBUTION. Indonesia: Kalimantan.

Scaphidium brendelli Fierroz-López

Scaphidium brendelli Fierroz-López, 2005: 31, Figs 22, 23, 109g, h. Holotype male, INBIO; type locality: Costa Rica: Est. Pitilla, 700 m, 9 km south Sta. Cecilia, Guanacaste.

DISTRIBUTION. Costa Rica; Nicaragua.

Scaphidium brunneopictum (Achard)

Hemiscaphium brunneopictum Achard, 1922b: 35. Holotype female, NMPC; type locality: Myanmar: Carin [?=Carin Chebà, 900–1100 m, approx. 19°13'N 96°35'E].

Löbl, 1992a: 498, Figs 24, 46, 67, 84 (characters, records, invalid lectotype designation)

Leschen & Löbl, 1995: 470 (transfer to *Scaphidium*)

DISTRIBUTION. India: "Ober Assam", Meghalaya, West Bengal (Darjeeling District); Myanmar; Nepal.

Scaphidium brunneum Hoshina & Morimoto

Scaphidium brunneum Hoshina & Morimoto, 1999: 87, Figs 1, 8, 11, 12, 17–19, 26. Holotype male, KUIC; type locality: Japan: Okinawa, Mt. Yonahadake, Kunigami Village.

Hoshina, 2001: 100, Figs 3, 10 (characters)

DISTRIBUTION. Japan: Okinawa.

Scaphidium carinense Achard

Scaphidium carinense Achard, 1920i: 239. Lectotype female, NMPC; type locality: Myanmar: Carin-Cheba [approx. 19°13'N, 96°25'E].

Pic, 1921a: 160 (characters, records)

Tang et Li, 2013: 174, 177, Figs 5–8 (characters, records)

Tang et al., 2014: 61, Figs 26, 27, 94–97, 153 (lectotype designation, characters, records, immature stage)

Tu & Tang, 2017: 595 (records)

DISTRIBUTION. China: Fujian, Guangxi, Hainan, Hubei, Sichuan, Yunnan; Myanmar.

Scaphidium castaneum Perty

Scaphidium castaneum Perty, 1830: 34, Pl. 7, Fig. 10. Syntypes, ZSMC; type locality: Brazil.

DISTRIBUTION. Brazil.

Scaphidium celebense **Pic**

Scaphidium celebense Pic, 1915e: 3. Syntypes, MNHN; type locality: Indonesia: Sulawesi, Tjamba.
Achard, 1920d: 126 (characters, records)
Ogawa, 2015: 52, Figs 4-2e, 4-3a, 4-4a, 4-5d, 4-8a, 4-10b (characters, records)
DISTRIBUTION. Indonesia: Sulawesi.

Scaphidium cerasinum **Oberthür**

Scaphidium cerasinum Oberthür, 1883: 11. Syntypes, MNHN; type localities: Brazil: Tomantins, Santo-Paulo-d'Olivença and Amazonas.
DISTRIBUTION. Brazil.

Scaphidium chapuisii **Gestro**

Scaphidium chapuisii Gestro, 1879b: 57. Syntypes, MCSN; type locality: Indonesia: Sumatra, Ajer Mantcior [near Padangpanjang].
Achard, 1920c: 52 (characters)
DISTRIBUTION. Indonesia: Sumatra.

Scaphidium cheesmanae **Löbl**

Scaphidium cheesmanae Löbl, 1975a: 371, Figs 1, 3. Holotype male, NHML; type locality: Indonesia: Western New Guinea, Mt. Lima, 3500 ft, Cyclops Mts.
DISTRIBUTION. Indonesia: Western New Guinea.

Scaphidium chinense **Li**

Scaphidium chinensis Li, 1992: 62. Type material unknown; type locality: China: Heilongjiang.
Li, 2015: 37, Fig. 1 (characters, as subspecies of *S. montivagum* Shirôzu & Morimoto)
DISTRIBUTION. China: Heilongjiang.

Scaphidium chujoi **Löbl**

Scaphidium (*Isoscaphidium*) *chujoi* Löbl, 1967c: 129, 2 Figs. Holotype male, KUIC; type locality: Japan: Mt. Tsurugi-san, Tokushima Pref.
Löbl, 1968c: 389, Fig. 5 (characters)
Morimoto, 1985: Pl. 45, Fig. 15 (characters)
DISTRIBUTION. Japan: Kyushu, Shikoku.

Scaphidium cinnamomeum Champion

Scaphidium cinnamomeum Champion, 1927: 270. Holotype female, NHML; type locality: India: W. Almora Division, Kumaon.
Löbl, 1986c: 343 (records)
Löbl, 1992a: 486, Figs 45, 63, 83 (characters, records)
DISTRIBUTION. India: Himachal Pradesh, Uttarakhand (Kumaon).

Scaphidium clathratum Achard

Scaphidium clathratum Achard, 1920j: 264. Syntypes, NMPC; type locality: India: Assam.
DISTRIBUTION. India: "Assam".

Scaphidium coerulans Löbl

Scaphidium coerulans Löbl, 1978c: 118, Fig. 5. Holotype male, BPBM; type locality: Indonesia: Vogelkopf, Lake Anggi Giji, 2000–2100 m, Western New Guinea.
DISTRIBUTION. Indonesia: Western New Guinea; Papua New Guinea.

Scaphidium collarti (Pic)

Scaphidiolum collarti Pic, 1928b: 35. Syntypes, MRAC/MNHN; type locality: Democratic Republic of the Congo: Tibo.
Leschen & Löbl, 1995: 453 (implicit transfer/synonymy of *Scaphidiolum*)
DISTRIBUTION. Democratic Republic of the Congo.

Scaphidium comes Löbl

Scaphidium comes Löbl, 1968c: 388, Figs 1, 2, 6, 7, 11. Holotype male, NHMB; type locality: North Korea: Pu Ryong [Purjong].
He et al., 2008b: 180, Fig. 2, 9 (characters, records: possibly based on misindentification)
He et al., 2008c: 59 (characters)
Tang et al., 2014: 51, Figs 5. 6, 62–65, 142–144 (characters, records, habitat)
DISTRIBUTION. China: Guangxi, Hainan, Hunan, Hubei, Zhejiang; North Korea.

Scaphidium compressum Achard

Scaphidium compressum Achard, 1915a: 556. Holotype female, NMPC; type locality: India: West Bengal, Barway.
Note. According to Achard (l.c.) the "type" repository should be "Musée de Bruxelles".
DISTRIBUTION. India: West Bengal.

Scaphidium confusum **(Pic)**

Scaphidiolum confusum Pic, 1926d: 46. Syntypes, MNHN; type locality: Vietnam: Tonkin, Hoa Binh [Lac Tho].
Leschen & Löbl, 1995: 470 (transfer to *Scaphidium*)
DISTRIBUTION. Vietnam.

Scaphidium conjunctum **Motschulsky**

Scaphidium conjunctum Motschulsky, 1860: 95. Syntypes, ?ZMUM, not traced; type locality: "continent Indien".
Löbl, 1997: 18 (misspelled *conjuctum*)
DISTRIBUTION. "continent Indien".

Scaphidium connexum **Tang, Li & He**

Scaphidium connexum Tang, Li & He, 2014: 78. Holotype male, SNUC; type locality: China: Zhejiang, Kaihua County, Gutianshan, 800 m.
DISTRIBUTION. China: Fujian, Guangxi, Zhejiang.

Scaphidium consimile **Achard**

Scaphidium consimile Achard, 1920c: 55. Syntypes, MNHN; type locality: Thailand: Lakhon, Isthmus of Kra.
DISTRIBUTION. Thailand.

Scaphidium consobrinum **Laporte**

Scaphidium consobrinum Laporte, 1840: 19. Syntypes, MNHN; type locality: Madagascar [Superbievelle].
DISTRIBUTION. Madagascar.

Scaphidium coomani **(Pic)**

Scaphidiolum coomani Pic, 1926a: 323. Lectotype female, MNHN; type locality: Vietnam: Tonkin, Hoa Binh [Lac Tho].
Löbl, 1992a: 495, Figs 20, 43, 64, 75 (lectotype designation, transfer to *Scaphidium*, characters, records)
Tang & Li, 2013: 178, Figs 9–10 (characters, records)
DISTRIBUTION. China: Yunnan; India: Meghalaya; Nepal; Vietnam.

Scaphidium costaricense **Fierroz-López**

Scaphidium costaricense Fierroz-López, 2005: 33, Figs 24, 25, 109i, j. Holotype female, INBIO; type locality: Costa Rica: Reserva Forestal, Los Santos, 1800 m, San José.
DISTRIBUTION. Costa Rica.

Scaphidium crassipes Löbl

Scaphidium crassipes Löbl, 2006: 28, Figs 3, 4. Holotype male, MHNG; type locality: Philippines: Todaya, Mindanao.

DISTRIBUTION. Philippines: Mindanao.

Scaphidium crypticum Tang, Li & He

Scaphidium crypticum Tang, Li & He, 2014: 56, Figs 13–16, 78–81, 151, 152. Holotype male, SNUC; type locality: China: Zhejiang, Longquan City, Fengyangshan, 1100 m.

DISTRIBUTION. China: Fujian, Guangxi, Jiangxi, Zhejiang.

Scaphidium cyanellum Oberthür

Scaphidium cyanellum Oberthür, 1883: 5. Holotype female, MNHN; type locality: "Inde boréale".

Champion, 1927: 271 (characters, records)

Löbl, 1992a: 488, Figs 42, 74 (characters, records)

DISTRIBUTION. India: Meghalaya, "North India"; Nepal.

Scaphidium cyanipenne Gestro

Scaphidium cyanipenne Gestro, 1879a: 559. Lectotype male, MCSN; type locality: Papua New Guinea: Fly River.

Achard, 1920d: 126 (characters, records)

Löbl, 1971a: 247 (characters, records)

Löbl, 1975a: 370 (lectotype fixed by inference, records)

DISTRIBUTION. New Britain, Papua New Guinea.

Scaphidium dalatense (Pic)

Scaphidiolum dalatense Pic, 1928c: 2. Syntypes, MNHN; type locality: Vietnam: Annam.

DISTRIBUTION. Vietnam.

Scaphidium decorsei Achard

Scaphidium decorsei Achard, 1920i: 239. Syntypes, MNHN; type locality: Madagascar: Ambowombé.

DISTRIBUTION. Madagascar.

Scaphidium delatouchei Achard

Scaphidium delatouchei Achard, 1920h: 210. Lectotype male, MNHN; type locality: China: Kouang-Toung [Guandong].

Löbl, 1999: 708, Fig. 26 (lectotype designation, characters, records)
Tang et al., 2014: 66, Figs 34–36, 102–105, 155 (characters, records)
DISTRIBUTION. China: Anhui, Hubei, Hunan, Guandong, Guangxi, Sichuan, Yunnan, Zhejiang.

***Scaphidium direptum* Tang & Li**
Scaphidium direptum Tang & Li, 2010b: 318, Figs 1, 4, 7–10. Holotype male, SNUC; type locality: China: Chebaling N.R., Shixing County, 365–500 m, Guandong.
Tang & Li, 2012: 188, Figs 7, 8 (characters)
Tang et al., 2014: 78, Figs 54, 55, 134–137 (characters, records)
DISTRIBUTION. China: Fujian, Guandong, Guangxi.

***Scaphidium discerptum* (Achard)**
Hemiscaphium discerptum Achard, 1922b: 33. Syntypes, NMPC/MCSN; type locality: Indonesia: Sumatra, Si Rambé [near Balige].
Leschen & Löbl, 1995: 470 (transfer to *Scaphidium*)
DISTRIBUTION. Indonesia: Sumatra.

***Scaphidium disclusum* (Achard)**
Scaphidiolum disclusum Achard, 1924a: 150. Syntypes, NHML; type locality: Indonesia: Java.
Leschen & Löbl, 1995: 470 (transfer to *Scaphidium*)
DISTRIBUTION. Indonesia: Java.

***Scaphidium discoidale* Pic**
Scaphidium discoidale Pic, 1920f: 23. Syntypes, MNHN; type locality: Madagascar.
DISTRIBUTION. Madagascar.

***Scaphidium discomaculatum* (Pic)**
Scaphidiolum discomaculatum Pic, 1954b: 33. Syntypes, MRAC/MNHN; type locality: Democratic Republic of the Congo: Lulua, Kapanga.
Leschen & Löbl, 1995: 470 (transfer to *Scaphidium*)
DISTRIBUTION. Democratic Republic of the Congo.

***Scaphidium disconotatum* Pic**
Scaphidium disconotatum Pic, 1915c: 36. Syntypes, MNHN; type locality: Borneo.
Achard, 1921b: 85 (records)
DISTRIBUTION. "Borneo"; Indonesia: Java.

Scaphidium distinctum Achard

Scaphidium distinctum Achard, 1916: 89. Lectotype male, NMPC; type locality: Autralia, New South Wales.

Löbl, 1976e: 289, Fig. 7 (lectotype designation, characters, records)

DISTRIBUTION. Australia: New South Wales, Queensland.

Scaphidium dohertyi Pic

Scaphidium dohertyi Pic, 1915f: 43. Syntypes, MNHN; type locality: Malaysia: Malacca.

Achard, 1920c: 51 (characters)

Achard, 1922b: 33 (transfer to *Hemiscaphium*)

Leschen & Löbl, 1995: 470 (transfer to *Scaphidium*)

DISTRIBUTION. West Malaysia.

Scaphidium donckieri Pic

Scaphidium donckieri Pic, 1917b: 3. Syntypes, MNHN; type locality: Madagascar: Fort Carnot.

DISTRIBUTION. Madagascar.

Scaphidium dureli (Achard)

Scaphidiolum dureli Achard, 1922c: 37. Syntypes males, MNHN/MHNG; type locality: India: West Bengal, "Bootan britanique: Palong" [Darjeeling District, Pedong].

Löbl, 1992a: 489 (transfer to *Scaphidium*, characters)

He et al., 2009: 483, Figs 2, 4, 6, 8, 11, 12 (characters, records)

DISTRIBUTION. China: Xizang; India: West Bengal.

Scaphidium egregium Achard

Scaphidium atripenne Pic, 1922a: 1. Syntypes, MNHN; type locality: Vietnam: Tonkin.

Scaphidium egregium Achard, 1922d: 261; replacement name for *Scaphidium atripenne* Pic, 1922 (nec *Scaphidium atripenne* Gestro, 1879).

Achard, 1922d: 261 (characters, records)

Rougemont, 1996: 12 (records)

Löbl, 1999: 708 (records)

Tang & Li, 2013: 180 (records)

DISTRIBUTION. China: Hainan, Hongkong; Singapore; Vietnam.

Scaphidium elisabethae (Pic)

Scaphidiolum elisabethae Pic, 1954b: 35. Holotype, MRAC; type locality: Democratic Republic of the Congo: Elisabethville [Lubumbashi].
Leschen & Löbl, 1995: 470 (transfer to *Scaphidium*)
DISTRIBUTION. Democratic Republic of the Congo.

Scaphidium ellenbergeri (Paulian)

Scaphidiolum ellenbergeri Paulian, 1951: 197. Holotype, MNHN; type locality: Gabon: Ogoué, Lambaréné.
Leschen & Löbl, 1995: 470 (transfer to *Scaphidium*)
DISTRIBUTION. Gabon.

Scaphidium elongatum Achard

Scaphidium elongatum Achard, 1915a: 558. Holotype, NMPC; type locality: India: Tamil Nadu, Chambaganor [Shembaganur].
Löbl, 1979a: 81 (records)
DISTRIBUTION. India: Tamil Nadu.

Scaphidium emarginatum Lewis

Scaphidium emarginatum Lewis, 1893: 291. Syntypes, NHML; type localities: Japan: Mountains in Kiushiu, at Chiuzenji and on Ontaki-san.
Achard, 1923: 103 (characters, records)
Miwa & Mitono, 1943: 522 (characters)
Nakane, 1955b: 53, Figs 1, 19, 20 (characters)
Hayashi et al., 1959: 428 (immature stages)
Nakane, 1963b: 79 (characters)
Shirôzu & Morimoto, 1963: 65 (characters, records)
Morimoto, 1985: Pl. 45, Fig. 24 (characters)
DISTRIBUTION. Japan: Hokkaido, Honshu, Kyushu, Shikoku.

Scaphidium exclamans Oberthür

Scaphidium exclamans Oberthür, 1883: 9. Syntypes, MNHN; type locality: Brazil: "Saint-Paul" [Sao Paulo].
Fierros-López, 2006b: 40, Figs 1b, 2b, 3b (characters, records)
DISTRIBUTION. Brazil; Paraguay.

Scaphidium exornatum Oberthür

Scaphidium exornatum Oberthür, 1883: 6. Lectotype male, MNHN; type locality: Australia: New South Wales, Clarence River.
Scaphidium australe Achard, 1916: 87. Lectotype female, NMPC; type locality: Australia: New South Wales.

Löbl, 1976e: 289, Fig. 9 (lectotype fixed by inference for *S. exornatum*, lectotype designation for *S. australe*, synonymy of *S. australe* with *S. exornatum*, characters, records)
Hawkeswood, 1989: 94 (records, fungal hosts)
Hawkeswood, 1990: 96 (records, fungal hosts)
DISTRIBUTION. Australia: New South Wales, Queensland.

Scaphidium fainanense **Pic**
Scaphidium fainanense Pic, 1915f: 43. Syntypes, MNHN; type locality: Taiwan [Tainan].
Miwa & Mitono, 1943: 528, 555 Fig. I (characters)
Hoshina, & Morimoto, 1999: 88, Fig. 7 (characters)
DISTRIBUTION. Taiwan.

Scaphidium fairmairei **Pic**
Scaphidium fairmairei Pic, 1920b: 4. Syntypes, MNHN; type locality: Madagascar [Diego Suarez].
Note. Pic (l.c.) credited the species epithet to Portevin.
DISTRIBUTION. Madagascar.

Scaphidium falsum **He, Tang & Li**
Scaphidium falsum He, Tang & Li, 2008a: 103, Figs 1–6. Holotype male, SNUC; type locality: China: Hongmiaohe, Zhashui County, 1.110 m, Shaanxi.
Tang & Li, 2013: 178 (records)
DISTRIBUTION. China: Beijing, Shaanxi.

Scaphidium fasciatomaculatum **Oberthür**
Scaphidium fasciatomaculatum Oberthür, 1883: 10. Syntypes, MNHN; type locality: Brazil: Ega; Amazones.
Fierros-López, 2006b: 40, Figs 2c, 3c (characters, records)
DISTRIBUTION. Brazil; Ecuador; Peru.

Scaphidium fasciatum **Laporte**
Scaphidium fasciatum Laporte, 1840: 19. Syntypes, MNHN; type locality: Madagascar.
DISTRIBUTION. Madagascar.

Scaphidium fascipenne **Reitter**
Scaphidium fascipenne Reitter, 1880a: 38. Syntypes, MNHN; type locality: Brazil.
DISTRIBUTION. Brazil.

***Scaphidium feai* Pic**

Scaphidium feai Pic, 1920f: 23. Syntypes, MCSN/MNHN; type locality: Myanmar: Palon, Pegu.
Pic, 1920g: 189 (characters)
Achard, 1920d: 123 (records)
Pic, 1921a: 159 (characters, records)
DISTRIBUTION. Myanmar.

***Scaphidium femorale* Lewis**

Scaphidium femorale Lewis, 1893: 292. Syntypes, NHML; type localities: Japan: "Main Island" [Honshu] and "Kiuschiu".
Achard, 1923: 105, 106 (characters, records)
Miwa & Mitono, 1943: 524 (characters, records)
Nakane, 1955a: 54, Figs 2, 3, 6 (characters)
Nakane, 1955b: 52, Figs 7–12, 18 (characters)
Nakane, 1963b: 78 (characters)
Shirôzu & Morimoto, 1963: 68 (characters, records)
Hisamatsu, 1977: 193 (records)
Morimoto, 1985: Pl. 45, Fig. 20 (characters)
DISTRIBUTION. Japan: Honshu, Kyushu, Shikoku.

***Scaphidium flavicorne* Löbl**

Scaphidium flavicorne Löbl, 2006: 30, Figs 5–7. Holotype male, SMNS; type locality: Philippines: 30 km NW of Maramag, Bagong Silang, 1700 m, Mindanao.
DISTRIBUTION. Philippines: Mindanao.

***Scaphidium flavofasciatum* Champion**

Scaphidium flavofasciatum Champion, 1913: 68, Figs 26, 27, 109k, l. Lectotype male, NHML; type locality: Mexico: Chilpancingo, Guerrero 4000 ft.
Fierros-López, 2005: 35 (lectotype designation, characters, records)
DISTRIBUTION. Mexico: Guerrero.

***Scaphidium flavomaculatum* Miwa & Mitono**

Scaphidium flavomaculatum Miwa & Mitono, 1943: 526, 555: Fig. B. Syntypes, TARI; type localities: Taiwan: Chiasien, Chishan, Kaohsiung; Henchun, Kueitzuchiao, Pingtung; Fushan, Taipei; Takengshan, Chiai.
Löbl, 1999: 708 (records)
DISTRIBUTION. China: Sichuan; Taiwan.

Scaphidium flohri Fierros-López

Scaphidium flohri Fierros-López, 2005: 37, Figs 28, 29, 110a, b. Holotype male, MHNG; type locality: Mexico: Tlanchinol, 43 km sw Huejutla, 1500 m, Hidalgo.

DISTRIBUTION. Mexico.

Scaphidium formosanum Pic

Scaphidium formosanum Pic, 1915c: 36. Syntypes, MNHN; type locality: Taiwan [Tainan].

Achard, 1921b: 84 (records)

Miwa & Mitono, 1943: 527, 528, 555, Fig. C (characters, records)

Löbl, 1999: 708 (characters, records)

Tang et al., 2014: 60, Figs 22–25, 90–93 (characters, records)

Tu & Tang, 2017: 595 (records)

DISTRIBUTION. China: Fujian, Guangdong, Guangxi, Hainan, Jianxi, Yunnan; Taiwan.

Scaphidium fossulatum Pic

Scaphidium fossulatum Pic, 1921a: 160. Syntypes, MCSN; type locality: Myanmar: Tenasserim, Kawkareet.

DISTRIBUTION. Myanmar.

Scaphidium frater He, Tang & Li

Scaphidium frates He, Tang & Li, 2008a: 106, Figs 7–12. Holotype male, SNUC; type locality: China: Qingling, Houzhenzi, Zhouzhi County, 1343 m, Shaanxi.

DISTRIBUTION. China: Shaanxi.

Scaphidium fraternum Achard

Scaphidium fraternum Achard, 1920c: 54. Syntypes, MNHN; type locality: Singapore.

DISTRIBUTION. Singapore.

Scaphidium fryi Achard

Scaphidium fryi Achard, 1920j: 264. Syntypes, NMPC; type locality: Myanmar.

Löbl, 1992a: 581, Fig. 11 (characters)

DISTRIBUTION. Myanmar.

***Scaphidium fukienense* Pic**

Scaphidium fukienense Pic, 1954c: 58, Figs 33, 34. Lectotype male, MNHN; type locality: China: Fujian, Kuatun.
Löbl, 1999: 710 (as *fukiense*, lectotype designation, characters)
Tang et al., 2014: 53, Figs 7, 8, 66–69 (characters, records)
DISTRIBUTION. China: Fujian.

***Scaphidium gabonicum* (Paulian)**

Scaphidiolum gabonicum Paulian, 1951: 198. Holotype, MNHN; type locality: Gabon: Ogoué, Sam Kita.
Leschen & Löbl, 1995: 471 (transfer to *Scaphidium*)
DISTRIBUTION. Gabon.

***Scaphidium geniculatum* Oberthür**

Scaphidium geniculatum Oberthür, 1883: 8. Syntypes, MNHN; type locality: Panama: Matachin.
Matthews, 1888: 162 (characters, records)
Fierros-López, 2005: 39, 42, Figs 30, 31, 110c, d (characters, records, fungal hosts)
DISTRIBUTION. Costa Rica; Guatemala; Honduras; Nicaragua; Panama.

***Scaphidium gestroi* Pic**

Scaphidium gestroi Pic, 1920f: 23. Syntypes, MCSN; type locality: Myanmar [Carin: Asciui Chebà, possibly misspelled Carin Asciuii-Ghecù, app. 19°41'N, 97°00'E].
Pic, 1921a: 159 (characters)
DISTRIBUTION. Myanmar.

***Scaphidium gibbosum* Pic**

Scaphidium gibbosum Pic, 1915e: 4. Syntypes, MNHN; type locality: Indonesia: Kalimantan, Martapoera [Riam Kanan].
DISTRIBUTION. Indonesia: Kalimantan.

***Scaphidium gounellei* Pic**

Scaphidium gounellei Pic, 1920b: 4. Syntypes, MNHN; type locality: Brazil [Caraça, Minas Geraez].
DISTRIBUTION. Brazil.

***Scaphidium grande* Gestro**

Scaphidium grande Gestro, 1879b: 50. Holotype male, MCSN; type locality: Malaysia: Sarawak.

Scaphidium grande var. *subannulatum* Pic, 1915e: 3. Syntypes, MNHN; type locality: Indonesia: Kalimantan, Pontianak.
Scaphidium grande var. *inimpressum* Pic, 1920g: 189. Syntypes, MNHN; type locality: Taiwan.
Scaphidiolum grande var. *melanopus* Achard, 1924d: 91. Syntypes, NMPC/NHML; type localities: India: Meghalaya, Khasi Hills; "Haut Laos".
Achard, 1920c: 56 (characters)
Achard, 1920d: 125 (records)
Achard, 1920f: 211 (records)
Achard, 1920h: 211 (characters)
Achard, 1921b: 84 (records)
Pic, 1920g: 189 (characters, records)
Achard, 1924d: 91 (*Scaphidiolum*) (characters)
Champion, 1927: 268 (characters, records)
Miwa & Mitono, 1943: 530 (characters, records)
Löbl, 1990b: 511 (records)
Löbl, 1992a: 488, Fig. 71 (synonymy of var. *inimpressum* and var. *subannulatum* with *S. grande*, characters, records)
He et al., 2008: 181, Figs 3, 10 (records)
Tang & Li, 2010a: 68, Figs 1–3, 14–18 (characters, records)
Tang et al., 2014: 53, Figs 9, 10, 70–73, 145–150 (characters, records, habitat, immature stage)
Tang et al., 2016b: 281, Fig. 7, 8, 10 (characters)

DISTRIBUTION. China: Chongqing, Guangdong, Guangxi, Guizhou, Hainan, Hunan, Sichuan, Yunnan, Zhejiang; East Malaysia: Sarawak, Sabah; India: Meghalaya, West Bengal (Darjeeling District); Indonesia: Kalimantan; Laos; Myanmar; Nepal; Taiwan; Thailand; Vietnam; West Malaysia.

Scaphidium grandidieri Achard

Scaphidium grandidieri Achard, 1920i: 240. Syntypes, MNHN; type locality: Madagascar.
DISTRIBUTION. Madagascar.

Scaphidium grouvellei Achard

Scaphidium grouvellei Achard, 1920d: 124. Syntypes males, MNHN; type locality: Indonesia: Sumatra, Palembang.
DISTRIBUTION. Indonesia: Sumatra.
Note. Pic, 1920f: 23 proposed a new name, *Scaphidium achardi*, for *S. grouvellei* Achard, nec Pic. I am not aware of an available name spelled *Scaphidium grouvellei* Pic and suppose Pic's confusion with *Scaphidium gounellei* Pic, 1920.

Scaphidium guanacaste **Fierros-López**

Scaphidium guanacaste Fierros-López, 2005: 44, Figs 32, 33, 110e-f. Holotype male, SEMC; type locality: Costa Rica: Estación Biológica Palo Verde, La Venada, 10 m, Guanacaste.

DISTRIBUTION. Costa Rica.

Scaphidium guillermogonzalezi **Fierros-López**

Scaphidium guillermogonzalezi Fierros-López, 2005: 46, Figs 34, 35, 110g, h. Holotype male, SEMC; type locality: Panama: 20 km north Gualaca, Finca La Suiza, Chiriquí.

DISTRIBUTION. Costa Rica; Panama.

Scaphidium gurung **Löbl**

Scaphidium gurung Löbl, 1992a: 493, Figs 14, 49, 56, 70. Holotype male, MHNG; type locality: Nepal: Ghoropani Pass, south slope, 2700 m, Parbat District.

Löbl, 2005: 178 (records)

DISTRIBUTION. Nepal.

Scaphidium harmandi **Achard**

Scaphidium harmandi Achard, 1920b: 125. Lectotype female, MNHN; type locality: India: Sikkim.

Achard, 1920h: 210 (characters)

Löbl, 1992a: 494, Figs 16, 40, 72 (lectotype designation, characters, records)

Löbl, 2005: 179 (characters, records)

DISTRIBUTION. India: Sikkim [or Darjeeling District]; Nepal.

Scaphidium hexaspilotum **(Achard)**

Scaphidiolum hexaspilotum Achard, 1924a: 152. Lectotype male, NHML; type locality: Sri Lanka: Kandy.

Löbl, 1971c: 941 (lectotype fixed by inference, transfer to *Scaphidium*, records)

DISTRIBUTION. Sri Lanka.

Scaphidium holzschuhi **Löbl**

Scaphidium holzschuhi Löbl, 1992a: 491, Figs 54, 78. Holotype male, MHNG; type locality: Nepal: Lamobagar, 1400 m, Arun Valley, Sankhuwasabha District.

DISTRIBUTION. Nepal.

***Scaphidium ifanense* (Pic)**

Scaphidiolum ifanense Pic, 1951b: 210. Syntypes, MNHN; type locality: Ivory Coast: Yapo.

Leschen & Löbl, 1995: 471 (transfer to *Scaphidium*)

DISTRIBUTION. Ivory Coast.

***Scaphidium ilanum* Löbl**

Scaphidium ilanum Löbl, 2006: 32, Figs 8, 9. Holotype male, ZMUB; type locality: Philippines: Socorro.

DISTRIBUTION. Philippines: Socorro.

***Scaphidium impictum* Boheman**

Scaphidium impictum Boheman, 1851: 557. Syntypes, NHRS; type locality: South Africa: "regione fluvii Gariepis".

DISTRIBUTION. South Africa.

***Scaphidium impuncticolle* Pic**

Scaphidium impuncticolle Pic, 1915f: 43. Lectotype, MNHN; type locality: Malaysia: Sabah, Kinabalu.

Pic, 1920g: 188 (characters)

Achard, 1922b: 33 (lectotype fixed by inference, transfer to *Hemiscaphium*, characters, records)

Leschen & Löbl, 1995: 471 (retransfer to *Scaphidium*)

DISTRIBUTION. Brunei; East Malaysia: Sabah.

***Scaphidium inagoense* Kimura**

Scaphidium inagoense Kimura, 2008: 160, Figs 9–15. Holotype male, KCMI; type locality: Japan: Inago-Yu, Koumi-machi, Nagano Pref.

DISTRIBUTION. Japan: Honshu, Shikoku.

***Scaphidium incisum* Lewis**

Scaphidium incisum Lewis, 1893: 294. Syntypes, NHML; type localities: Japan: Miyanoshita, Nikko, Mayebashi.

Achard, 1923: 107, 108 (characters, records)

Miwa & Mitono, 1943: 532; (characters)

Nakane, 1955b: 51, Fig. 5 (characters)

Nakane, 1963b: 78 (characters)

Shirôzu & Morimoto, 1963: 60 (characters, records)

Löbl, 1968c: 387, Figs 8, 12 (characters)

Morimoto, 1985: Pl. 45, Fig. 17 (characters)

DISTRIBUTION. Japan: Hokkaido, Honshu, Kyushu.

Scaphidium incrassatum Achard

Scaphidium incrassatum Achard, 1920j: 263. Syntypes, NMPC; type locality: Myanmar [Burma].

Löbl, 1992a: 485, Fig. 12. (characters, records

DISTRIBUTION. India: "Assam"; Myanmar.

Scaphidium indicum Löbl

Scaphidium indicum Löbl, 1979a: 82, Figs 1, 2. Holotype male, MHNG; type locality: India: Kaikatty, Neliumpatty Hills, 900 m, Kerala.

DISTRIBUTION. India: Kerala, Tamil Nadu.

Scaphidium inexspectatum Löbl

Scaphidium inexspectatum Löbl, 1999: 713, Figs 35, 36. Holotype male, MHNG; type locality: China: Gaoliong Mts, 1500–2500 m, Yunnan.

DISTRIBUTION. China: Yunnan.

Scaphidium inflexitibiale Tang & Li

Scaphidium inflexitibiale Tang & Li, 2010a: 72, Figs 10, 11, 31–34. Holotype male, SNUC; type locality: China: Hekou, Yaoshan, 1345 m, Yunnan.

Tang, Tu & Li, 2016b: 282 (records)

DISTRIBUTION. China: Yunnan; Laos.

Scaphidium innotatum (Pic)

Scaphidiolum innotatum Pic, 1940a: 2. Syntypes, MNHN; type locality: Vietnam: Tonkin.

Leschen & Löbl, 1995: 471 (transfer to *Scaphidium*)

DISTRIBUTION. Vietnam.

Scaphidium inornatum Gestro

Scaphidium inornatum Gestro, 1879b: 55. Holotype, MCSN; type locality: Malaysia: Sarawak.

DISTRIBUTION. East Malaysia: Sarawak.

Scaphidium interruptum Fairmaire

Scaphidium interruptum Fairmaire, 1897: 368. Syntypes, MNHN; type locality: Madagascar.

DISTRIBUTION. Madagascar.

Scaphidium irregulare Pic

Scaphidium (*Cribroscaphium*) *irregulare* Pic, 1920e: 93. Syntypes, MCSN; type locality: Indonesia: Sumatra, Siboga.

Achard, 1922a: 11, 13 (*Cribroscaphium*, characters)
Leschen & Löbl, 1995: 471 (transfer to *Scaphidium*)
DISTRIBUTION. Indonesia: Sumatra.

***Scaphidium jacobsoni* Achard**
Scaphidium jacobsoni Achard, 1921b: 85. Syntypes, NBCL/NMPC; type locality: Indonesia: Java, Goenoeng, Oengaran [Gn. Ungaran].
DISTRIBUTION. Indonesia: Java.

***Scaphidium japonum* Reitter**
Scaphidium japonum Reitter, 1877: 369. Syntypes, MNHN; type locality: Japan.
Scaphidium longipes Lewis, 1893: 292. Syntypes, NHML; type localities: Japan: Higo, Kiga, Miyanoshita.
Lewis, 1893: 292 (misspelled *S. japonicum*, records)
Achard, 1923: 104 (characters, records)
Miwa & Mitono, 1943: 522 (characters, records)
Nakane, 1955b: 52, Figs 2, 17, 21–23 (characters)
Chûjô, 1961: 5 (records)
Nakane, 1963b: 79 (characters)
Shizôzu & Morimoto, 1963: 71 (synonymy of *S. longipes* with *S. japonum*, characters, records)
Morimoto, 1985: Pl. 45, Fig. 25 (characters)
DISTRIBUTION. Japan: Honshu, Kyushu, Oshima Islands, Sado, Shikoku, Tsushima.

***Scaphidium javanum* Pic**
Scaphidium javanum Pic, 1915d: 40. Syntypes, MNHN; type locality: Indonesia: Java [Pengarengan, W. Java].
DISTRIBUTION. Indonesia: Java.

***Scaphidium jinmingi* Tang, Li & He**
Scaphidium jinmingi Tang, Li & He, 2014: 49, Figs 1, 1, 58–61. Holotype male, SNUC; type locality: China: Lin'an City, West Tianmushan, 1400 m.
DISTRIBUTION. China: Anhui, Chongqing, Zhejiang.

***Scaphidium jizuense* Löbl**
Scaphidium jizuense Löbl, 1999: 713. Holotype male, MHNG; type locality: China: Jizu Mts, 2800 m, Yunnan.
DISTRIBUTION. China: Yunnan.

***Scaphidium klapperichi* Pic**

Scaphidium klapperichi Pic, 1954c: 57. Lectotype male, NHRS; type locality: China: Fujian, Kuatun.
Löbl, 1999: 710, Fig. 25 (lectotype designation, characters)
Tang et al., 2014: 73, Figs 44, 45, 164, 165 (characters, habitat)
Tu & Tang, 2017: 595 (records)
DISTRIBUTION. China: Fujian, Zhejiang.

***Scaphidium kubani* Löbl**

Scaphidium kubani Löbl, 1999: 714, Figs 37, 38. Holotype male, MHNG; type locality: China: Gaoligong Mts, 1500–2500 m, Yunnan.
DISTRIBUTION. China: Yunnan.

***Scaphidium kumejimaense* Hoshina & Maruyama**

Scaphidium kumejimaense Hoshina & Maruyama, 1999: 480. Holotype male, KUIC; type locality: Japan: Kumejima Island, forest along Shirase river, Ryukyus.
Hoshina, 2001: 102, Figs 2, 9, 19, 21, 22 (characters, records)
DISTRIBUTION. Japan: Ryukyus.

***Scaphidium kurbatovi* Löbl**

Scaphidium kurbatovi Löbl, 1999: 717, Figs 30, 47, 48. Holotype male, MHNG; type locality: China: Wolong Nat. Res., 1500 m, Sichuan.
DISTRIBUTION. China: Sichuan.

***Scaphidium kurosawai* Löbl**

Scaphidium kurosawai Löbl, 2006: 34, Figs 10, 11. Holotype male, MHNG; type locality: Philippines: Ifugao Prov., Mt. Polis, 1900 m, Luzon.
DISTRIBUTION. Philippines: Luzon.

***Scaphidium lafertei* Pic**

Scaphidium lafertei Pic, 1920f: 23. Syntypes, MNHN; type locality: Madagascar.
DISTRIBUTION. Madagascar.

***Scaphidium laosense* (Pic)**

Scaphidiolum laosense Pic, 1928c: 1. Syntypes, MNHN; type locality: Laos [Bin Sab].
Leschen & Löbl, 1995: 471 (transfer to *Scaphidium*)
DISTRIBUTION. Laos.

Scaphidium lateflavum **(Pic)**

Scaphidiolum lateflavum Pic, 1928c: 1. Syntypes, MNHN; type locality: Vietnam: Dalat.

Leschen & Löbl, 1995: 471 (transfer to *Scaphidium*)

Distribution. Vietnam.

Scaphidium laxum **Tang & Li**

Scaphidium laxum Tang & Li, 2010a: 72, Figs 8, 9, 27–30. Holotype male, SNUC; type locality: China: Yunnan, Nabanhe N.P., Benggangxinzhai, 1750 m.

Distribution. China: Yunnan.

Scaphidium leleupi leleupi **(Pic)**

Scaphidiolum leleupi Pic, 1954b: 34. Syntypes, MRAC; type locality: Democratic Republic of the Congo, Kundelungu Massif.

Leschen & Löbl, 1995: 471 (transfer to *Scaphidium*)

Distribution. Democratic Republic of the Congo.

Scaphidium leleupi atropygum **(Pic)**

Scaphidiolum leleupi var. *atropygum* Pic, 1954b: 35. Holotype, MRAC; type locality: Democratic Republic of the Congo, Mayidi.

Distribution. Democratic Republic of the Congo.

Scaphidium lescheni **Fierros-López**

Scaphidium lescheni Fierros-López, 2005: 49, Figs 36, 37, 110i, j. Holotype male, SEMC; type locality: Nicaragua: Hotel Selva Negra, Matagalpa.

Distribution. Honduras; Nicaragua

Scaphidium lesnei **Achard**

Scaphidium lesnei Achard, 1920c: 52. Syntypes, MNHN; type locality: Thailand: Isthmus of Kra.

Achard, 1922d: 262 (characters)

Distribution. Thailand.

Scaphidium lewisi **(Achard)**

Scaphidiolum lewisi Achard, 1923: 108. Syntypes, NHML; type locality: Japan.

Nakane, 1955a: 56 (characters, records)

Leschen & Löbl, 1995: 471 (transfer to *Scaphidium*)

Distribution. Japan: Kyushu.

Scaphidium lineaticolle **Matthews**

Scaphidium lineaticolle Matthews, 1888: 163, Pl. 4, Fig. 5. Holotype, NHML; type locality: Mexico: Córdoba, Veracruz.

Fierros-López, 2005: 52, Figs 38, 39, 110k-l (characters, records)

DISTRIBUTION. Mexico.

Scaphidium lineatipes **(Pic)**

Scaphidiolum lineatipes Pic, 1925b: 8. Syntypes, ?MRAC; type locality: Congo.

Paulian, 1951: 198 (*Scaphidiolum*, characters, records)

Leschen & Löbl, 1995: 471 (transfer to *Scaphidium*)

DISTRIBUTION. Democratic Republic of the Congo; Gabon.

Scaphidium linwenhsini **Tang & Li**

Scaphidium linwenhsini Tang & Li, 2013: 174, Figs 1–4, 11–14. Holotype male, SNUC; type locality: China: Sichuan, Baiyun Temple, Qichengshan, 1650 m.

DISTRIBUTION. China: Chongqing, Guangxi, Sichuan.

Scaphidium liui **Tang & Li**

Scaphidium liui Tang & Li, 2010a: 75, Figs 39–42. Holotype male, SEM; type locality: China: Motuo County, Yadong, Xizang, 1250 m.

DISTRIBUTION. China: Xizang.

Scaphidium loebli **Fierros-López**

Scaphidium loebli Fierros-López, 2005: 54, Figs 40, 41, 111a, b. Holotype male, CZUG; type locality: Mexico: San José de los Laureles, 1768 m, Tlayacapan, Morelos.

DISTRIBUTION. Mexico.

Scaphidium longicolle **Pic**

Scaphidium longicolle Pic, 1915c: 36. Syntypes, MNHN; type locality: Borneo.

[*Scaphidium longicolle* var. *bicoloripes* Pic, 1948a: 12. Syntypes, MNHN; type locality: Borneo – Deemed unavailable, due to sympatry of two varieties with the nominate taxon].

[*Scaphidium longicolle* var. *kudatense* Pic, 1948a: 12. Syntypes, MNHN; type locality: Borneo – Deemed unavailable, due to sympatry of two varieties with the nominate taxon].

Scaphidium plagatum Achard, 1920d: 124. Holotype female, MNHN; type locality: Indonesia: Java, Meuwen Bay.

Pic, 1920g: 189 (characters)

Achard, 1920c: 54 (characters)

Achard, 1922d: 262 (characters, records, downgraded *S. plagatum* to variety and implicitly synonym with *S. longicolle*)

DISTRIBUTION. Brunei; Indonesia: Java, Kalimantan.

Scaphidium longithorax longithorax Pic

Scaphidium longithorax Pic, 1916c: 17. Syntypes, MNHN; type locality: Malaysia: Sabah, Kinabalu.

Achard, 1924d: 91 (transfer to *Scaphidiolum*, records)

Leschen & Löbl, 1995: 471 (transfer to *Scaphidium*)

DISTRIBUTION. Brunei; East Malaysia: Sabah; Indonesia: Kalimantan; West Malaysia.

Scaphidium longithorax nigriventre Achard

Scaphidiolum longithorax var. *nigriventre* Achard, 1924d: 91. Syntypes, NMPC/NHML; type locality: Singapore.

DISTRIBUTION. Singapore.

Scaphidium longum Tang & Li

Scaphidium longum Tang & Li, 2010a: 70. Holotype male, SNUC; type locality: China: Limu Shan, Hainan, 800 m.

DISTRIBUTION. China: Hainan.

Scaphidium lucidum Achard

Scaphidium lucidum Achard, 1915a: 557. Holotype, NMPC; type locality: Colombia, Cauca.

DISTRIBUTION. Colombia.

Scaphidium lunare Löbl

Scaphidium lunare Löbl, 1999: 717, Figs 28, 43, 44. Holotype male, MHNG; type locality: China: Heishui, 35 km north of Lijing, Yunnan.

DISTRIBUTION. China: Yunnan.

Scaphidium lunatum lunatum Motschulsky

Scaphidium lunatum Motschulsky, 1860: 94. Syntypes, ZMUM; type locality: "continent Indien".

Pic, 1921a: 158 (misspelled *lunulatum*, records)

Achard, 1924d: 91 (transfer to *Scaphidiolum*, characters)

Löbl, 1990b: 510 (as *Scaphidium*, records)

DISTRIBUTION. Myanmar; Thailand.

Scaphidium lunatum bioculatum Achard
Scaphidiolum lunatum var. *bioculatum* Achard, 1924d: 91. Syntypes, NMPC/MNHN; type locality: Myanmar: Tenasserim.
DISTRIBUTION. Myanmar.

Scaphidium lunatum inconjunctum Pic
Scaphidium lunulatum [sic] var. *inconjunctum* Pic, 1921a: 158. Syntypes, MCSN; type locality: Myanmar: Carin Ghecù, 1300–1400 m [approx. 19°41'N 97°00'E].
DISTRIBUTION. Myanmar.

Scaphidium lunatum rufithorax Pic
Scaphidium lunulatum [sic] var. *rufithorax* Pic, 1921a: 159. Syntypes, MCSN; type locality: Myanmar: Carin Chebà, 900–1100 m [approx. 19°13'N 96°35'E].
DISTRIBUTION. Myanmar.

Scaphidium luteomaculatum (Pic)
Scaphidiolum luteomaculatum Pic, 1923d: 194. Syntypes, MNHN; type locality: Indonesia: Sumatra.
Achard, 1920d: 130 (records)
Leschen & Löbl, 1995: 472 (transfer to *Scaphidium*)
DISTRIBUTION. Indonesia: Sumatra.

Scaphidium luzonicum (Achard)
Scaphidiolum luzonicum Achard, 1924a: 152. Lectotype male, NHML; type locality: Philippines: Luzon.
Leschen & Löbl, 1995: 472 (transfer to *Scaphidium*)
Löbl, 2006: 35, Figs 12, 13 (lectotype designation, characters)
DISTRIBUTION. Philippines: Luzon.

Scaphidium macuilimaculatum Fierros-López
Scaphidium macuilimaculatum Fierros-López, 2005: 57, Figs 42, 43, 111c, d. Holotype male, SEMC; type locality: Honduras: 28.2 km west Tegucigalpa, 1630 m, Francisco Morazán.
DISTRIBUTION. Honduras; Nicaragua.

Scaphidium maculaticeps (Pic)
Scaphidiolum maculaticeps Pic, 1923d: 194. Syntypes, MNHN; type locality: Indonesia: Sumatra, Sibolangit [Sibaulangit].
Leschen & Löbl, 1995: 472 (transfer to *Scaphidium*)
DISTRIBUTION. Indonesia: Sumatra.

Scaphidium madecassum Pic

Scaphidium madecassum Pic, 1917c: 3. Syntypes, MNHN; type locality: Madagascar [Tananarive].

DISTRIBUTION. Madagascar.

Scaphidium madurense Achard

Scaphidium madurense Achard, 1915a: 555. Lectotype male, NMPC; type locality: India: Tamil Nadu, Chambaganor [Shembaganur].

Löbl, 1979a: 81 (lectotype designation, records)

DISTRIBUTION. India: Tamil Nadu.

Scaphidium malaccanum Pic

Scaphidium malaccanum Pic, 1915e: 2. Syntypes, MNHN; type locality: Malaysia: Malacca.

Achard, 1920c: 50 (characters)

DISTRIBUTION. West Malaysia.

Scaphidium mangenoti (Paulian)

Scaphidiolum mangenoti Paulian, 1951: 199. Holotype, MNHN; type locality: Ivory Coast: Azaguié.

Leschen & Löbl, 1995: 472 (transfer to *Scaphidium*)

DISTRIBUTION. Ivory Coast.

Scaphidium marginale Reitter

Scaphidium marginale Reitter, 1880a: 38. Syntypes, MNHN; type locality: India: Calcutta [Kolkata].

DISTRIBUTION. India: West Bengal.

Scaphidium marginatum Matthews

Scaphidium marginatum Matthews, 1888: 160, Pl. IV, Fig. 1. Lectotype male, NHML; type locality: Guatemala: Capetillo, Sacatepequez.

Fierros-López, 2005: 60, Figs 44, 45, 111e, f (lectotype designation, characters, records)

DISTRIBUTION. Guatemala.

Scaphidium martapuranum Pic

Scaphidium martapuranum Pic, 1916c: 18. Syntypes, MNHN; type locality: Indonesia: Kalimantan, Martapura.

DISTRIBUTION. Indonesia: Kalimantan.

***Scaphidium mastersii* Macleay**

Scaphidium mastersii Macleay, 1871: 156. Syntypes, ?AMSC, not traced; type locality: Australia: Gayndah.

Löbl, 1976e: 294 (misspelled *mastersi*, of doubtful identity, possibly synonym with *S. notatum*)

Löbl, 1997: 27 (misspelled *mastersi*)

DISTRIBUTION. Australia: Queensland.

***Scaphidium matthewsi* Csiki**

Scaphidium unicolor Matthew, 1888: 161. Lectotype male, NHML; type locality: Panama: Volcán Chiriquí.

Scaphidium matthewsi Csiki, 1904: 85; replacement name for *Scaphidium unicolor* Matthew, 1888: 161 (nec *Scaphidium unicolor* Laporte, 1840).

Fierros-López, 2005: 62, 64, Figs 46, 47, 111g, h (lectotype designation, characters, records, fungal hosts)

DISTRIBUTION. Costa Rica; Panama.

***Scaphidium mauroi* Fierros-López**

Scaphidium mauroi Fierros-López, 2005: 66, Figs 48, 49, 111i, j. Holotype male, SEMC; type locality: Guatemala: 3.5 km SE La Unión, 1500 m, Zacapa.

DISTRIBUTION. Guatemala.

***Scaphidium maynei* (Pic)**

Scaphidiolum maynei Pic, 1954b: 35. Holotype, MRAC; type locality: Democratic Republic of the Congo: Kunzulu.

Leschen & Löbl, 1995: 472 (transfer to *Scaphidium*)

DISTRIBUTION. Democratic Republic of the Congo.

***Scaphidium medionigrum* Pic**

Scaphidium medionigrum Pic, 1915c: 36. Syntypes, MNHN; type locality: West Malaysia: Malacca, Perak.

Achard, 1920c: 54 (in synonymy with *S. sundaicum* [sic] Gestro

Pic, 1920g: 188 (resurrected)

DISTRIBUTION. West Malaysia.

***Scaphidium melanogaster* Löbl**

Scaphidium melanogaster Löbl, 1992a: 487, Figs 66. 82. Holotype male, MHNG; type locality: India: Vashisht Baths north Manali, 1990 m, Himachal Pradesh.

DISTRIBUTION. India: Himachal Pradesh, Uttarakhand; Nepal.

***Scaphidium melli* Löbl**

Scaphidium melli Löbl, 1972a: 115. Holotype female, ZMUB; type locality: China: Chao-chow-fu [Fengyi, Dali Zhou], 2300 m.
Löbl, 1999: 710 (characters, records)

DISTRIBUTION. China: Yunnan.

***Scaphidium metallescens* Gestro**

Scaphidium metallescens Gestro, 1879a: 560. Holotype female, MCSN; type locality: Papua New Guinea: Fly River.
Löbl, 1971a: 247 (records)
Löbl, 1975a: 370 (records)

DISTRIBUTION. New Britain, Papua New Guinea.

***Scaphidium minutum* Pic**

Scaphidium minutum Pic, 1920b: 3. Syntypes, MNHN; type locality: Malaysia: Sabah, Kinabalu.

DISTRIBUTION. East Malaysia: Sabah.

***Scaphidium monteithi* Löbl**

Scaphidium monteithi Löbl, 1976e: 291, Figs 2, 2a, 10. Holotype male, UQBA; type locality: Australia: Mt Spec via Paluma.

DISTRIBUTION. Australia: Queensland.

***Scaphidium montivagum* Shirôzu & Morimoto**

Scaphidium montivagum Shirôzu & Morimoto, 1963: 70. Holotype male, KUEC; type locality: Japan: Masutomi, Yamanashi Pref.
Morimoto, 1985: Pl. 45, Fig. 23 (characters)
Li, 1992: 62 (records)
[*Scaphidium montivagum heilongiangensis* Li, L. Zhang & X. Zhang, 2015: 47, unavailable]

DISTRIBUTION. China: Heilongjiang, Jilin; Japan: Honshu.

***Scaphidium morimotoi* Löbl**

Scaphidium takahashii Shirôzu & Morimoto, 1963: 64. Holotype male, KUEC; type locality: Japan: Yôrô-shima, Amami.

Scaphidium morimotoi Löbl, 1982b: 101; replacement name for *Scaphidium takahashii* Shirôzu & Morimoto, 1963: 64 (nec *Scaphidium takahashii* Miwa & Mitono, 1943).

Scaphidium amamiense Hoshina & Morimoto, 1999: 90, Figs 2, 4, 9, 13, 14, 20–22, 27. Holotype male, KUIC; type locality: Japan: Amami-ôshima, Mt. Yuwandake.

Löbl, 1982b: 102, Fig. 1 (characters, records)

Morimoto, 1985: 255 (as *Scaphidium* sp.)

Hoshina, 2001: 100, Fig. 6 (characters)

Ogawa et al., 2016: 70, Figs 3a-d, 4a-j, 5 a-d (synonymy of *S. amamiense* with *S. morimotoi*, characters, records)

DISTRIBUTION. Japan: Amami Islands, Kakeroma-jima, Tokunoshima, Yoro-shima.

Scaphidium multinotatum **Pic**

Scaphidium multinotatum Pic, 1921a: 161. Syntypes, MCSN/MNHN; type locality: Myanmar: Carin Chebà, 900–1100 m [approx. 19°13'N 96°35'E].

DISTRIBUTION. Myanmar.

Scaphidium multipunctatum multipunctatum **(Pic)**

Scaphidiolum multipunctatum Pic, 1928b: 34. Syntypes, MRAC; type locality: Democratic Republic of the Congo: Tibo.

[*Scaphidiolum multipunctatum* var. *atricentre* Pic, 1928b: 34. Syntypes, MRAC; type locality: Democratic Republic of the Congo: Tibo. Note: Sympatric with the nominate subspecies, deemed infrasubspecific]

[*Scaphidiolum multipunctatum* var. *prolongatum* Pic, 1928b: 35. Syntypes, MRAC/MNHN; type locality: Democratic Republic of the Congo: Tibo. Note: Sympatric with the nominate subspecies, deemed infrasubspecific; listed as species in Leschen & Löbl, 1995: 472]

DISTRIBUTION. Democratic Republic of the Congo.

Scaphidium multipunctatum luluanum **(Pic)**

Scaphidiolum multipunctatum var. *luluanum* Pic, 1954b: 33. Syntypes, MRAC/MNHN; type locality: Democratic Republic of the Congo: Lulua, Kapanga.

Leschen & Löbl, 1995: 472 (transfer to *Scaphidium*)

DISTRIBUTION. Democratic Republic of the Congo.

Scaphidium negrito **Heller**

Scaphidium negrito Heller, 1917: 42. Lectotype male, SMTD; type locality: Philippines: Palawan, Porto Princesa.

Löbl, 2006: 38, Figs 14, 15 (lectotype designation, characters)

DISTRIBUTION. Philippines: Palawan.

Scaphidium nepalense Löbl

Scaphidium nepalense Löbl, 1992a: 494, Figs 13, 50, 57, 69. Holotype male, MHNG; type locality: Nepal: Phulcoki, 2550 m, Patan District.
Löbl & Ogawa, 2016a: Fig. 5 (characters)
Distribution. Nepal.

Scaphidium newtoni Fierros-López

Scaphidium newtoni Fierros-López, 2005: 71, Figs 52, 53, 112a, b. Holotype male, CZUG; type locality: Mexico: Cañón del Rio Metlac, Fortín de las Flores, 975 m, Veracruz.
Distribution. Mexico.

Scaphidium nigripes Guérin-Méneville

Scaphidium nigripes Guérin-Méneville, 1834: Pl. 17, Fig. 14. Syntypes, MNHN. [note: issued October-December, 1834, see Bousquet, 2016]
Scaphidium mexicanum Laporte, 1840: 19. Syntypes, MNHN, not traced; type locality: Mexico.
Scaphidium nigripes Chevrolat, 1844: 62 [text]; type locality: Mexico.
Matthews, 1888: 161, Pl. 4, Fig. 2 (synonymy of *S. nigripes* with *S. mexicanum*, characters, records)
Fierros-López, 1988: 36, Fig. 1 (as *S. mexicanum*, characters, records, fungal hosts)
Fierros-López, 2005: 68, 69, Figs 50, 51, 111k-l (as *S. mexicanum*, characters, records, fungal hosts)
Distribution. Costa Rica; Mexico.

Scaphidium nigrocinctulum Oberthür

Scaphidium nigrocinctulum Oberthür, 1883: 7. Syntypes, MNHN/NBCL; type locality: India: Andaman Islands.
Achard, 1920d: 123 (characters)
Miwa & Mitono, 1943: 527 (characters, records)
Distribution. India: Andaman Islands; Myanmar; Taiwan.

Scaphidium nigromaculatum Reitter

Scaphidium nigromaculatum Reitter, 1880b: 170. Holotype female, ZMUB; type locality: Sri Lanka.
Scaphidium nigromaculatum Reitter var. *effigiatum* Achard, 1922d: 261. Lectotype female, NHML; type locality: Sri Lanka: Dikoya, 1300–1400 m.
Löbl, 1971c: 941 (lectotype for var. *effigiatum* fixed by inference, synonymy with *S. nigromaculatum*, records)
Distribution. Sri Lanka.

Scaphidium nigrosuturale **Pic**

Scaphidium nigrosuturale Pic, 1920f: 23. Syntypes, MNHN; type locality: Malaysia: Sabah, Kinabalu.

DISTRIBUTION. East Malaysia: Sabah.

Scaphidium nigrotibiale **Fierros-López**

Scaphidium nigrotibiale Fierros-López, 2005: 73, Figs 54, 55, 112c-d. Holotype male, SEMC; type locality: Costa Rica: Reserva Biológia Lomas Barbudal, 17 m, Guanacaste.

DISTRIBUTION. Costa Rica; Panama.

Scaphidium nigrum **Laporte**

Scaphidium nigrum Laporte, 1840: 19. Syntypes, MNHN; type locality: Madagascar [Antongil Bay].

DISTRIBUTION. Madagascar.

Scaphidium nopillohuan **Fierros-López**

Scaphidium nopillohuan Fierros-López, 2005: 76, Figs 56, 57, 112e-f. Holotype male, SEMC; type locality: Honduras: Jardín Botánico Lacertilla, Tela, 10 m, Atlántida.

DISTRIBUTION. Honduras.

Scaphidium notaticolle **Pic**

Scaphidium notaticolle Pic, 1915b: 30. Syntypes, MNHN; type locality: Madagascar: Ambositra.

DISTRIBUTION. Madagascar.

Scaphidium notatum **(Pic)**

Scaphidiolum notatum Pic, 1923b: 16. Lectotype male, MNHN; type locality: Australia: Western Australia, Swan River.

Löbl, 1976b: 288, Fig. 5 (lectotype fixed by inference, transfer to *Scaphidium*, characters, records)

DISTRIBUTION. Australia: Queensland, Western Australia.

Scaphidium ocellatum ocellatum **Achard**

Scaphidium ocellatum Achard, 1920j: 264. Syntypes, MNHN; type locality: India: "Assam".

DISTRIBUTION. India: "Assam".

Scaphidium ocellatum birmanicum Achard

Scaphidium ocellatum var. *birmanica* Achard, 1920j: 264. Syntypes, MNHN; type locality: Birmanie.

DISTRIBUTION. Myanmar.

Scaphidium ocelotl Fierros-López

Scaphidium ocelotl Fierros-López, 2005: 78, Figs 58, 59, 112g, h. Holotype male, SEMC; type locality: Costa Rica: La Selva, 3.2 km SE Puerto Viejo, Heredia.

DISTRIBUTION. Costa Rica; Mexico; Panama.

Scaphidium oculare (Pic)

Scaphidiolum oculare Pic, 1923b: 16. Syntypes, MNHN; type locality: Madagascar.
Leschen & Löbl, 1995: 472 (transfer to *Scaphidium*)

DISTRIBUTION. Madagascar.

Scaphidium okinawaense Hoshina & Morimoto

Scaphidium okinawaense Hoshina & Morimoto, 1999: 91, Figs 3, 5, 10, 15, 16, 23–25, 28. Holotype male, KUIC; type locality: Japan: Okinawa, Mt. Yonohadake.
Hoshina, 2001: 100, Fig. 7 (characters)

DISTRIBUTION. Japan: Ryukyus: Okinawa.

Scaphidium omemaculatum Fierros-López

Scaphidium omemaculatum Fierros-López, 2005: 81, Figs 60, 61, 112i, j. Holotype male, SEMC; type locality: Costa Rica: Monte Verde, 1400 m, Puntarenas.

DISTRIBUTION. Costa Rica.

Scaphidium optabile (Lewis)

Scaphium optabile Lewis, 1893: 290. Holotype female, NHML; type locality: Japan: Ichiuchi, in Higo.
Achard, 1923: 97, 98 (transfer to *Parascaphium*, characters, records)
Nakane, 1955a: 54, Figs 1, 8 (as *Parascaphium*, characters, records)
Nakane, 1963b: 78 (as *Parascaphium*, characters)
Löbl, 1968c: 386, Figs 3, 4, 9, 10 (transfer to *Scaphidium*, characters, records)
Morimoto, 1985: Pl. 45, Fig. 14 (characters)
Hwang & Ahn, 2001: 372 (characters, records)

DISTRIBUTION. Japan: Kyushu; South Korea.

Scaphidium orbiculosum Reitter

Scaphidium orbiculosus Reitter, 1880a: 40. Syntypes, MNHN; type locality: Borneo.
Achard, 1922d: 262 (characters)
DISTRIBUTION. "Borneo".

Scaphidium ornatum Casey

Scaphidium ornatum Casey, 1900: 56. Syntypes, USNM; type locality: USA: ?Colorado.
DISTRIBUTION. United States: ?Colorado.

Scaphidium overlaeti (Pic)

Scaphidiolum overlaeti Pic, 1954b: 34. Syntypes, MRAC/MNHN; type localities: Democratic Republic of the Congo: Lulua, Kapanga; riv. Kasai-Lumeni; Kafakumba.
Leschen & Löbl, 1995: 472 (transfer to *Scaphidium*)
DISTRIBUTION. Democratic Republic of the Congo.

Scaphidium pallidum He, Tang & Li

Scaphidium pallidum He, Tang & Li, 2008a: 107, Figs 13–18. Holotype male, SNUC; type locality: China: East Daba-Shan, Chongqing, 2,039 m.
Tang & Li, 2013: 180 (records, misspelled *pallidium*)
DISTRIBUTION. China: Chongqing, Guizhou, Sichuan.

Scaphidium palonense Achard

Scaphidium palonense Achard, 1920i: 239. Syntypes, MNHN/NMPC; type locality: Myanmar: Palon [Tikekee, Pegu].
Pic, 1921a: 159 (characters)
DISTRIBUTION. Myanmar.

Scaphidium panamense Fierros-López

Scaphidium panamense Fierros-López, 2005: 83, Figs 62, 63, 112k, l. Holotype male, SEMC; type locality: Panama: Parque Nacional Soberanía, Pipeline km 6.1, 80 m, Colón.
DISTRIBUTION. Panama.

Scaphidium pantherinum Oberthür

Scaphidium pantherinum Oberthür, 1883: 9. Syntypes, MNHN; type locality: Brazil: Rio Negro.
DISTRIBUTION. Brazil.

Scaphidium papuanum Löbl

Scaphidium papuanum Löbl, 1975a: 372, Figs 2, 4. Holotype male, ZMUB; type locality: Papua New Guinea: Sattelberg [Sattelburg].

Löbl, 1978c: 114 (characters, records)

DISTRIBUTION. Papua New Guinea.

Scaphidium pardale pardale Laporte

Scaphidium pardalis Laporte, 1840: 19. Syntypes, MNHN; type locality: French Guiana: Cayenne.

Oberthür, 1883: 11 (records)

Achard, 1921b: 86 (records)

DISTRIBUTION. Brazil; French Guiana.

Scaphidium pardale nigripenne Oberthür

Scaphidium pardale var. *nigripenne* Oberthür, 1883: 11. Syntypes, MNHN; type locality: Brazil: Manès.

DISTRIBUTION. Brazil.

Scaphidium patinoi Oberthür

Scaphidium patinoi Oberthür, 1883: 7. Syntypes, MNHN; type locality: Colombia: Manizales.

DISTRIBUTION. Colombia.

Scaphidium pauliani Leschen & Löbl

Scaphidiolum sulcatum Paulian, 1951: 196. Holotype male, MNHN; type locality: Gabon: Ogoué, Lambaréné.

Scaphidium pauliani Leschen & Löbl, 1995: 472; replacement name for *Scaphidiolum sulcatum* Paulian, 1951 (nec *Scaphidium sulcatum* Pic, 1915).

DISTRIBUTION. Gabon.

Scaphidium pech Fierros-López

Scaphidium pech Fierros-López, 2005: 85, Figs 64, 65, 113a, b. Holotype male, SEMC; type locality: Honduras: La Muralla, 14 km north La Unión, 1450 m, Olancho.

DISTRIBUTION. Guatemala; Honduras.

Scaphidium peckorum Fierros-López

Scaphidium peckorum Fierros-López, 2005: 87, Figs 66, 67, 113c, d. Holotype male, FMNH; type locality: Honduras: Confradía, 25 km north N.P. Cursuco, 1550 m, Cortés.

DISTRIBUTION. Honduras.

Scaphidium peninsulare **Achard**

Scaphidium peninsulare Achard, 1920c: 51. Holotype male, MNHN; type locality: Malaysia: Perak.
Pic, 1920g: 188 (characters, as variety of *S. dohertyi* Pic)
Achard, 1922b: 31, 33 (transfer to *Hemiscaphium*, records)
Leschen & Löbl, 1995: 473 (transfer to *Scaphidium*)

DISTRIBUTION. Indonesia: Sumatra, Kalimantan; West Malaysia.

Scaphidium peraffine **Oberthür**

Scaphidium peraffine Oberthür, 1883: 6. Syntypes, MNHN; type locality: Colombia.

DISTRIBUTION. Colombia.

Scaphidium perezrodriguezae **Fierros-López**

Scaphidium perezrodriguezae Fierros-López, 2005: 89, Figs 68, 69, 113e, f. Holotype male, SEMC; type locality: Panama: 20 km north Gualaca, Finca La Suiza, Chiriquí.

DISTRIBUTION. Costa Rica; Panama.

Scaphidium perpulchrum **Csiki**

Scaphidium perpulchrum Csiki, 1909: 340. Syntypes, HNHM/NMPC; type locality: Vietnam: Tonkin, Mts Mauson [Maoson].
Achard, 1920h: 212 (characters)
Tang & Li, 2010b: 319 (characters)

DISTRIBUTION. Vietnam.

Scaphidium philippense **Reitter**

Scaphidium philippense Reitter, 1880a: 39. Syntypes, MNHN; type locality: Philippines.
Heller, 1917: 42, 45 (characters, records, misspelled *philippinense*)
Löbl, 2006: 40, Figs 18–20 (doubtful identity, characters, records)

DISTRIBUTION. Philippines: Luzon.

Scaphidium phungi **Pic**

Scaphidium phungi Pic, 1923a: 4. Syntypes, MNHN; type locality: Vietnam: Tonkin [Lac Tho].

DISTRIBUTION. Vietnam.

Scaphidium picconii picconii **Gestro**

Scaphidium picconii Gestro, 1879b: 52. Syntypes, MCSN; type locality: Indonesia: Sumatra, Ajer Mantcior [near Padangpanjang].

DISTRIBUTION. Indonesia: Sumatra.

Scaphidium picconii sexmaculatum Reitter

Scaphidium picconii var. *sexmaculatum* Reitter, 1889: 7. Syntypes, NBCL; type locality: Indonesia: Sumatra, Tambang-Salida.

DISTRIBUTION. Indonesia: Sumatra.

Scaphidium piceoapicale (Pic)

Scaphidiolum piceoapicale Pic, 1940a: 2. Syntypes, MNHN; type locality: Vietnam: Tonkin [Hoa Binh].

Leschen & Löbl, 1995: 473 (transfer to *Scaphidium*)

DISTRIBUTION. Vietnam.

Scaphidium piceum Melsheimer

Scaphidium piceum Melsheimer, 1846: 103, Syntypes, MCZC; type locality: USA: Pennsylvania.

Scaphidium amplum Casey, 1900: 56. Syntypes, USNM; type locality: USA: Indiana.

LeConte, 1860: 322 (characters)

Reitter, 1880a: 36 (characters, as variety of *S. quadriguttatum*)

Casey, 1893: 513 (characters, as variety of *S. quadriguttatum*)

Casey, 1900: 56 (characters, records)

Blatchley, 1910: 492, Fig. 174 (*characters,* as variety of *S. quadriguttatum*, synonymy of *S. amplum* with *S. piceum*)

Leonard, 1928: 314 (records, as variety of *S. quadriguttatum*)

Brimley, 1938: 149 (as subspecies of *S. quadriguttatum*, records)

Kirk, 1969: 30 (as subspecies of *S. quadriguttatum*, records)

Campbell, 1991: 125 (records)

Downie & Arnett, 1996: 365 (characters, as variety of *S. quadriguttatum*)

Poole & Gentili, 1996: 413 (*S. piceum* and *S. amplum* as synonyms of *S. quadriguttatum*)

Newton et al., 2000: Fig. 4.22 (characters)

Sikes, 2004: 105 (as synonym of *S. quadriguttatum* Say)

DISTRIBUTION. Canada: Ontario; United States: Connecticut, Maine, New Hampshire, New York, North Carolina, Pennsylvania, Rhode Island, South Carolina, "Rhode Island to Indiana and Iowa".

Scaphidium pilarae Fierros-López

Scaphidium pilarae Fierros-López, 2005: 91, Figs 70, 71, 113g, h. Holotype male, SEMC; type locality: Panama: Estación Biológica Cana, Serranía del Pirre, Darien, 800 m.

DISTRIBUTION. Panama.

Scaphidium politum **Fairmaire**

Scaphidium politum Fairmaire, 1899: 468. Syntypes, MNHN; type locality: Madagascar: Betsiboka Valley.

DISTRIBUTION. Madagascar.

Scaphidium pulchellum **Reitter**

Scaphidium pulchellum Reitter, 1880a: 40. Syntypes, MNHN; type locality: Madagascar.

DISTRIBUTION. Madagascar.

Scaphidium punctaticolle **Pic**

Scaphidium punctaticolle Pic, 1923a: 4. Syntypes, MNHN; type locality: Vietnam: Tonkin [Lac Tho].

DISTRIBUTION. Vietnam.

Scaphidium punctatostriatum **Kolbe**

Scaphidium punctatostriatum Kolbe, 1897: 88. Syntypes, ZNUB; type locality: Usamara, Derema.

DISTRIBUTION. Tanzania.

Scaphidium punctatum **Laporte**

Scaphidium punctatum Laporte, 1840: 19. Syntypes, MNHN; type locality: Madagascar.

DISTRIBUTION. Madagascar.

Scaphidium punctipenne **Macleay**

Scaphidium punctipenne Macleay, 1871: 156. Lectotype male, ANIC; type locality: Australia: Gayndah.

Scaphidium coronatum Reitter, 1880: 40. Lectotype male, MNHN; type locality: Nova Hollandia.

Scaphidium thoracicum Achard, 1916: 88. Lectotype male, NMPC; type locality: Australia: New South Wales.

Löbl, 1976e: 290, Figs 1, 8 (lectotype fixations by inference for *S. punctipenne* and *S. coronatum*, lectotype designation for *S. thoracicum*; synonymy of *S. coronatum* and *S. thoracicum* with *S. punctipenne*, characters, records)

Hawkeswood, 1989: 94 (records, fungal hosts)

Hawkeswood, 1990: 97 (records, fungal hosts)

DISTRIBUTION. Australia: New South Wales, Queensland.

***Scaphidium pygidiale pygidiale* Pic**

Scaphidium pygidiale Pic, 1917b: 3. Syntypes, MNHN; type locality: Madagascar.

DISTRIBUTION. Madagascar.

***Scaphidium pygidiale bicoloricolle* Pic**

Scaphidium pygidiale var. *bicoloricolle* Pic, 1917b: 3. Syntypes, MNHN; type locality: Madagascar.

DISTRIBUTION. Madagascar.

***Scaphidium quadriguttatum* Say**

Scaphidium quadriguttatum Say, 1823: 198. Type material not traceable; type locality: United States.

Scaphidium quadripustulatum Say, 1823: 198 [nec *Scaphidium quadripustulatum* (Fabricius, 1775)]. Type material not traceable; type locality: United States.

Scaphidium quadrinotatum Laporte, 1840: 19. Syntypes, MNHN; type locality: North America.

Scaphidium obliteratum LeConte, 1860: 322. Holotype, MCZC; type locality: USA: Evansville, Indiana.

LeConte, 1860: 322) (characters, records)

Reitter, 1880a: 36 (synonymy of *S. quadrinotatum* with *S. quadriguttatum*, *S. obliteratum* as variety of *S. quadriguttatum*, characters)

Casey, 1893: 514 (*S. obliteratum*, as valid species, characters, records)

Bowditch, 1896: 5 (*S. quadripustulatum*, as var. of *S. quadriguttatum*, records)

Casey, 1900: 56 (*S. obliteratum* as valid species, characters)

Blatchley, 1910: 492 (*S. obliteratum* as variety of *S. quadriguttatum*, characters)

Weiss & West, 1920: 5 (fungal hosts)

Achard, 1921b: 86 (records)

Leonard, 1928: 313 (records)

Brimley, 1938: 149 (*obliteratum* and *quadripustulatum* as subspecies of *S. quadriguttatum*, records)

Kirk, 1969: 30 (records)

Kirk, 1970: 26 (records)

Newton, 1984: 317 (fungal hosts)

Downie & Arnett, 1996: 365 (records)

Campbell, 1991: 125 (*S. obliteratum* as valid species, records)

Downie & Arnett, 1996: 365 (*S. obliteratum* as variety of *S. quadriguttatum*, characters)

Poole & Gentili 1996: 413 (*S. obliteratum* as synonym of *S. quadripustulatum*)

Sikes, 2004: 104 (records)

Majka et al., 2011: 88 (distribution)
Webster et al., 2012: 242 (records, habitat)
Evans, 2014: 132 (characters, records)

DISTRIBUTION. Canada: New Brunswick, Nova Scotia, Ontario, Quebec; United States: Connecticut, Indiana, Iowa, Kansas, Maine, Massachussetts, New Hampshire, New Jersey, New York, North Carolina, Rhode Island, South Carolina, Vermont, Texas.

***Scaphidium quadrillum quadrillum* Fairmaire**

Scaphidium quadrillum Fairmaire, 1898b: 394. Syntypes, MNHN; type locality: Madagascar: Suberbieville [in Maevatanana District].

DISTRIBUTION. Madagascar.

***Scaphidium quadrillum biconjunctum* Pic**

Scaphidium quadrillum var. *biconjunctum* Pic, 1920b: 4. Syntypes, MNHN; type locality: Madagascar [Antongil].

DISTRIBUTION. Madagascar.

***Scaphidium quadrimaculatum* Olivier**

Scaphidium quadrimaculatum Olivier, 1790: 20: 4, pl. 1. Syntypes, MNHN, not traced; type locality: France: environs de Paris.

Fabricius, 1792: 509 (characters)
Panzer, 1792: 1 (characters)
Panzer, 1793: 17 (characters)
Herbst, 1793: 132, pl. 49 (characters)
Paykull, 1800: 338 (characters)
Marsham, 1802: 233 (characters)
Latreille, 1804: 247 (characters)
Gyllenhal, 1808: 186 (characters)
Duftschmid, 1825: 70 (characters)
Stephens, 1830: 2 (characters, records)
Guérin-Méneville, 1834: pl. 17, Fig. 15 (characters)
Laporte, 1840: 18 (characters)
Heer, 1841: 372 (characters, records)
Erichson, 1845: 5 (characters)
Fairmaire & Laboulbène, 1855: 342 (characters, records)
Gutfleisch, 1859: 222 (characters)
Thomson, 1862: 126 (characters)
Sahlberg, 1889: 80 (records)
Heyden, 1880: 88 (records)

Fowler, 1889: 346 (characters)
Paulino de Oliveira, 1895: 124 (records)
Stierlin, 1900: 489 (records)
Everts, 1903: 445 (characters)
Reitter, 1909: 276, pl. 65(characters)
Jakobson, 1910: 636 (records)
Fuente, 1924: 43 (records)
Vitale, 1929: 110 (records)
Roubal, 1930: 297 (records)
Paulian, 1943: 147 (larva)
Scheerpeltz & Höfler, 1948: 149 (fungal hosts, habitat, ecology)
Horion, 1949: 254 (records, habitat)
Novak, 1952: 74 (records)
Benick, 1952: 46 (fungal hosts)
Rehfous, 1955: 19 (fungal hosts)
Tamanini, 1955: 12 (records)
Palm, 1959: 223 (fungal hosts, habitat)
Koch & Lucht, 1962: 29 (records)
Löbl, 1965g: 732 (records)
Kasule, 1966: 266, Figs 14–16 (immature stages)
Kasule, 1968: 117, Figs 4–6 (immature stages)
Löbl, 1968c: 389 (characters, records)
Tamanini, 1969a: 486 (records)
Tamanini, 1969b: 353, Fis +F-I (characters, fungal hosts)
Löbl, 1970a: 10, Figs 4, 7 (characters, records)
Tamanini, 1970: 7, Figs 1B, 2F-I (characters, fungal hosts)
Franz, 1970: 266 (records)
Freude, 1971: 345 (characters, records)
Zachariassen, 1973: 335 (records)
Nuss, 1975: 109 (fungal hosts)
Burakowski et al., 1978: 233 (records, habitat)
Löbl, 1982g: 47 (records)
Iablokoff-Khnzorian, 1985: 137 (records)
Angelini, 1986: 55 (records)
Chandra & Srivaramakrishnan, 1986: 512 (Indian records based on misidentification)
Kompantsev & Pototskaya, 1987: 91 (larva)
Merkl, 1987: 113 (records)
Merkl, 1996: 261 (records)
Nikitsky et al., 1996: 24 (fungal hosts)

Krasutskij, 1996a: 96 (records, fungal hosts)
Krasutskij, 1997a: 307 (habitat, fungal hosts)
Nikitsky et al., 1998: 7 (fungal hosts)
Kofler, 1998: 651 (fungal hosts)
Jeremías & Pérez De-Gregorio, 2003: 57 (records)
Nikitsky & Schigel, 2004: 8, 13 (fungal hosts)
Tronquet, 2006: 79 (records)
Mateleshko, 2005: 128 (fungal hosts)
Byk et al., 2006: 337 (ecology, records)
Borowski, 2006: 54–58 (records, fungal hosts)
Nikitsky et al., 2008: 120 (records, habitat, fungal hosts)
Guéorguiev & Ljubomirov, 2009: 247 (records)
Tsurikov, 2009: 107 (records)
Delwaide & Thieren, 2010: 5 (records)
Vinogradov et al., 2010: 15 (records, fungal hosts)
Zamotajlov & Nikitsky, 2010: 105 (records, fungal hosts)
Grosso-Silva & Soares-Vieira, 2011: 4 (records)
Chillo, 2012: 45 (records, habitat, fungal hosts)
Löbl, 2012: 202 (characters)
Torrella Allegue, 2013: 44 (records, habitat, fungal hosts)
Viñolas et al., 2014: 46 (records)
Hamed & Vancl, 2016: 79 (records)
Löbl & Ogawa, 2016a: 163 (records)
Nikitsky et al., 2016: 138 (records, fungal hosts)
Teofilova, 2017: 77 (records)
Vlasov & Nikitsky, 2017: 4 (records, fungal hosts)

Distribution. Albania; Austria; Belarus; Belgium; Bosnia and Herzegovina; Bulgaria; Croatia; Cyprus; Czech Republik; Denmark, Estonia; Finland; France; Germany; Great Britain; Greece; Hungary; Israel; Italy; Latvia; Lithuania; Luxemburg; Makedonia; Moldova; Montenegro; Netherlands; Norway; Poland; Portugal; Romania; Russia: European, Altai, Siberia; Serbia; Slovakia; Slovenia; Spain; Sweden; Switzerland; Turkey; Ukraine.

Scaphidium quadriplagatum **Achard**

Scaphidium quadriplagatum Achard, 1915b: 290. Syntypes, NMPC; type locality: Malaysia: Sabah, Kinabalu.
Pic, 1916a: 3 (misspelled *quadriplagiatum*, characters)
Achard, 1920d: 125 (characters)
Achard, 1922d: 125 (characters, records)

Distribution. East Malaysia: Sabah; Indonesia: Kalimantan; Brunei.

Scaphidium quadripustulatum (Fabricius)

Sphaeridium quadripustulatum Fabricius, 1775: 67. Lectotype male, NHML; type locality: Australia: "Nova Hollandia".

Scaphidium bimaculatum Macleay, 1871: 119. Lectotype female, ANIC; type locality: Australia: Port Denison.

Olivier, 1790: 20: 4, pl. 1 (transfer to *Scaphidium*)

Löbl, 1976e: 287, Fig. 4 (lectotype fixed by inference for *S. quadripustulatum* and lectotype designation for *S. bimaculatum*, synonymy of *S. bimaculatum* with *S. quadripustulatum*, characters, records)

DISTRIBUTION. Australia: Queensland, Victoria.

Scaphidium quinquemaculatum Pic

Scaphidium 5-maculatum Pic, 1915c: 35. Syntypes, MNHN; type locality: French Guiana, Cayenne.

DISTRIBUTION. French Guiana.

Scaphidium reductum (Pic)

Scaphidiolum reductum Pic, 1954a: 12. Syntypes, MNHN; type locality: Congo.

Leschen & Löbl, 1995: 473 (transfer to *Scaphidium*)

DISTRIBUTION. "Congo".

Scaphidium reitteri Lewis

Scaphidium reitteri Lewis, 1879: 460. Syntypes, NHML; type locality: Japan: Hiogo, near Maiyasan temple.

Scaphidium insulare Achard, 1922c: 38. Syntypes, NMPC, MNHN; type localities: Japan: Kyushu: Kagoshima; Oshima Island.

Lewis, 1893: 293 (records)

Achard, 1923: 106 (characters, records)

Miwa & Mitono, 1943: 524 (characters, records)

Nakane, 1955b: 51, Figs 3, 24, 25 (characters, records)

Nakane, 1963b: 78 (characters)

Shirôzu & Morimota, 1963: 62 (synonymy of *S. insulare* with *S. reitteri*, characters, records)

Morimoto, 1985: Pl. 45, Fig. 18 (characters)

Hoshina & Morimoto, 1999: 88, Figs 6, 29 (characters, records)

Hoshina, 2001: 100, Figs 5, 14–16, 18 (characters, records)

Hoshina, 2015: 46 (records, fungal hosts)

DISTRIBUTION. Japan: Honshu, Kyushu, Ryukyus, Shikoku, Tsushima.

Scaphidium reni **Tang & Li**

Scaphidium reni Tang & Li, 2010a: 73, Figs 12, 13, 35–38. Holotype male, SNUC; type locality: China: Guizhou, Leigonshan N.P., Lianhuaping, 1550–1680 m.

DISTRIBUTION. China: Guizhou.

Scaphidium robustum **Tang, Li & He**

Scaphidium robustum Tang, Li & He, 2014: 70, Figs 40, 41, 110–113. Holotype male, SNUC; type locality: China: Fujian, Wuyishan City, Guadun Vil., 1200–1500 m.

DISTRIBUTION. China: Chongqing, Fujian, Guangxi, Guizhou, Yunnan.

Scaphidium rochaloredoae **Fierros-López**

Scaphidium rochaloredoae Fierros-López, 2005: 93, Figs 72, 73, 113i, j. Holotype male, SEMC; type locality: Panama: La Fortuna Hydro trail, 1150 m, Chiriquí.

DISTRIBUTION. Costa Rica; Panama.

Scaphidium rosenbergi **Pic**

Scaphidium rosenbergi Pic, 1956a: 175. Holotype, MNHN; type locality: Ecuador: Paramba, 3500 m.

DISTRIBUTION. Ecuador.

Scaphidium rouyeri **Pic**

Scaphidium rouyeri Pic, 1915f: 43. Syntypes, MNHN; type locality: Indonesia: Sumatra [Pajucombo].

DISTRIBUTION. Indonesia: Sumatra.

Scaphidium rubicundum **Reitter**

Scaphidium rubicundum Reitter, 1880a: 37. Syntypes, MNHN; type locality: Colombia: Cartagena.

DISTRIBUTION. Colombia.

Scaphidium rubricolle **(Pic)**

Scaphidiolum rubricolle Pic, 1951a: 6. Syntypes, MNHN; type locality: Madagascar: Antongil Bay.

Leschen & Löbl, 1995: 473 (transfer to *Scaphidium*)

DISTRIBUTION. Madagascar.

Scaphidium rubritarse **Pic**

Scaphidium rubritarse Pic, 1915c: 36. Syntypes, ?MNHN, not traced; type locality: Indonesia: Java [Pengarengan, W. Java].

Löbl, 1992a: 496, Figs 19, 41, 62, 73 (characters, records)

DISTRIBUTION. Indonesia: Java; Nepal.

Scaphidium ruficolor Pic

Scaphidium ruficolor Pic, 1915c: 36. Syntypes, MNHN; type locality: Malaysia: Malacca, Perak.
Achard, 1920c: 54 (as aberration of *S. sundaicum* [sic] Gestro
Pic, 1920g: 188 (resurrected)
DISTRIBUTION. West Malaysia.

Scaphidium ruficorne Fairmaire

Scaphidium ruficorne Fairmaire, 1898c: 465. Syntypes, MNHN; type locality: Madagascar: Suberbieville [in Maevatanana District].
DISTRIBUTION. Madagascar.

Scaphidium rufipenne Pic

Scaphidium rufipenne Pic, 1917a: 3. Syntypes, MNHN; type locality: Madagascar: Tananarive.
DISTRIBUTION. Madagascar.

Scaphidium rufipes Pic

Scaphidium rufipes Pic, 1922a: 1. Syntypes, MNHN; type locality: Vietnam: Tonkin [Lac Tho].
DISTRIBUTION. Vietnam.

Scaphidium rufitarse rufitarse Achard

Scaphidium rufitarse Achard, 1920h: 211. Syntypes, NMPC; type locality: Laos.
DISTRIBUTION. Laos.

Scaphidium rufitarse modiglianii Achard

Scaphidium rufitarse var. *modiglianii* Pic, 1920e: 94. Syntypes, MCSN; type locality: Indonesia: Sumatra, Si Rambé [near Balige].
DISTRIBUTION. Indonesia: Sumatra.

Scaphidium rufofemorale Löbl

Scaphidium rufofemorale Löbl, 2006: 42, Figs 16, 17. Holotype male, MHNG; type locality: Philippines: Sagada, 1550 m, Luzon.
DISTRIBUTION. Philippines: Luzon.

Scaphidium rufofemoratum Pic

Scaphidium rufofemoratum Pic, 1921c: 5. Syntypes, MNHN; type locality: Indonesia: Sumatra [Balimbigan = ?Balimbing, 100°03'E, 0°07'N].
DISTRIBUTION. Indonesia: Sumatra.

Scaphidium rufonotatum **(Pic)**
Scaphidiolum rufonotatum Pic, 1928c: 1. Syntypes, MNHN; type locality: Vietnam: Annam, Dalat.
Leschen & Löbl, 1995: 473 (transfer to *Scaphidium*)
DISTRIBUTION. Vietnam.

Scaphidium rufopygum **Lewis**
Scaphidium rufopygum Lewis, 1893: 293. Syntypes, NHML; type localities: Japan: Yuyama, Kiga, Nikko.
Achard, 1923: 107 (characters, records)
Miwa & Mitono, 1943: 531 (characters, records)
Nakane, 1955b: 52, Figs 4, 6 (characters)
Nakane, 1963b: 78 (characters)
Shirôzu & Morimoto, 1963: 61 (characters, records)
Löbl, 1968c (characters)
Morimoto, 1985: Pl. 45, Fig. 16 (characters)
DISTRIBUTION. Japan: Honshu, Kyushu, Shikoku.

Scaphidium rufum **Brancsik**
Scaphidium rufum Brancsik, 1891: 220. Syntypes, FMNH; type locality: Madagascar: Nossibé.
DISTRIBUTION. Madagascar.

Scaphidium rugatum **Löbl**
Scaphidium rugatum Löbl, 1976b: 317. Holotype male, NBCL; type locality: Indonesia: Western New Guinea, Arabu River, 1750 m.
DISTRIBUTION. Indonesia: Western New Guinea.

Scaphidium sakura **Hoshina**
Scaphidium sakura Hoshina, 2001: 101, Figs 1, 4, 8, 11, 17. Holotype male, KMNH; type locality: Japan: Okinawa Island, Chinufiku-Rindo, Kunigami Village.
DISTRIBUTION. Japan: Ryukyus: Okinawa, Kume-jima.

Scaphidium salvazai **(Pic)**
Scaphidiolum salvazai Pic, 1928c: 1. Syntypes, MNHN; type locality: Laos: N. Mo.
Leschen & Löbl, 1995: 473 (transfer to *Scaphidium*)
DISTRIBUTION. Laos.

Scaphidium sauteri **Miwa & Mitono**
Scaphidium sauteri Miwa & Mitono, 1943: 529, Fig. A. Syntypes, TARI; type localities: Taiwan: Henchun, Tapanliah; Kueitzuchiao, Pingtung.

Tang et al., 2014: 60, Figs 20, 21, 86–89 (characters, records)
Tu & Tang, 2017: 595 (records)
DISTRIBUTION. China: Anhui, Fujian, Guangdong, Jiangxi, Zhejiang; Taiwan.

Scaphidium schuelkei **Löbl**
Scaphidium schuelkei Löbl, 1999: 718, Figs 45, 46. Holotype male, MHNG; type locality: China: Shennongjia Nat. Res., 2000–2200 m, Hubei.
DISTRIBUTION. China: Hubei, Shaanxi.

Scaphidium semiflavum **Gestro**
Scaphidium semiflavum Gestro, 1879b: 51. Syntypes, MCSN; type locality: Indonesia: Sumatra, Mt. Singalan.
DISTRIBUTION. Indonesia: Sumatra.

Scaphidium semilimbatum **Pic**
Scaphidium semilimbatum Pic, 1917c: 3. Lectotype male, MNHN; type locality: India: West Bengal, Kurseong, Darjeeling District.
Löbl, 1992a: 495, Fig. 18 (lectotype designation, characters, records)
Note. Accordung to Pic (l.c.) the patria would be "Chine".
DISTRIBUTION. India: West Bengal (Darjeeling District); Nepal.

Scaphidium septemmaculatum **(Pic)**
Scaphidiolum septemmaculatum Pic, 1940a: 2. Syntypes, MNHN; type locality: Vietnam: Tonkin [Phuyen Binh].
Leschen & Löbl, 1995: 474 (transfer to *Scaphidium*)
DISTRIBUTION. Vietnam.

Scaphidium septemnotatum **Champion**
Scaphidium septemnotatum Champion, 1927: 270. Holotype male, NHML; type locality: India: West Bengal, Gopaldhara.
Löbl, 1992a: 497, Figs 21, 44, 65, 80 (characters, records)
DISTRIBUTION. India: West Bengal (Darjeeling District); Nepal.

Scaphidium seriatum **Heller**
Scaphidium seriatum Heller, 1917: 44. Lectotype male, SMTD; type locality: Philippines: Luzon, Mt. Makiling.
Hoshina, 2001: 105, Figs 20, 23, 24 (characters)
Löbl, 2006: 44, Figs 24–26 (lectotype designation, characters)
DISTRIBUTION. Philippines: Luzon.

***Scaphidium shelfordi* (Achard)**

Hemiscaphium shelfordi Achard, 1922b: 34. Holotype male, NMPC; type locality: Malaysia: Sarawak.

Leschen & Löbl, 1995: 474 (transfer to *Scaphidium*)

DISTRIBUTION. East Malaysia: Sarawak.

***Scaphidium shibatai* Kimura**

Scaphidium shibatai Kimura, 1987: 9, Figs 1–4. Holotype male, OMNH; type locality: Taiwan: Mt. Lishan, Taichung Hsien.

DISTRIBUTION. Taiwan.

***Scaphidium shirakii* Miwa & Mitono**

Scaphidium shirakii Miwa & Mitono, 1943: 525, 555: Fig. E. Syntypes, TARI; type locality: Niitaka, Nantou.

DISTRIBUTION. Taiwan.

***Scaphidium sichuanum* Löbl**

Scaphidium sichuanum Löbl, 1999: 720, Figs 32, 49, 50. Holotype male, MHNG; type locality: China: south Xichang, Mt. Luoji, 2300–2400 m, Sichuan.

DISTRIBUTION. China: Sichuan.

***Scaphidium sikorai* (Pic)**

Ascaphidium sikorai Pic, 1915a: 24. Syntypes, MNHN; type locality: Madagascar, Annanarivo.

Scaphidium latissimum Achard, 1915b: 290. Syntypes, NMPC; type locality: Madagascar: Tananarive.

Pic, 1916a: 3 (synonymy of *S. latissimum* with *S. sikorai*)

Leschen & Löbl, 1995: 474 (transfer to *Scaphidium*)

DISTRIBUTION. Madagascar.

***Scaphidium simile* Laporte**

Scaphidium simile Laporte, 1840: 19. Syntypes, MNHN; type locality: Madagascar.

DISTRIBUTION. Madagascar.

***Scaphidium simplicicolle* (Pic)**

Hemiscaphium simplicicolle Pic, 1923b: 17. Syntypes, MNHN; type locality: Borneo.

Leschen & Löbl, 1995: 474 (transfer to *Scaphidium*)

DISTRIBUTION. "Borneo".

***Scaphidium sinense* Pic**

Scaphidium sinense Pic, 1954c: 57. Holotype female, NHRS; type locality: China: Fujian, Kuatun.

Löbl, 1992a: 495, Fig. 17 (characters)
Löbl, 1999: 711, Fig. 23 (characters)
He et al., 2008c: 59, Figs 2, 3, 7, 11, 16, 19, 22 (characters, records)
Tang et al., 2014: 64, Figs 28–33, 98–101, 154 (characters, records, immature stages)
Tu & Tang, 2017: 575 (records)

DISTRIBUTION. China: Fujian, Guangxi, Hunan, Jiangxi, Zhejiang.

***Scaphidium sinuatum* Csiki**

Scaphidium sinuatum Csiki, 1924: 32. Syntypes, HNHM; type locality: Taiwan: Fuhosho, Kosempo.

Miwa & Mitono, 1943: 594, 555 Fig. F (characters, records)

DISTRIBUTION. Taiwan.

***Scaphidium solukhumbu* Löbl & Ogawa**

Scaphidium solukhumbu Löbl & Ogawa, 2016a: 159, Figs 1, 3, 4. Holotype male, SMNS; type locality: Nepal: Solukhumbu, Junbesi, 2700 m.

DISTRIBUTION. Nepal.

***Scaphidium sondaicum* Gestro**

Scaphidium sondaicum Gestro, 1879b: 54. Syntypes, MCSN; type locality: Indonesia: Sumatra, Ajer Mantcior [near Padangpanjang].

Csiki, 1910: 8 (records)
Achard, 1920c: 53 (misspelled *sundaicum*)
Pic, 1920g: 188 (notes)
Ogawa, 2015: 54, Figs 4-2f, 4-3b, 4-4k, 4-7a-c (characters, records)

DISTRIBUTION. Indonesia: Sumatra.

***Scaphidium spinatum* Tang & Li**

Scaphidium spinatum Tang & Li, 2010a: 70, Figs 6, 7, 23–26. Holotype male, HBUM; type locality: China: Anhui, Yuexi County, Yaoluoping.

Tang et al., 2014: 54, Figs 11, 12, 74–77, 163 (characters, records, immature stages)

DISTRIBUTION. China: Anhui.

***Scaphidium stigmatinotum* Löbl**

Scaphidium stigmatinotum Löbl, 1999: 719, Fig. 31. Holotype female, MHNG; type locality: China: Yizu Mts [=Jizushan], 2800 m, Yunnan.
He et al., 2008c: 60, Figs 4, 8, 12, 13, 17, 20, 23 (characters, records)
Tang et al., 2014: 73, Figs 45, 47, 122–127 (characters, records)
Tang & Li, 2013: 179 (records)
DISTRIBUTION. China: Anhui, Fujian, Guangxi, Guangdong, Hunan, Jiangsu, Shaanxi, Yunnan, Zhejiang.

***Scaphidium striatipenne striatipenne* Gestro**

Scaphidium striatipenne Gestro, 1879b: 55. Holotype female, MCSN; type locality: Indonesia: Java, Cibodas.
Achard, 1922b: 34 (transfer to *Hemiscaphium*, characters, records)
Leschen & Löbl, 1995: 474 (transfer to *Scaphidium*)
DISTRIBUTION. Indonesia: Java, Sumatra.

***Scaphidium striatipenne ornatipenne* Achard**

Hemiscaphium striatipenne var. *ornatipenne* Achard, 1922b: 34. Holotype female, NMPC; type locality: Indonesia: Java, Preanger.
Leschen & Löbl, 1995: 474 (implicit transfer to *Scaphidium*)
DISTRIBUTION. Indonesia: Java.

***Scaphidium striatum* Pic**

Scaphidium striatum Pic, 1920e: 94. Syntypes, MCSN/MNHN; type locality: Indonesia: Java, Buitenzorg [Bogor].
Achard, 1922b: 32 (transfer to *Hemiscaphium*)
Leschen & Löbl, 1995: 474 (transfer to *Scaphidium*)
DISTRIBUTION. Indonesia: Java.

***Scaphidium suarezicum* Pic**

Scaphidium suarezicum Pic, 1920f: 23. Syntypes, MNHN; type locality: Madagascar: Diego-Suarez.
DISTRIBUTION. Madagascar.

***Scaphidium subdepressum* Pic**

Scaphidium subdepressum Pic, 1921c: 5. Syntypes, MNHN; type locality: Indonesia: Sumatra [Brastagi].
Pic, 1923b: 16 (transfer to *Falsoascaphidium*, characters)
Leschen & Löbl, 1995: 474 (transfer to *Scaphidium*)
DISTRIBUTION. Indonesia: Sumatra.

***Scaphidium subelongatum* Pic**

Scaphidium subelongatum Pic, 1915f: 44. Syntypes, MNHN; type locality: Malaysia: Sabah [Kinabalu].
Achard, 1920d: 124 (characters, records)
DISTRIBUTION. "Borneo"; East Malaysia: Sabah; Indonesia: Java.

***Scaphidium subpunctatum* (Pic)**

Scaphidiolum subpunctatum Pic, 1951a: 5. Syntypes, MNHN; type locality: Madagascar, Diego-Suarez.
Leschen & Löbl, 1995: 474 (transfer to *Scaphidium*)
DISTRIBUTION. Madagascar.

***Scaphidium sulcaticolle* (Pic)**

Hemiscaphium sulcaticolle Pic, 1923b: 16. Syntypes, MNHN; type locality: Borneo [Riam Kanan, Martapura].
Leschen & Löbl, 1995: 474 (transfer to *Scaphidium*)
DISTRIBUTION. Indonesia: Kalimantan.

***Scaphidium sulcatum* Pic**

Scaphidium sulcatum Pic, 1915f: 43. Syntypes, MNHN; type locality: Malaysia: Banguey [Banggi Is.].
Achard, 1920d: 126 (records)
Achard, 1922b: 32 (transfer to *Hemiscaphium*)
Leschen & Löbl, 1995: 474 (transfer to *Scaphidium*)
DISTRIBUTION. East Malaysia: Sabah.

***Scaphidium sulcipenne* Gestro**

Scaphidium sulcipenne Gestro, 1879b: 57. Holotype female, MCSN; type locality: Indonesia: Sumatra, Ajer Mantcior [near Padangpanjang].
Achard, 1922b: 32 (transfer to *Hemiscaphium*, characters, records)
Leschen and Löbl, 1995: 474 (transferred to *Scaphidium*)
DISTRIBUTION. Indonesia: Sumatra; East Malaysia: Sarawak, West Malayasia: Penang.

***Scaphidium sylhetense* Achard**

Scaphidium sylhetense Achard, 1920j: 263. Lectotype male, NHML; type locality: Bangladesh: Sylhet.
Champion, 1927: 269 (lectotype designation, characters)
Löbl, 1986c: 343 (records)
Löbl, 1992a: 490, Figs 15, 53, 59, 79 (characters, records)

DISTRIBUTION. Bangladesh; Bhutan; India: Uttarakhand (Kumaon, Garhwal), West Bengal (Darjeeling District); Nepal.

***Scaphidium taichii* Kimura**

Scaphidium taichii Kimura, 2008: 158, Figs 1–8. Holotype male, KCMI; type locality: Japan: Miyanoura, Yaku Is., Kagoshima Pref.

DISTRIBUTION. Japan: Yaku-shima.

***Scaphidium takahashii* Miwa & Mitono**

Scaphidium takahashii Miwa & Mitono, 1943: 529, 555: Fig. D. Type material not located; type locality: Taiwan: Taipingshan, Ilan.

DISTRIBUTION. Taiwan.

***Scaphidium takemurai* Nakane**

Scaphidium takemurai Nakane, 1956: A 160. Holotype female, HUSJ; type locality: Japan: Kyushu, Kirishima.
Nakane, 1963b: 79 (characters)
Shirôzu & Morimoto, 1963: 73 (characters, records)
Morimoto, 1985: Pl. 45, Fig. 21 (characters)

DISTRIBUTION. Japan: Kyushu.

***Scaphidium tecuani* Fierros-López**

Scaphidium tecuani Fierros-López, 2005: 95, Figs 74, 75b 113k, l. Holotype male, EMEC; type locality: Mexico: Estación de Biología Los Tuxtlas, 1350 m, Catemaco, Veracruz.

DISTRIBUTION. Mexico.

***Scaphidium testaceum* Reitter**

Scaphidium testaceum Reitter, 1880a: 38. Syntypes, MNHN; type locality: Brazil.

DISTRIBUTION. Brazil.

***Scaphidium teuhtimaculatum* Fierros-López**

Scaphidium teuhtimaculatum Fierros-López, 2005: 98, Figs 76, 77, 114a, b. Holotype male, SEMC; type locality: Panama: 20 km north Gualaca, Finca La Suiza, Chiriquí.

DISTRIBUTION. Costa Rica; Panama.

***Scaphidium thakali* Löbl**

Scaphidium thakali Löbl, 1992a: 490, Figs 52, 61, 78. Holotype male, MHNG; type locality: Nepal: Lete, 2550 m, Mustang District.

DISTRIBUTION. Nepal.

Scaphidium thayerae **Fierros-López**

Scaphidium thayerae Fierros-López, 2005: 101, Figs 78, 79, 114c, d. Holotype male, FMNH; type locality: Mexico: Rd to Yuvila 3.3 mi east jet Mex 175, 2468 ft, Oaxaca.

DISTRIBUTION. Mexico.

Scaphidium theni **Fierros-López**

Scaphidium theni Fierros-López, 2005: 103, Figs 80, 81, 114[e], f. Holotype male, CZUG; type locality: Mexico: Dos Amates, Catemaco, Veracruz.

DISTRIBUTION. Mexico.

Scaphidium thomasi **(Pic)**

Scaphidiolum thomasi Pic, 1926b: 3. Syntypes, MNHN; type locality: Philippines: Luzon [St. Thomas].

Leschen & Löbl, 1995: 474 (transfer to *Scaphidium*)

Löbl, 2006: 45, Figs 27, 28 (characters, records)

DISTRIBUTION. Philippines: Luzon.

Scaphidium tibiale **Fierros-López**

Scaphidium tibiale Fierros-López, 2005: 105, Figs 82, 83, 114g, h. Holotype male, SEMC; type locality: Costa Rica: Estación Biológica Cacao, 1050 m, Guanacaste.

DISTRIBUTION. Costa Rica.

Scaphidium tiboense **(Pic)**

Scaphidiolum tiboense Pic, 1928b: 35. Syntypes, MRAC; type locality: Democratic Republic of the Congo: Tibo.

DISTRIBUTION. Democratic Republic of the Congo.

Scaphidium tlatlauhqui **Fierros-López**

Scaphidium tlatlauhqui Fierros-López, 2005: 107, Figs 84, 85, 114i, j. Holotype male, CZUG; type locality: Mexico: Arroyo Tacubaya, Casimiro Castillo, Jalisco.

DISTRIBUTION. Guatemala; Honduras; Mexico.

Scaphidium tlileuac **Fierros-López**

Scaphidium tlileuac Fierros-López, 2005: 110, Figs 86, 87, 114k, l. Holotype male, SEMC; type locality: Costa Rica: Estación Biológica Palo Verde, Sendero Natural la Venada, 10 m, Guanacaste.

DISTRIBUTION. Costa Rica; Nicaragua.

Scaphidium tlilmetztli Fierros-López

Scaphidium tlilmetztli Fierros-López, 2005: 113, Figs 88, 89, 115a, b. Holotype male, EMEC; type locality: Mexico: Simojovel, Chiapas.

DISTRIBUTION. Costa Rica; Mexico.

Scaphidium transversale Matthews

Scaphidium transversale Matthews, 1888: 164, Pl. 4, Fig. 6. Holotype male, NHML; type locality: Panama: Caldera in Chiriquí.

Fierros-López, 2005: 116, Figs 90, 91, 115c, d (characters, records)

DISTRIBUTION. Costa Rica; Panama.

Scaphidium tricolor Achard

Scaphidium tricolor Achard, 1920b: 126. Lectotype male, MNHN; type locality: Indonesia: Sumatra, Palembang.

Achard, 1922b: 34 (lectotype fixed by inference, transfer to *Hemiscaphium*, characters, records)

Leschen & Löbl, 1995: 474 (transfer to *Scaphidium*)

DISTRIBUTION. Indonesia: Java, Sumatra.

Scaphidium trimaculatum Löbl

Scaphidium trimaculatum Löbl, 1976e: 294, Figs 3, 3a, 11. Holotype male, QMBA; type locality: Australia: Queensland, Cooper Creek, 21 km north of Daintree River.

DISTRIBUTION. Australia: Queensland.

Scaphidium trinotatum Pic

Scaphidium trinotatum Pic, 1920f: 23. Syntypes, MNHN; type locality: Malaysia: Sabah, Kinabalu.

DISTRIBUTION. East Malaysia: Sabah.

Scaphidium tuberculipes (Löbl)

Scaphidiolum tuberculipes Löbl, 1972b: 45, Figs 1–4. Holotype male, ZMUB; type locality: Philippines: Bucas Sorocco.

Leschen & Löbl, 1995: 474 (transfer to *Scaphidium*)

Löbl, 2006: 46 (characters, records)

DISTRIBUTION. Philippines: Bucas, Siargao.

Scaphidium tzinti Fierros-López

Scaphidium tzinti Fierros-López, 2005: 118; Figs 92, 93, 115e, f. Holotype male, SEMC; type locality: Honduras: 7.6 km north Guaimaca, 820 m, Francisco Morazán.

DISTRIBUTION. Guatemala; Honduras; Mexico.

Scaphidium uinduri Fierros-López

Scaphidium uinduri Fierros-López, 2005: 121, Figs 94, 95, 115g, h. Holotype male, CZUG; type locality: Mexico: Arroyo Tacubaya, 600 m, Casimiro Castillo, Jalisco.

DISTRIBUTION. Mexico.

Scaphidium undulatum Pic

Scaphidium undulatum Pic, 1915c: 36. Syntypes, MNHN; type locality: Brazil.

DISTRIBUTION. Brazil.

Scaphidium unifasciatum Pic

Scaphidium unifasciatum Pic, 1916c: 17. Syntypes, MNHN; type locality: China: Yunnan [Fou].

Achard, 1920h: 210 (characters)

Löbl, 1999: 711, Fig. 22 (characters, records)

DISTRIBUTION. China: Yunnan.

Scaphidium vagefasciatum Pic

Scaphidium vagefasciatum Pic, 1920b: 4. Syntypes, MNHN; type locality: Madagascar [Diego Suarez].

Note. Pic (l.c.) credited the species epithet to Portevin.

DISTRIBUTION. Madagascar.

Scaphidium variabile Matthews

Scaphidium variabile Matthews, 1888: 162, Pl. 4, Fig. 3. Lectotype male, NHML; type locality: Panama: Volcán Chiriquí.

Fierros-López, 2005: 124, Figs 96, 97, 115i, j (lectotype designation, characters, records)

DISTRIBUTION. Costa Rica, Panama.

Scaphidium variegatum Pic

Scaphidium variegatum Pic, 1915c: 35. Syntypes, MNHN; type locality: French Guiana: Cayenne.

[*Scaphidium variegatum* var. *portevini* Pic, 1920b: 4. Syntypes, MNHN; type locality: French Guiana: Cayenne. Note: Sympatric with the nominate subspecies, deemed infrasubspecific]

DISTRIBUTION. French Guiana.

Scaphidium varifasciatum Tang, Li & He

Scaphidium varifasciatum Tang, Li & He, 2014: 58, Figs 17–19, 82–85. Holotype male, SNUC; type locality: China: Zhejiang, Lin'an city, West Tianmushan, 1000 m.

DISTRIBUTION. China: Anhui, Zhejiang.

Scaphidium vernicatum (Pic)

Scaphium vernicatum Pic, 1954c: 57. Lectotype male, NHRS; type locality: China: Kuatun, Fujian.

Löbl, 1992a: 482 (transfer to *Scaphidium*)

Löbl, 1999: 711, Fig. 24 (lectotype designation, characters, records)

Tang & Li, 2010b: 319, Figs 2, 5, 11, 12 (characters)

Tang et al., 2014: 76, Figs 50, 51, 130–133 (characters, records)

Tu & Tang, 2017: 596 (records)

DISTRIBUTION. China: Fujian, Jiangxi.

Scaphidium vicinum Pic

Scaphidium vicinum Pic, 1915e: 2. Syntypes, MNHN; type locality: Taiwan: Fainan.

Scaphidium castaneicolor Csiki, 1924: 32. Syntypes, HNHM; type localities: Taiwan: Fuhosho, Kosempo.

Scaphidium longipenne Achard, 1921b: 85. Syntypes, NBCL; type localities: Taiwan: Fuhosho and Kosempo.

Miwa & Mitono, 1943: 530 (synonymy of *S. castaneicolor* and *S. longipenne* with *S. vicinum*, characters, records)

Tu & Tang, 2017: 595 (records)

DISTRIBUTION. China: Fujian; Taiwan.

Scaphidium violaceipenne Pic

Scaphidium violaceipenne Pic, 1927: 10. Syntypes, MNHN; type locality: Vietnam: Tonkin, Chapa.

DISTRIBUTION. Vietnam.

Scaphidium viride Löbl

Scaphidium viride Löbl, 1978c: 117, Fig. 4. Holotype male, BPBM; type locality: Papua New Guinea: Mt. Missim, 1600–2000 m, Morobe District.

DISTRIBUTION. Papua New Guinea.

Scaphidium vitalisi Achard

Scaphidium vitalisi Achard, 1920h: 211. Syntypes, NMPC; type locality: Laos: Upper Mekong, Vien Poukha [Vien Pou Kha].

DISTRIBUTION. Laos.

Scaphidium vittipenne Oberthür

Scaphidium vittipenne Oberthür, 1883: 12. Syntypes, MNHN; type localities: Peru: Pebas and Iquitos.

DISTRIBUTION. Peru.

***Scaphidium waterstradti* Pic**

Scaphidium waterstradti Pic, 1915e: 4. Syntypes, MNHN; type locality: Malayasia: Banguey [Banggi Island].

DISTRIBUTION. East Malayasia: Sabah, Banggi Island.

***Scaphidium wuyongxiangi* He, Tang & Li**

Scaphidium wuyongxiangi He, Tang & Li, 2008c: 57, Figs 1, 5, 9, 14, 18, 21. Holotype male, SNUC; type locality: China: Zhejiang Prov., Lin'an City, Tianmushan.
Tang & Li, 2013: 178 (records)
Tang et al., 2014: 74, Figs 48, 49, 126–129, 166 (characters, records)

DISTRIBUTION. China: Anhui, Fujian, Jiangxi, Sichuan, Zhejiang.

***Scaphidium xicaltetl* Fierros-López**

Scaphidium xicaltetl Fierros-López, 2005: 126, Figs 98, 99, 115k, l. Holotype male, SEMC; type locality: Costa Rica: Reserva Biológica San Ramón, 810 m, Alajuela.

DISTRIBUTION. Costa Rica.

***Scaphidium xolotl* Fierros-López**

Scaphidium xolotl Fierros-López, 2005: 129, Figs 100, 101, 116a, b. Holotype male, CC-UAEH; type locality: Mexico: Huejutla, Río Atlapexco, 120 m, Hidalgo.

DISTRIBUTION. Mexico.

***Scaphidium yasumatsui* Shirôzu & Morimoto**

Scaphidium yasumatsui Shirôzu & Morimoto, 1963: 67. Holotype male, ?KUIC; type locality: Japan: Mt. Hiko, Fukuoka Pref.
Hisamatsu, 1977: 193 (records)
Morimoto, 1985: Pl. 45, Fig. 19 (characters)

DISTRIBUTION. Japan: Kyushu, Shikoku.

***Scaphidium yayactic* Fierros-López**

Scaphidium yayactic Fierros-López, 2005: 131, Figs 102, 103, 116c, d. Holotype male, CZUG; type locality: Mexico: Racho La Peña, Reserva de la Biosfera La Michilia, Durango.

DISTRIBUTION. Mexico.

***Scaphidium yeilineatum* Fierros-López**

Scaphidium yeilineatum Fierros-López, 2005: 133, Figs 104, 105, 116e, f. Holotype female, FMNH; type locality: Mexico: east slope of Nevado de Colima, 6800 ft, Jalisco.

DISTRIBUTION. Mexico.

***Scaphidium yeti* Löbl & Ogawa**

Scaphidium yeti Löbl & Ogawa, 2016a: 161, Figs 2, 6, 7. Holotype male, NMEC; type locality: Nepal: Rolwaling Himal, upp. Simigau village, 2600 m.

DISTRIBUTION. Nepal.

***Scaphidium yinziweii* Tang & Li**

Scaphidium yinziweii Tang & Li, 2012: 186, Figs 1, 2, 9–12. Holotype male, SNUC; type locality: China: Yunnan, Nabanhe National Reserve, Benggang Hani, 1700 m.

DISTRIBUTION. China: Yunnan.

***Scaphidium yocupitziae* Fierros-López**

Scaphidium yocupitziae Fierros-López, 2005: 135, Figs 106, 107, 116g, h. Holotype male, SEMC; type locality: Panama: Estación Biológica Cana, Serranía del Pirre, 800 m, Darién.

DISTRIBUTION. Costa Rica; Panama.

***Scaphidium yunnanum* Fairmaire**

Scaphidium yunnanum Fairmaire, 1886: 318. Lectotype female, MNHN; type locality: China: Yunnan.

Achard, 1920d: 124 (note about type material)

Achard, 1920h: 209 (characters)

Löbl, 1999 (lectotype designation, characters)

DISTRIBUTION. China: Sichuan, Yunnan.

***Scaphidium yuzhizhoui* Tang, Tu & Li**

Scaphidium yuzhizhoui Tang, Tu & Li, 2016b: 279, Figs 1–6, 9, 11. Holotype male, SNUC; type locality: China: Guangdong, Ruyuan County, Nanling N.R., 1050 m.

DISTRIBUTION. China: Guangdong.

***Scaphidium zhoushuni* He, Tang & Li**

Scaphidium zhoushuni He, Tang & Li, 2009: 481, Figs 1, 3, 5, 7, 9, 10. Holotype male, SNUC; type locality: China: Huashuping, Hu County, Shaanxi, 1700 m.

Tang & Li, 2012: 187, Fig. 3, 4 (characters)

DISTRIBUTION. China: Chongqing, Shaanxi.

SCAPHISOMATINI Casey

Scaphisomini Casey, 1893: 511. Type genus: *Scaphisoma* Leach, 1815.

Baeoceridiitae Achard, 1924b: 39. Type genus: *Baeoceridium* Reitter, 1889. Included in Scaphisomatini: Löbl, 1997.
Baeoceritae Achard, 1924b: 30. Type genus: *Baeocera* Erichson, 1845. Included in Scaphisomatini: Löbl, 1969b.
Cyparellini Achard, 1924b: 28. Type genus: *Cyparella* Achard, 1924 (= *Baeocera* Erichson, 1845).
Heteroscaphini Achard, 1914: 395. Type genus: *Heteroscapha* Achard, 1914 (= *Bironium* Csiki, 1909). Included in Scaphisomatini: Löbl, 1997.
Scaphicomitae Achard, 1924b: 31. Type genus: *Scaphicoma* Motschulsky, 1863. Included in Toxidiini: Löbl, 1971c.
Scaphisomidae Tamanini, 1969c: 136. Type genus: *Scaphisoma* Leach, 1815.
Sciatrophitae Achard, 1924b: 30. Type genus: *Sciatrophes* Blackburn, 1903 (= *Baeocera* Erichson, 1845).
Toxidiini Achard, 1924b: 31. Type genus: *Toxidium* LeConte, 1860. Included in Scaphisomatini: Löbl, 1997.
Toxidiitae; Achard, 1924b: 31.

Reitter, 1908: 31 (key to East African genera)
Achard, 1924b: 28 (characters)
Casey, 1893: 511 (characters)
Ganglbauer, 1899: 341 (characters)
Pic, 1928b: 35 (key to taxa of the Democratic Republic of the Congo)
Tamanini, 1969c: 131 (as Scaphisomini, characters)
Löbl, 1970a: 10 (as Scaphisominae, characters)
Löbl, 1970c: 729 (key to Palaearctic genera)
Löbl, 1971c: 942, 985 (as Scaphisomini and Toxiini respectively, keys to Sri Lankan genera)
Löbl, 1987d: 856 (synonymy of Cyparellini with Scaphisomatini)
Löbl, 1992a: 476, Figs 3–8 (prothoracic corbiculum)
Newton & Thayer, 1992: 63, 64 (type genera of the family group taxa, correction of stem formation)
Leschen & Löbl, 1995: 447 (characters)
Newton et al., 2000: 375 (characters)
Löbl & Leschen, 2003b: 18 (characters)
Leschen & Löbl, 2005: 3 (phylogeny, characters, mycophagy, termitophily, key to genera)

Afroscaphium Löbl

Afroscaphium Löbl, 1989b: 277. Type species: *Afroscaphium palpale* Löbl, 1898; by original designation. Gender: neuter.

***Afroscaphium palpale* Löbl**

Afroscaphium palpale Löbl, 1989b: 279, Figs 2, 3. Holotype male, MRAC; type locality: Democratic Republic of the Congo: riv. Gombe, terr. Ikela, Tschuapa.

DISTRIBUTION. Democratic Republic of the Congo.

***Afroscaphium striatulum* Löbl**

Afroscaphium striatulum Löbl, 1989b: 280, Fig. 1. Holotype female, MRAC; type locality: Democratic Republic of the Congo: riv. Gombe, terr. Ikela, Tschuapa. Gender: feminine.

DISTRIBUTION. Democratic Republic of the Congo.

Alexidia Reitter

Alexidia Reitter, 1880a: 43. Type species: *Alexidia rogenhoferi* Reitter, 1880; by monotypy. Gender: feminine.

Löbl & Leschen, 2003a: 316 (characters, key to species)

***Alexidia carltoni* Löbl & Leschen**

Alexidia carltoni Löbl & Leschen, 2003a: 318, Figs 8, 10. Holotype male, SEMC; type locality: Ecuador: Azuay 50 km NW Cuenca, 2470 m.

DISTRIBUTION. Ecuador.

***Alexidia dybasi* Löbl & Leschen**

Alexidia dybasi Löbl & Leschen, 2003a: 322, Figs 11–14. Holotype male, FMNH; type locality: Panama: El Valle, traile to Las Minas, Cocle Prov.

DISTRIBUTION. Panama.

***Alexidia plaumanni* Löbl & Leschen**

Alexidia plaumanni Löbl & Leschen, 2003a: 319, Figs 1, 3–7. Holotype male, MHNG; type locality: Brazil: Santa Catharina.

DISTRIBUTION. Brazil.

***Alexidia rogenhoferi* Reitter**

Alexidia rogenhoferi Reitter, 1880a: 44. Syntypes, MNHN; type locality: Colombia: New Granada.

Löbl & Leschen, 2003a: 317 (characters)

DISTRIBUTION. Colombia.

Amalocera Erichson

Amalocera Erichson, 1845: 4. Type species: *Amalocera picta* Erichson, 1845; by monotypy. Gender: feminine.

[*Omalocera* Reitter, 1880a: 43 misspelling]
Lacordaire, 1854: 240 (characters)
Löbl, 1974d: 39 (characters, key to species)

Amalocera basipennis **Löbl**
Amalocera basipennis Löbl, 1974d: 43, Figs 8, 9, 13. Holotype male, MNHN; type locality: Brazil: Ega.
DISTRIBUTION. Brazil.

Amalocera dentifera **Löbl**
Amalocera dentifera Löbl, 1974d: 40, Figs 1–3, 10. Holotype male, MHNG; type locality: Brazil: Santa Catarina, Nova Teutonia, 300–500 m.
DISTRIBUTION. Brazil.

Amalocera paulistana **Achard**
Amalocera paulistana Achard, 1922c: 42. Syntypes, NMPC; type locality: Brazil: near San Paulo.
Löbl, 1974d: 40, Figs 6, 7, 11 (characters)
DISTRIBUTION. Brazil.

Amalocera picta **Erichson**
Amalocera picta Erichson, 1845: 4, Lectotype female, ZMUB; type locality: Brazil.
Löbl, 1974d: 39 (lectotype fixed by inference)
DISTRIBUTION. Brazil.

Amalocera tibialis **Löbl**
Amalocera tibialis Löbl, 1974d: 41, Figs 4, 5, 12. Holotype male, MNHN; type locality: Brazil: Caraça.
DISTRIBUTION. Brazil.

Amaloceromorpha Pic

Amaloceromorpha Pic, 1920f: 23. Type species: *Amaloceromorpha rufa* Pic, 1920; by monotypy. Gender: feminine.
Pic, 1921a: 164 (characters)

Amaloceromorpha rufa **Pic**
Amaloceromorpha rufa Pic, 1920f: 23. Syntypes, MCSN; type locality: Myanmar: Tenasserim, Meetan.
Pic, 1921a: 164 (characters)
DISTRIBUTION. Myanmar.

Baeocera Erichson

Baeocera Erichson, 1845: 4. Type species: *Baeocera falsata* Achard, 1920; ICZN, 1982: 175, Oppinion 1221: Official List of Specific Names in Zoology, name number 2163. Gender: feminine.

Sciatrophes Blackburn, 1903: 100. Type species: *Sciatrophes latens* Blackburn, 1903; by monotypy. Gender: masculine.

Cyparella Achard, 1924b: 28. Type species: *Scaphisoma rufoguttatum* Fairmaire, 1898; by original designation. Gender: feminine.

Amaloceroschema Löbl, 1967a: 1 (as subgenus). Type species: *Baeocera freudei* Löbl, 1967; by original designation. Gender: neuter.

Eubaeocera Cornell, 1967: 2, Fig. 1. Type species: *Baeocera abdominalis* Casey, 1900; by original designation. Gender: feminine.

Lacordaire, 1854: 240 (characters)
LeConte & Horn, 1883: 111 (characters)
Reitter, 1880a: 44 (key to species)
Reitter, 1886: 7 (characters)
Matthews, 1888: 168 (characters, key to Central American and Mexican species)
Casey, 1893: 515 (characters, key to species of the United States)
Reitter, 1899: 158 (key to European species)
Ganglbauer, 1899: 345 (characters)
Casey, 1900: 57 (key to species of the United States)
Blatchley, 1910: 493 (key to Indiana species)
Achard, 1923: 117 (characters)
Cornell, 1967: 2, 4 (*Baeocera* Erichson in synonymy with *Cyparium*, *Baeocera* sensu auct. replaced by *Eubaeocera*, key to species of the United States)
Löbl, 1969b: 328 (*Eubaeocera*, characters, nomenclature, key to Palaearctic species)
Tamanini, 1969c: 133, Figs 24, 28, 30, 33, 38, 42 (*Eubaeocera*, charactes)
Löbl, 1971c: 942 (*Eubaeocera*, key to Sri Lankan species, species groups)
Löbl, 1975a: 374 (*Eubaeocera*, key to New Guinean species)
Löbl, 1976c: 207 (nomenclature, species groups)
Löbl, 1976d: 777 (nomenclature)
Löbl, 1977e: 5 (synonymy of *Eubaeocera* with *Sciatrophes*, key to Australian species)
Löbl, 1977h: 101 (nomenclature, type species)
Löbl, 1979a: 86 (key to South Indian species, species groups)
Lawrence & Newton, 1980: 132, 137, Figs 4, 7, 8 (morphology, larva, hosts)
Löbl, 1980a: 93 (key to Taiwanese species)

Löbl, 1980c: 380 (characters, key to Fijian species)
Löbl, 1981a: 350 (key to New Caledonian species)
ICZN, 1982: 175 (fixation of type species)
Löbl, 1983b: 162 (key to Chilean species)
Newton, 1984: 319 (fungal and slime mould hosts)
Löbl, 1984a: 62 (key to Northeast Indian and Bhutanese species, species groups)
Löbl, 1984b: 182 (key to Japanese species)
Iablokoff-Khnzorian, 1985: 138 (characters, key to Russian species)
Löbl, 1987a: 86 (key to Bornean species)
Löbl, 1987d: 856 (synonymy of *Cyparella* with *Baeocera*)
Lawrence, 1989: 10, 12 (immature stages, feeding)
Newton & Stephenson, 1990: 204 (hosts)
Löbl, 1990b: 515 (key to Thai species)
Löbl, 1992a: 504 (species groups, key to Himalayan species)
Leschen, 1993: 73 (larval mandible)
Löbl & Stephan, 1993: 678 (characters, key to species of America north of Mexico, species groups)
Löbl, 1997: 45 (implicite synonymy of *Amaloceromorpha* with *Baeocera*)
Löbl, 1999: 722 (key to Chinese species)
Löbl, 2002a: 2 (key to New Guinean species)
Löbl & Leschen, 2003b: 27 (characters, key to New Zealand species)
Fierros-López, 2010: 206 (key to Mexican species)
Hoshina et al., 2011: 45 (key to Korean species)
Löbl, 2012a: 129 (key to Taiwanese species)
Löbl, 2012d: 352 (key to Philippine species)
Ogawa & Löbl, 2013: 302 (characters, key to Japanese species)
Löbl, 2014a: 55 (key to Maluku Islands species)
Ogawa, 2015: 104 (characters, key to Sulawesi species)
Löbl, 2015a: 76 (key to Lesser Sunda species)

Baeocera abdominalis **Casey**

Baeocera abdominalis Casey, 1900: 58. Lectotype male, USNM; type locality: USA: Tyngsboro, Massachusetts.
Cornell, 1967: 6, Figs 2E, 4F, N (lectotype fixed by inference, transfer to *Eubaeocera*, characters, records)
Löbl & Stephan, 1993: 706, Figs 37, 125, 126 (characters, records)

DISTRIBUTION. United States: Georgia, Illinois, Maine, Maryland, Massachusetts, Mississippi, New Jersey, North Carolina, Texas, Virginia.

Baeocera abnormalis Nakane

Baeocera abnormalis Nakane, 1963a: 22. Holotype female, HUSJ; type locality: Japan: Honshu, Daisen, Tottori.
Nakane, 1963b: 79 (characters)
Löbl, 1969b: 337, Figs 20, 21 (transfer to *Eubaeocera*, characters, records)
Löbl, 1984b: 188, Figs 10, 11 (characters, records)
Morimoto, 1985: 258 (characters)
Ogawa & Löbl, 2013: 317, Figs 5, 14, 27, 52, 59, 65, 77left, 82 (characters, records)

DISTRIBUTION. Japan: Honshu, Kyushu, Shikoku.

Baeocera abrupta Löbl & Leschen

Baeocera abrupta Löbl & Leschen, 2003b: 29, Figs 11, 37–39, Map 1. Holotype male, NZAC; type locality: New Zealand: AK, Lynfield, Tropicana Drive.

DISTRIBUTION. New Zealand: North Island.

Baeocera actuosa (Broun)

Scaphisoma actuosa Broun, 1881: 664. Lectotype, NHML; type locality: New Zealand: Parua [Whangarei Harbour].
Kuschel, 1990: 43 (as *Scaphisoma*, habitat)
Newton, 1996: 159 (transfer to *Scaphoxium*)
Löbl & Leschen, 2003b: 30, 46, Figs 10, 34–36, 85, 96, 103, 105, 109 (lectotype designation, transfer to *Baeocera*, fungal and slime mould hosts)
Löbl & Leschen, 2005: 37, Figs 59, 60 (characters)

DISTRIBUTION. New Zealand: North Island.

Baeocera africana Löbl

Baeocera africana Löbl, 1987d: 853, Figs 18–20. Holotype male, MHNG; type locality: Zimbabwe: Melsetter, Umtali, 1700 m.

DISTRIBUTION. Kenya; Rwanda; Zimbabwe.

Baeocera agostii Löbl

Baeocera agostii Löbl, 2014a: 50, Figs 1–3. Holotype male, MHNG; type locality: Indonesia: Maluku Islands, Yamdena Isl., 22 km north Saumlaki.

DISTRIBUTION. Indonesia: Maluku Islands: Yamdena Island.

Baeocera alesi Löbl

Baeocera alesi Löbl, 2012a: 119, Figs 1–3. Holotype male, MHNG; type locality: Taiwan: Meifeng, 2310 m, Nantou Co.

DISTRIBUTION. Taiwan.

Baeocera aliena Löbl

Baeocera aliena Löbl, 2012a: 120, Figs 4–7. Holotype male, MHNG; type locality: Taiwan: Takuanshan Forest, 1650 m, Taoyuan Co.

DISTRIBUTION. Taiwan.

Baeocera alishana Löbl

Baeocera alishana Löbl, 2012a: 122, Figs 8–10. Holotype male, MHNG; type locality: Taiwan: Alishan Nat. Scenic Area, ca 2350 m, road no. 18 km 102, old Lulin Tree Track, Nantou Co.

DISTRIBUTION. Taiwan.

Baeocera alternans (Löbl)

Sciatrophes alternans Löbl, 1977e: 8, Figs 5, 6. Holotype male, SAMA; type locality: Australia: near Brisbane.

DISTRIBUTION. Australia: Queensland.

Baeocera alticola Löbl

Baeocera alticola Löbl, 2012d: 378, Figs 63–66. Holotype male, MHNG; type locality: Philippines: Pacay, 2400 m, near Sayangan, Luzon.

DISTRIBUTION. Philippines: Luzon.

Baeocera amicula Löbl & Stephan

Baeocera amicula Löbl & Stephan, 1993: 709, Figs 39, 134, 135. Holotype male, CNCI; type locality: USA: Sugarloaf Key, Kitchings, Monroe Co.

DISTRIBUTION. United States: Florida.

Baeocera anchorifera Löbl

Baeocera anchorifera Löbl, 2012a: 123, Figs 11–15. Holotype male, MHNG; type locality: Taiwan: road from Tengchih to Chuyunshan, 1400 m, Kaohsiung Co.

DISTRIBUTION. Taiwan.

Baeocera apicalis LeConte

Baeocera apicalis LeConte, 1860: 323. Lectotype male, MCZC; type locality: USA: Southern States.

Scaphisoma distincta Blatchley, 1910: 496. Lectotype, PURC; type locality: USA: Putnam Co., Indiana.

Casey, 1893: 518 (characters)
Casey, 1900: 57 (characters)
Blatchley, 1910: 494 (as *Scaphisoma distincta*, characters, records)

Blatchley, 1930: 35 (lectotype designation for *Scaphisoma distincta*)
Brimley, 1938: 149 (records)
Cornell, 1967: 7, Figs 2F, 4F, L (lectotype fixed by inference for *B. apicalis*, transfer to *Eubaeocera*, characters, records)
Löbl, 1987b: 317 (synonymy of *Scaphisoma distincta* with *B. apicalis*)
Löbl & Stephan, 1993: 682, Figs 17, 58, 59 (characters, records, habitat)
Downie & Arnett, 1996: 366 (as *Scaphisoma distinctum*, characters)
Majka et al., 2011: 88 (distribution)

DISTRIBUTION. Canada: Manitoba, New Brunswick, Nova Scotia, Ontario, Quebec; United States: Arkansas, Connecticut, Florida, Georgia, Illinois, Indiana, Kansas, Kentucky, Maine, Maryland, Massachussets, Michigan, Minnesota, Mississippi, Missouri, Nebraska, New Hampshire, New Jersey, New York, North Carolina, Ohio, Oklahoma, Pennsylvania, Rhode Island, Tennessee, Texas, Vermont, Virginia, West Virginia, Wisconsin.

***Baeocera argentina argentina* Pic**

Baeocera argentinum [sic] Pic, 1916c: 19. Syntypes, MNHN; type locality: Argentina: [Buenos Aires].

DISTRIBUTION. Argentina.

***Baeocera argentina tucumana* Pic**

Baeocera argentina var. *tucumana* Pic, 1920i: 50. Syntypes, MNHN; type locality: Argentina: Tucuman

DISTRIBUTION. Argentina.

***Baeocera atricollis* Pic**

Baeocera atricollis Pic, 1920b: 3. Lectotype male, MNHN; type locality: Chili.
Löbl, 1983b: 167, Figs 11, 12 (lectotype designation, characters, records)

DISTRIBUTION. Chile.

***Baeocera australica* (Löbl)**

Sciatrophes australica Löbl, 1977e: 13, Figs 11, 12. Holotype male, SAMA; type locality: Australia: Windsor, NSW.

DISTRIBUTION. Australia: New South Wales, Queensland.

***Baeocera bacchusi* (Löbl)**

Eubaeocera bacchusi Löbl, 1975a: 379, Figs 13, 14. Holotype male, NHML; type locality: Papua New Guinea: Mt. Kainde, 8000 ft., Morobe District.

DISTRIBUTION. Papua New Guinea.

***Baeocera badia* Löbl**

Baeocera badia Löbl, 2015a: 77, Figs 1–5. Holotype male, MHNG; type locality: Indonesia: Timor, Camplong, 250 m.

DISTRIBUTION. Indonesia: Timor.

***Baeocera baliensis* Löbl**

Baeocera baliensis Löbl, 2015a: 77, Figs 6–8. Holotype male, MHNG; type locality: Indonesia: Bali, Lake Tamblingan, ca 1300 m.

DISTRIBUTION. Indonesia: Bali.

***Baeocera barbara* Löbl**

Baeocera barbara Löbl, 1990b: 532, Figs 35–37. Holotype male, MHNG; type locality: Thailand: Khao Yai Nat. Park, near Headquarters, 750–850 m.

DISTRIBUTION. Thailand.

***Baeocera barda* Löbl**

Baeocera barda Löbl, 2015a: 79, Figs 9–13. Holotype male, MHNG; type locality: Indonesia: Lombok, Pusuk Pass, 300 m.

DISTRIBUTION. Indonesia: Lombok.

***Baeocera basalis* Löbl**

Baeocera basalis Löbl, 2015a: 79, Figs 14–19. Holotype male, MHNG; type locality: Indonesia: Bali, Mt. Agung above Besakih Temple, 1000–1100 m.

DISTRIBUTION. Indonesia: Bali.

***Baeocera batukoqensis* Löbl**

Baeocera batukoqensis Löbl, 2015a: 81, Figs 20–24. Holotype male, MHNG; type locality: Indonesia: Lombok, Batu Koq, 500 m.

DISTRIBUTION. Indonesia: Lombok.

***Baeocera beata* Löbl**

Baeocera beata Löbl, 2015a: 83, Figs 25–30. Holotype male, MHNG; type locality: Indonesia: Timor, between Soe and Kapan, 1000 m.

DISTRIBUTION. Indonesia: Timor.

***Baeocera bella* Löbl**

Baeocera bella Löbl, 2015a: 85, Figs 31–33. Holotype male, MHNG; type locality: Indonesia: Timor, Camplong, 250 m.

DISTRIBUTION. Indonesia: Timor.

Baeocera bengalensis Löbl

Baeocera bengalensis Löbl, 1984a: 78, Fig. 28. Holotype male, MHNG; type locality: India: West Bengal, Mahanadi, near Kurseong, 1200 m.

DISTRIBUTION. India: West Bengal (Darjeeling District).

Baeocera benolivia Löbl & Leschen

Baeocera benolivia Löbl & Leschen, 2003b: 31, 47, Figs 60–63, Map 3. Holotype male, NZAC; type locality: New Zealand: South Island, BR, Capleston.

DISTRIBUTION. New Zealand: South Island.

Baeocera bicolor bicolor Achard

Baeocera bicolor Achard, 1920e: 351. Syntypes, MACN/NMPC; type locality: Argentina: Buenos Aires.

Pic, 1920i: 50 (as variety of *B. argentina* Pic)

DISTRIBUTION. Argentina.

Baeocera bicolor diluta Achard

Baeocera bicolor Achard var. *diluta* Achard, 1920e: 352. Syntypes, MBRA; type locality: Argentina: Buenos Aires.

DISTRIBUTION. Argentina.

Baeocera bicolorata Löbl

Eubaeocera bicolor Löbl, 1972b: 80, Figs 5–9. Holotype male, ZMUB; type locality: Philippines: Luzon, "Heightspe".

Baeocera bicolorata Löbl, 1979a: 86; replacement name for *Eubaeocera bicolor* Löbl, 1972 (nec *Baeocera bicolor* Achard, 1920).

Löbl, 2012d: 354 (characters, records)

DISTRIBUTION. Philippines: Luzon.

Baeocera bifurcata Löbl

Baeocera bifurcata Löbl, 2015a: 85, Figs 34–39. Holotype male, MHNG; type locality: Indonesia: Bali, Lake Buyan, ca 1200 m.

Newton, 2017: 17 (first reviser action)

DISTRIBUTION. Indonesia: Bali.

Baeocera bifurcilla Löbl

Baeocera bifurcilla Löbl, 2015a: 87, Figs 40–45. Holotype male, MHNG; type locality: Indonesia: Bali, Yehbuah, 250 m.

DISTRIBUTION. Indonesia: Bali.

Baeocera biroi (Löbl)

Eubaeocera biroi Löbl, 1975a: 375, Figs 7, 8. Holotype male, HNHM; type locality: Papua New Guinea: Friedrich-Wilhelms-Hafen [Madang], Astrolabe Bay.

DISTRIBUTION. Papua New Guinea.

Baeocera bironis (Pic)

Scaphosoma bironis Pic, 1956b: 73. Lectotype male, HNHM; type locality: Papua New Guinea: Friedrich-Wilhelms-Hafen [Madang], Astrolabe Bay.

Löbl, 1975a: 381, Figs 15, 16 (lectotype designation, transfer to *Eubaeocera*, characters, records)

DISTRIBUTION. Papua New Guinea.

Baeocera boettcheri (Löbl)

Eubaeocera boettcheri Löbl, 1972b: 83, Figs 13, 14. Holotype male, ZMUB; type locality: Philippines: Luzon, Imugan.

Löbl, 2012d: 359, Figs 17, 18 (characters)

DISTRIBUTION. Philippines: Luzon.

Baeocera bogotensis Reitter

Baeocera bogotensis Reitter, 1880a: 45. Syntypes, MNHN; type locality: Colombia: Bogotá.

DISTRIBUTION. Colombia.

Baeocera bona Löbl

Baeocera bona Löbl, 2015a: 89, Figs 46–50. Holotype male, MHNG; type locality: Indonesia: Lombok, Batu Koq, 500 m.

DISTRIBUTION. Indonesia: Lombok.

Baeocera borealis Löbl & Stephan

Baeocera borealis Löbl & Stephan, 1993: 689, Figs 84, 85. Holotype male, USNM; type locality: United States: North Dakota, Mirror Pool, Richland Co.

DISTRIBUTION. United States: North Dakota.

Baeocera bournei Löbl

Baeocera bournei Löbl, 1980b: 221, Figs 1, 2. Holotype male, MHNG; type locality: Papua New Guinea: New Ireland, Limbin, Lelet Plateau, 900 m.

DISTRIBUTION. Papua New Guinea: New Ireland.

Baeocera bremeri **Löbl**

Baeocera bremeri Löbl, 1990b: 517, Figs 5, 6. Holotype male, MHNG; type locality: Thailand: Khon Kaen.

DISTRIBUTION. Thailand.

Baeocera breveapicalis **(Pic)**

Scaphosoma breveapicale Pic, 1926d: 45. Lectotype male, MNHN; type locality: Vietnam: Tonkin, Lac Tho.

Löbl, 1975b: 269, Fig. 1 (lectotype fixed by inference, transfer to *Eubaeocera,* characters)

Löbl, 1999: 726, Fig. 63 (characters, records)

DISTRIBUTION. China: Guangxi, Yunnan; Vietnam.

Baeocera brevicornis **(Löbl)**

Eubaeocera brevicornis Löbl, 1971c: 948, Figs 6, 7. Holotype male, MHNG; type locality: Sri Lanka: Palatupana, at entry of Yala Nat. Park.

Löbl, 1979a: 89 (records)

Löbl, 1992a: 511 (records)

DISTRIBUTION. India: Kerala; Nepal; Sri Lanka.

Baeocera brevis **Löbl**

Baeocera brevis Löbl, 2015a: 91, Figs 51–54. Holotype male, MHNG; type locality: Indonesia: Bali, Mt. Batukaru near Luhur temple, 500–700 m.

DISTRIBUTION. Indonesia: Bali.

Baeocera breviuscula **Löbl**

Baeocera breviuscula Löbl, 2015a: 91, Figs 55–59. Holotype male, MHNG; type locality: Indonesia: Bali, Badingkayu, 300–500 m.

DISTRIBUTION. Indonesia: Bali, Lombok.

Baeocera bruchi **Pic**

Baeocera bruchi Pic, 1928a: 3. Syntypes, MACN/MNHN; type locality: Argentina: San Pedro Cola lasTucuman.

DISTRIBUTION. Argentina.

Baeocera brunnea **(Löbl)**

Eubaeocera brunnea Löbl, 1972b: 82, Figs 10–12. Holotype male, ZMUB; type locality: Philippines: Dapa.

Löbl, 2012d: 355, Figs 4, 5 (characters, records)

DISTRIBUTION. Philippines: Luzon, Leyte, Siargao.

Baeocera caliginosa Löbl

Baeocera caliginosa Löbl, 1984b: 186, Figs 8, 9. Holotype male, MCZC; type locality: Japan: Yonaha-dake, Okinawa.
Hoshina, 2010: 16 (records)
Löbl, 2012a: 124, Figs 16–19 (characters, records)
Ogawa & Löbl, 2013: 318, Figs 6, 15, 28, 53, 58, 66, 76left (characters, records)
DISTRIBUTION. Japan: Kyushu, Ryukyus; Taiwan.

Baeocera callida Löbl

Baeocera callida Löbl, 1986c: 354, Figs 6–8. Holotype male, MHNG; type locality: India: Uttarakhand, Bhim Tal, 1500 m, Kumaon.
Löbl, 1992a: 515 (records)
Löbl, 1999: 725 (records)
DISTRIBUTION. China: Yunnan; India: Himachal Pradesh, Uttarakhand (Kumaon); Nepal.

Baeocera carinata (Löbl)

Eubaeocera carinata Löbl, 1975b: 275, Figs 5, 6. Holotype male, MHNG; type locality: Indonesia: East Sumatra.
DISTRIBUTION. Indonesia: Sumatra.

Baeocera cekalovici Löbl

Baeocera cekalovici Löbl, 1983b: 164, Fig. 8. Holotype male, MHNG; type locality: Chile: Pinares, Concepción.
DISTRIBUTION. Chile.

Baeocera cerbera (Cornell)

Eubaeocera cerbera Cornell, 1967: 13, Figs 3E, I, 4V. Holotype male, UAIC; type locality: USA: Davis Mountains, Jeff Davis Co., Texas.
Löbl & Stephan, 1993: 699, Figs 102, 103 (characters, records)
DISTRIBUTION. United States: Arizona, Texas.

Baeocera ceylonensis (Löbl)

Eubaeocera ceylonensis Löbl, 1971c: 957, Figs 16, 17, 20. Holotype male, MHNG; type locality: Sri Lanka: Mihintale.
Löbl, 1979a: 90 (records)
DISTRIBUTION. India: Kerala, Tamil Nadu; Sri Lanka.

Baeocera charybda (Cornell)

Eubaeocera charybda Cornell, 1967: 9, Figs 2J, 4C, P. Holotype male, INHS; type locality: USA: Putnam, at Lake Senachwine, Illinois.

Lawrence & Newton, 1980: 137 (slime mould hosts)
Löbl & Stephan, 1993: 689, Figs 75–77 (characters, records, fungal hosts)

DISTRIBUTION. United States: Arkansas, Illinois, Iowa, Kentucky, Maryland, Massachusetts, Michigan, Mississippi, New Hampshire, North Carolina, Ohio, Oklahoma, Virginia, Wisconsin.

Baeocera chilensis Reitter

Baeocera chilensis Reitter, 1880a: 45. Lectotype male, MNHN; type locality: Chili.
Löbl, 1983b: 162, Figs 1, 2 (lectotype designation, characters, records)

DISTRIBUTION. Chile.

Baeocera chisosa Löbl & Stephan

Baeocera chisosa Löbl & Stephan, 1993: 683, Figs 35, 55, 67. Holotype male, USNM; type locality: USA: Texas, Koppe's Bridge, 5 mi SE College Sta, Brazos Co.

DISTRIBUTION. United States: Texas.

Baeocera choi Hoshina & Park

Baeocera choi Hoshina & Park, 2011: 47, Figs 1, 4, 7–9. Holotype male, CNUI; type locality: South Korea: Gacheon-ri, Sumnam-myeon, Uju-gun, Ulsan City.

DISTRIBUTION. South Korea.

Baeocera coalita Löbl

Baeocera coalita Löbl, 2003a: 64, Figs 1–3. Holotype male, ZMUB; type locality: China: W. Hubei, Daba Shan, valley 8 km NW Muyuping, 1550–1650 m.

DISTRIBUTION. China: Hubei, Shaanxi.

Baeocera compacta Löbl & Stephan

Baeocera compacta Löbl & Stephan, 1993: 694, Figs 55, 90, 91. Holotype male, CNCI; type locality: USA: Florida, Long Key, Leyton, Monroe Co.

DISTRIBUTION. United States: Florida.

Baeocera congener Casey

Baeocera congener Casey, 1893: 517. Lectotype male, USNM; type locality: USA: New York.
Blatchley, 1910: 493 (characters)
Brimley, 1938: 149 (records)
Cornell, 1967: 11, Figs 2A, 3H, 4U (lectotype fixed by inference, transfer to *Eubaeocera*, characters, records)
Kirk, 1969: 30 (records)

Löbl & Stephan, 1993: 700, Figs 18–20, 104, 105 (characters, records)
Majka et al., 2011: 88 (distribution)

DISTRIBUTION. Canada: Manitoba, New Brunswick, Nova Scotia, Ontario, Quevec, Saskatchewan; United States: Alabama, Arkansas, Arizona, District of Columbia, Connecticut, Florida, Georgia, Illinois, Indiana, Iowa, Kansas, Louisiana, Massachusetts, Maine, Michigan, Missouri, Montana, Nebraska, New Hampshire, New Mexico, New York, North Carolina, North Dakota, Oklahoma, Pennsylvania, Rhode Island, South Carolina, Tennessee, Texas, Vermont, Virginia, Wyoming.

Baeocera convexa **(Pic)**

Amalocera convexa Pic, 1920f: 24. Lectotype female, MNHN; type locality: Indonesia: Java, Gn. Gedeh, NW Pranger.
Löbl, 1975b: 270 (lectotype fixed by inference, transfer to *Eubaeocera*)
Löbl, 1986b: 89 (characters)

DISTRIBUTION. Indonesia: Java.

Baeocera coomani **Pic**

Baeocera coomani Pic, 1923c: 269. Syntypes, MNHN; type locality: Vietnam: Tonkin, Lac Tho.

DISTRIBUTION. Vietnam.

Baeocera cooteri **Löbl**

Baeocera cooteri Löbl, 1999: 729, Figs 54, 55. Holotype male, MHNG; type locality: China: Lin'an County, West Tianmu shan, Nat. r., ca 500 m, Zhejiang.
Löbl, 2003a: 61 (records)
Löbl, 2012: 125 (records, habitat)

DISTRIBUTION. China: Anhui, Fujian, Hong Hong, Jiangxi, Zhejiang; Taiwan.

Baeocera cribrata **Löbl**

Baeocera cribrata Löbl, 1992a: 519, Figs 106–108. Holotype male, SMNS; type locality: Nepal: Paniporua, 2300 m, Panthar District.

DISTRIBUTION. Nepal.

Baeocera crinita **Löbl**

Baeocera crinita Löbl, 1992a: 519, Figs 103–105. Holotype male, MHNG; type locality: Nepal: south Mangsingma, 2800 m, Sankhuwasabha District.

DISTRIBUTION. Nepal.

Baeocera cuccodoroi **Löbl**

Baeocera cuccodoroi Löbl, 2002a: 17, Figs 30–32. Holotype male, MHNG; type locality: Papua New Guinea: Morobe District, Biaru Rd., Mt. Saredomo, 2450 m.

DISTRIBUTION. Papua New Guinea.

Baeocera curta **Löbl**

Baeocera curta Löbl, 2002a: 15, Figs 28, 29. Holotype male, MHNG; type locality: Papua New Guinea: road Bulolo – Wau.

DISTRIBUTION. Papua New Guinea.

Baeocera curtula **Achard**

Baeocera curtula Achard, 1923: 117. Lectotype female, NHML; type locality: Japan: Nagasaki.
Miwa & Mitono, 1943: 545 (characters)
Nakane, 1955b: 53 (characters)
Löbl, 1969b: 340, Fig. 27 (as *Eubaeocera*, lectotype fixed by inference, characters, records)
Löbl, 1984b: 184, Figs 6, 7 (characters, records)
Morimoto, 1985: 257 (characters)
Ogawa & Löbl, 2013: 319, Figs 7, 16, 29, 54, 67 (characters, records)

DISTRIBUTION. Japan: Kyushu, Honshu, Hokkaido, Shikoku.

Baeocera danielae **Löbl**

Baeocera danielae Löbl, 2012d: 359, Figs 19–22. Holotype male, MHNG; type locality: Philippines: Mt. Banahaw above Kinabuhayan, Cristalino trail, 600–700 m, Luzon.

DISTRIBUTION. Philippines: Luzon.

Baeocera deflexa **Casey**

Baeocera deflexa Casey, 1893: 517. Lectotype, USNM; type locality: USA: Boston Neck, Rhode Island.
Blatchley, 1910: 494 (characters, records)
Cornell, 1967: 9, Figs 3B, C, 4T (lectotype fixed by inference, transfer to *Eubaeocera*, characters, records)
Löbl & Stephan, 1993: 697, Figs 21–23, 96, 97 (characters, records)

DISTRIBUTION. Canada: New Brunswick, Nova Scotia, Ontario, Quebec; United States: Arkansas, Colorado, Connecticut, Kentucky, Illinois, Indiana, Louisiana, Maine, Massachusetts, New Hampshire, New Jersey, New York, North Carolina, Ohio, Oklahoma, Pennsylvania, Rhode Island, South Carolina, Tennessee, Texas, Vermont, Virginia, Wisconsin.

***Baeocera deharvengi* Löbl**

Baeocera deharvengi Löbl, 1990b: 530, Figs 30–32. Holotype male, MHNG; type locality: Thailand: Khao Yai Nat. Park, near Headquarters, 750–850 m.

DISTRIBUTION. Thailand.

***Baeocera dentipes* Löbl**

Baeocera dentipes Löbl, 1986c: 356, Figs 9–11. Holotype male, ZMUB; type locality: India: Uttarakhand, Mussoorie.

DISTRIBUTION. India: Uttarakhand (Garhwal).

***Baeocera derougemonti* Löbl**

Baeocera derougemonti Löbl, 1983a: 285, Figs 1, 2. Holotype male, MHNG; type locality: Indonesia: Sulawesi, Makale.

Ogawa, 2015: 105, Figs 4-20a, b (characters, records)

DISTRIBUTION. Indonesia: Sulawesi.

***Baeocera dilutior* Löbl**

Eubaeocera diluta Löbl, 1972b: 85, Figs 15, 16. Holotype male, ZMUB; type locality: Philippines: Luzon, Bambang, Nueva Vizcaya.

Baeocera dilutior Löbl, 1979a: 86; replacement name for *Eubaeocera diluta* Löbl, 1972 (nec *Baeocera diluta* Achard, 1920).

Löbl, 2012d: 376, Figs 59–62 (characters, records)

DISTRIBUTION. Philippines: Luzon, Leyte.

***Baeocera discolor* Casey**

Baeocera discolor Casey, 1900: 58. Syntype female, USNM; type locality: USA: Michigan.

Cornell, 1967: 1 (nomen dubium)

Löbl & Stephan, 1993: 676 (unknown identity)

DISTRIBUTION. United States: Michigan.

***Baeocera doriai* (Pic)**

Amalocera doriai Pic, 1920e: 95. Lectotype male, MCSN; type locality: East Malaysia: Sarawak.

Löbl, 1982d: 790, Figs 5–7 (lectotype designation, transfer to *Baeocera*, characters)

DISTRIBUTION. East Malaysia: Sarawak.

***Baeocera dufaui dufaui* Pic**

Baeocera dufaui Pic, 1920b: 3. Syntype, MNHN; type locality: Guadeloupe.

DISTRIBUTION. Guadeloupe.

Baeocera dufaui tricolor **Pic**

Baeocera dufaui var. *tricolor* Pic, 1920b: 3. Syntype, MNHN; type locality: Guadeloupe.

DISTRIBUTION. Guadeloupe.

Baeocera dufaui unicolor **Pic**

Baeocera dufaui var. *unicolor* Pic, 1920b: 3. Syntype, MNHN; type locality: Guadeloupe.

DISTRIBUTION. Guadeloupe.

Baeocera dugdalei **Löbl**

Baeocera dugdalei Löbl, 1981a: 352, Figs 4, 5. Holotype male, NZAC; type locality: New Caledonia: Mt. Koghis, 650 m.

DISTRIBUTION. New Caledonia.

Baeocera egena **(Löbl)**

Eubaeocera egena Löbl, 1975a: 375, Figs 5, 6. Holotype male, HNHM; type locality: Papua New Guinea: Friedrich-Wilhems-Hafen [Madang], Astrolabe Bay.

DISTRIBUTION. Papua New Guinea.

Baeocera eichelbaumi **(Reitter)**

Scaphosoma eichelbaumi Reitter, 1908: 33. Lectotype male, NHMW; type locality: Tanzania: Amani.

Löbl, 1989b: 280, Figs 4, 5 (lectotype designation, transfer to *Baeocera*, characters)

DISTRIBUTION. Tanzania.

Baeocera elenae **Löbl & Leschen**

Baeocera elenae Löbl & Leschen, 2003b: 32, 47, Figs 64–66, Map 4. Holotype male, NZAC; type locality: New Zealand: South Island, BR, L. Rotoiti, Nelson Lakes National Park.

DISTRIBUTION. New Zealand: South Island.

Baeocera elongata **Löbl & Stephan**

Baeocera elongata Löbl & Stephan, 1993: 710, Figs 48, 49, 136, 137. Holotype male, OSUC; type locality: USA: Colossal Cave Park, Pima Co., Arizona.

DISTRIBUTION. United States: Arizona, Texas.

Baeocera epipleuralis Löbl & Leschen

Baeocera epipleuralis Löbl & Leschen, 2003b: 32, 47, Figs 12, 40–42, Map 5. Holotype male, NZAC; type locality: New Zealand: NN, Mile Creek, 30 km SW Collingwood.

DISTRIBUTION. New Zealand: North and South Islands.

Baeocera errabunda Löbl

Baeocera errabunda Löbl, 1992a: 524, Figs 123–125. Holotype male, MHNG; type locality: Nepal: Goru Dzure Dara, eastern slope, 3350 m, Sankhuwasabha District.

DISTRIBUTION. Nepal.

Baeocera erroris Löbl

Baeocera erroris Löbl, 1990b: 528, Figs 26, 27. Holotype male, MHNG; type locality: Thailand: Khao Sabap Nat. Park, hills NE Phliu Waterfalls, 150–300 m.

DISTRIBUTION. Thailand.

Baeocera eurydice (Cornell)

Eubaeocera eurydice Cornell, 1967: 8, Figs 2G, 4K, S. Holotype male, USNM; type locality: USA: Santa Rita Mountains.
Löbl & Stephan, 1993: 684, Figs 1, 2, 62, 63 (characters, records, habitat)

DISTRIBUTION. United States: Arizona.

Baeocera excelsa Löbl

Baeocera excelsa Löbl, 1986c: 356, Figs 12, 13. Holotype male, MHNG; type locality: India: Uttarakhand, Bhim Tal, 1500 m, Kumaon.
Löbl, 1992a: 515 (records)

DISTRIBUTION. India: Uttarakhand (Kumaon); Nepal.

Baeocera falsata Achard

Scaphidium concolor sensu Erichson, 1845: 4. Lectotype male, USNM; type locality: USA: Pennsylvania.

Baeocera falsata Achard, 1920k: 307; replacement name for *Scaphidium concolor* sensu Erichson (nec *Scaphidium concolor* Fabricius, 1801).
Casey, 1893: 516 (as *B. concolor*, characters)
Blatchley, 1910: 493, Fig. 175 (as *B. concolor*, characters)
Brimley, 1938: 149 (as *B. concolor*, records)

Löbl, 1976d: 777 (as senior synonym of *E. youngi*, characters)
ICZN, 1982: 175, Opinion 1221: lectotype designation (name number 2813 on the Official List of Specific Names in Zoology)
Löbl, 1987b: 315, Figs 1, 2 (as senior synonym of *E. youngi*, characters)
Löbl & Stephan, 1993: 701, Figs 24–26, 106, 107 (characters, records, habitat)
Majka et al., 2011: 88 (distribution)
Ogawa & Löbl, 2013: 314, Fig. 73 (characters)
DISTRIBUTION. Canada: Ontario, Quebec; United States: Kentucky, Maine, New Hampshire, New York, North Carolina, Oklahoma, South Carolina, Virginia.

***Baeocera flagellata* (Löbl)**

Sciatrophes flagellata Löbl, 1976c: 207. Figs 1, 4, 7, 8. Holotype male, CNCI; type locality: United States: Arizona, Pinaleno Mts., 6100 ft, Wet Cn., Graham Co.
Löbl & Stephan, 1993: 697, Figs 27, 28, 98, 99 (characters, records)
DISTRIBUTION. Mexico; United States: Arizona.

***Baeocera flagrans* Löbl**

Baeocera flagrans Löbl, 2002a: 8, Figs 13–15. Holotype male, MHNG; type locality: Papua New Guinea: Morobe District, Biaru Rd., Mt. Koloring, 2200 m.
DISTRIBUTION. Papua New Guinea.

***Baeocera formosana* Löbl**

Baeocera formosana Löbl, 1980a: 97, Figs 11, 12. Holotype male, HNHM; type locality: Taiwan: Pilam.
DISTRIBUTION. Taiwan.

***Baeocera fortepunctata* Löbl**

Baeocera fortepunctata Löbl, 2002a: 6, Figs 7–9. Holotype male, MHNG; type locality: Papua New Guinea: Road from Kaindi (Wau).
DISTRIBUTION. Papua New Guinea.

***Baeocera fortis* Löbl**

Baeocera fortis Löbl, 2012d: 379, Figs 67–70. Holotype male, MHNG; type locality: Philippines: Los Banos, Luzon.
DISTRIBUTION. Philippines: Luzon.

***Baeocera franzi* (Löbl)**

Eubaeocera franzi Löbl, 1973c: 158, Figs 3, 4. Holotype male, NHMW; type locality: Thailand: forest above Kachong Reserve Station.
Löbl, 1990b: 523 (records)

Löbl, 1999: 725 (records)
Löbl, 2003a: 62 (records)

DISTRIBUTION. China: Fujian, Hubei, Jiangsu, Shaanxi, Sichuan, Yunnan; Thailand.

***Baeocera frater* (Löbl)**

Eubaeocera frater Löbl, 1969b: 334, Figs 16, 17. Holotype male, MHNG; type locality: Japan: Omogo Valley, Ehime Pref., Shikoku.
Morimoto, 1985: 258 (characters)
Ogawa & Löbl, 2013: 303, Figs 2, 11, 24, 49, 62 (characters, records)

DISTRIBUTION. Japan: Honshu, Shikoku.

***Baeocera freudei* Löbl**

Baeocera (*Amaloceroschema*) *freudei* Löbl, 1967a: 1, Figs 1, 2. Holotype female, ZSMC; type locality: Brazil: Tupurucuara on Rio Negro/Amazonas.

DISTRIBUTION. Brazil.

***Baeocera freyi* Löbl**

Baeocera freyi Löbl, 1966b: 129, Figs 1–3. Holotype male, NHMB; type locality: North Korea: "Pu Ryong" [Purjong].
Löbl, 1968b: 421 (records)
Löbl, 1969b: 338, Figs 24, 25 (as *Eubaeocera*, characters. records)
Iablokoff-Khnzorian, 1985: 138 (records)
Löbl, 1999: 725 (characters, records)
Hoshina et al., 2011: 49, Figs 5, 10–12 (characters, records)

DISTRIBUTION. China: Shaanxi; North and South Korea; Russia: Far East.

***Baeocera frigida* (Löbl)**

Eubaeocera frigida Löbl, 1971c: 952, Figs 10, 11. Holotype male, MHNG; type locality: Sri Lanka: Horton Plains.

DISTRIBUTION. Sri Lanka.

***Baeocera galapagoensis* Löbl**

Baeocera galapagoensis Löbl, 1977b: 249, Figs 1–6. Holotype male, NHMW; type locality: Galapagos: Cumbre, Pinzón, cloud zone.

DISTRIBUTION. Ecuador: Galapagos Is.

***Baeocera gerardi* (Pic)**

Scaphosoma gerardi Pic, 1928b: 43. Holotype male, MRAC; type locality: Democratic Republic of the Congo: Manyema, Tengo Katanta.

Löbl, 1987d: 846, Figs 10, 11 (transfer to *Baeocera*, characters)
DISTRIBUTION. Democratic Republic of the Congo.

Baeocera germaini **Pic**
Baeocera chilensis Reitter var. *germaini* Pic, 1920b: 4. Lectotype female, MNHN; type locality: Chili.
Löbl, 1983b: 164, Figs 3–7 (lectotype designation, raised to species rank, characters, records)
DISTRIBUTION. Chile.

Baeocera gilloghyi **(Löbl)**
Eubaeocera gilloghyi Löbl, 1973c: 170, Figs 21, 23. Holotype male, MHNG; type locality: Vietnam: Nui Ba Den, Tay Ninh.
Löbl, 1984a: 81 (characters, records)
Löbl, 1999: 725 (records)
DISTRIBUTION. China: Yunnan; India: Assam, Meghalaya; Vietnam.

Baeocera globosa **(Pic)**
Scaphosoma globosum Pic, 1926b: 1. Lectotype female, MNHN; type locality: Philippines: Luzon, Balbalan.
Löbl, 1971a: 248 (lectotype designation, transfer to *Eubaeocera*, characters)
Löbl, 1972b: 101 (as *Eubaeocera*, characters)
Löbl, 2012d: 361, Figs 23–26 (characters, records)
DISTRIBUTION. Philippines: Luzon.

Baeocera gnava **Löbl**
Baeocera gnava Löbl, 1980c: 384, Figs 11, 12. Holotype male, NZAC; type locality: Fiji: Nandarivatu ridge, 1000 m, Viti Levu.
DISTRIBUTION. Fiji.

Baeocera gracilis **(Löbl)**
Sciatrophes gracilis Löbl, 1977e: 11, Figs 9, 10. Holotype male, SAMA; type locality: Australia: Queensland, near Brisbane.
DISTRIBUTION. Australia: Queensland.

Baeocera gutierrezberaudi **Fierros-López**
Baeocera gutierrezberaudi Fierros-López, 2010: 204, Figs 7–12. Holotype male, CZUG; type locality: Mexico: El Floripondio, km 4 RMO Las Viboras, San Gabriel, Jalisco.
DISTRIBUTION. Mexico: Jalisco.

Baeocera gyrinoides Reitter

Baeocera gyrinoides Reitter, 1880a: 46. Lectotype female, MNHN; type locality: Mexico: Teapa.
Löbl, 1992b: 380 (lectotype designation, characters)
DISTRIBUTION. Mexico.

Baeocera hamata Löbl & Stephan

Baeocera hamata Löbl & Stephan, 1993: 711, Figs 56, 138, 139. Holotype male, CMNC; type locality: USA: Texas, Welder Wildlife Refuge, 17 km NE Sinton.
DISTRIBUTION. United States: Texas.

Baeocera hamifer Löbl

Baeocera hamifer Löbl, 1977a: 256, Figs 7, 8. Holotype male, NHMB; type locality: India: West Bengal, Tiger Hill, Darjeeling District.
Löbl, 1984a: 80 (records)
Löbl, 1992a: 516 (as *hamifera*, records)
DISTRIBUTION. India: West Bengal (Darjeeling District); Nepal.

Baeocera hammondi Löbl

Baeocera hammondi Löbl, 1984c: 994, Figs 1, 2. Holotype male, NHML; type locality: China: Shaanxi Nan Wutai, ca 20 mi south Xiang.
Löbl, 1999: 725 (records)
DISTRIBUTION. China: Shaanxi.

Baeocera hesperia Löbl & Stephan

Baeocera hesperia Löbl & Stephan, 1993: 691, Figs 9, 10, 86, 87. Holotype male, USNM; type locality: USA: Ouray, 7500–8000 ft, Colorado.
Löbl, 1997: 54 (misspelled *hesperida*)
DISTRIBUTION. United States: Arizona, Colorado.

Baeocera hillaryi Löbl & Leschen

Baeocera hillaryi Löbl & Leschen, 2003b: 33, 48, Fig. 54–56, Map 6. Holotype male, NZAC; type locality: New Zealand: NN, Oparara R., Karamea.
DISTRIBUTION. New Zealand: South Island.

Baeocera huashana Löbl

Baeocera huashana Löbl, 1999: 728, Fig. 53. Holotype male, MHNG; type locality: China: Shaanxi: Mt. Huashan, 1700 m, east Xian.
DISTRIBUTION. China: Shaanxi.

Baeocera humeralis Fall

Baeocera humeralis Fall, 1910: 116. Holotype female, MCZC; type locality: USA: Tacoma, Washington.

Hatch, 1957: 281, Pl. 36, Fig. 8 (characters, records)

Cornell, 1967: 13, Figs 2N, 4B (transfer to *Eubaeocera*, characters, records)

Löbl & Stephan, 1993: 695, Figs 14, 15, 92, 93 (characters, records)

DISTRIBUTION. Canada: British Columbia, Manitoba; United States: Arizona, Colorado, Idaho, New Hampshire, New Mexico, Washington.

Baeocera hygrophila Löbl

Baeocera hygrophila Löbl, 1984a: 74, Figs 21, 22. Holotype male, MHNG; type locality: India: Meghalaya, below Cherrapunjee, 1200 m, Khasi Hills.

DISTRIBUTION. India: Meghalaya.

Baeocera hypomeralis Löbl

Baeocera hypomeralis Löbl, 2012d: 363, Figs 27–30. Holotype male, MHNG; type locality: Philippines: Mt. Banahaw above Kinabuhayan, Cristalino trail, 600–700 m, Luzon.

DISTRIBUTION. Philippines: Leyte, Luzon, Mindanao.

Baeocera ignobilis Löbl

Baeocera ignobilis Löbl, 1980c. 386, Figs 13, 14. Holotype male, NZAC; type locality: Fiji: Ndelaikoro, 800 m, Vanua Levu.

DISTRIBUTION. Fiji.

Baeocera impunctata Löbl & Stephan

Baeocera impunctata Löbl & Stephan, 1993: 702, Figs 54, 114, 115. Holotype male, OSUC; type locality: USA: Pantano Wash 3 mi north Mt. View, Pima Co., Arizona.

DISTRIBUTION. United States: Arizona, New Mexico.

Baeocera inaequicornis Champion

Baeocera inaequicornis Champion, 1927: 274. Holotype male, NHML; type locality: India: Uttarakhand W. Almora, Kumaon.

Löbl, 1971c: 943 (as *Eubaeocera*, invalid lectotype designation, characters)

DISTRIBUTION. India: Uttarakhand (Kumaon).

Baeocera incisa (Löbl)

Eubaeocera incisa Löbl, 1973c: 161, Figs 7, 8. Holotype male, MHNG; type locality: East Malaysia: Sabah, Umas near Tawau.

Löbl, 1984a: 69, Fig. 13 (characters)

DISTRIBUTION. East Malaysia: Sabah.

Baeocera inculta Löbl

Baeocera inculta Löbl, 1984a: 73, Figs 18, 19. Holotype male, MHNG; type locality: India: West Bengal, 13 km north Ghoom, via Bijanbari, Darjeeling District.
DISTRIBUTION. India: Meghalaya, Sikkim, West Bengal (Darjeeling District).

Baeocera indistincta Löbl & Stephan

Baeocera indistincta Löbl & Stephan, 1993: 690, Figs 82, 83. Holotype male, UNHC; type locality: USA: 1 mi SW Durham, Strafford Co., New Hampshire.
DISTRIBUTION. Canada: New Brunswick, Ontario; United States: Alabama, Arkansas, Illinois, Indiana, Maryland, Mississippi, New Jersey, New Hampshire, New York, Ohio, Oklahoma.

Baeocera inermis Löbl

Baeocera inermis Löbl, 1988a: 373, Figs 1, 2. Holotype male, MHNG; type locality: India: Himachal Pradesh, Dhangiara, Junee Valley, 1800 m.
Newton & Stephenson, 1990: 204 (slime mould hosts)
DISTRIBUTION. India: Himachal Pradesh.

Baeocera inexspectata Löbl & Stephan

Baeocera inexspectata Löbl & Stephan, 1993: 702, Figs 108–110. Holotype male, CNCI; type locality: Canada: Saskatoon, Saskatchewan.
Webster et al., 2012: 243 (records, habitat)
DISTRIBUTION. Canada: New Brunswick, Saskatchewan.

Baeocera innocua Löbl

Baeocera innocua Löbl, 1990b: 537, Figs 42–46. Holotype male, MHNG; type locality: Thailand: Khao Yai Nat. Park, hills east Heo Suwat Waterfalls, 800 m.
DISTRIBUTION. Thailand.

Baeocera insolita Löbl

Baeocera insolita Löbl, 1990b: 531, Figs 33, 34. Holotype male, MHNG; type locality: Thailand: Khao Yai Nat. Park, near Headquarters, 750–850 m.
Löbl, 1997: 55 (misspelled *insolida*)
DISTRIBUTION. Thailand.

Baeocera insperata (Löbl)

Eubaeocera insperata Löbl, 1975a: 382, Figs 17, 18. Holotype male, HNHM; type locality: Papua New Guinea: Stephansort, Astrolabe Bay.
DISTRIBUTION. Papua New Guinea.

Baeocera irregularis Champion

Baeocera irregularis Champion, 1913: 70. Holotype female, NHML; type locality: Mexico: Omiltene in Guerrero, 8000 ft.
Löbl, 1976c: 209, Figs 3, 6, 11, 12 (as *Sciatrophes*, characters, records)
Löbl & Stephan, 1993: 696, Figs 33, 94, 95 (characters)
DISTRIBUTION. Mexico; United States: Arizona.

Baeocera jankodadai Löbl

Baeocera jankodadai Löbl, 2012d: 364, Figs 31–34. Holotype male, MHNG; type locality: Philippines: San Rafael, 300 m, Palawan.
DISTRIBUTION. Philippines: Palawan.

Baeocera jeani Löbl

Baeocera jeani Löbl, 2012d: 366, Figs 35–38. Holotype male, MHNG; type locality: Philippines: Mt. Makiling, 450–550 m, Luzon.
DISTRIBUTION. Philippines: Luzon.

Baeocera jejuna Löbl

Baeocera jejuna Löbl, 2012d: 368, Figs 39–42. Holotype male, MHNG; type locality: Philippines: Conception, 50 m, Palawan.
DISTRIBUTION. Philippines: Palawan.

Baeocera kaibesara Löbl

Baeocera kaibesara Löbl, 2014a: 53, Figs 8, 9. Holotype male, MHNG; type locality: Indonesia: Maluku Islands, Kai Besar, G. Turkan, 300 m.
DISTRIBUTION. Indonesia: Maluku Islands: Kai Besar.

Baeocera kapfereri Reitter

Baeocera kapfereri Reitter, 1915: 42. Holotype female, HNHM; type locality: Tunisia: Ain Draham.
Löbl, 1969b: 337, Figs 3, 22, 23, 30 (as *Eubaeocera*, characters, records)
Tamanini, 1969b: 358, Figs 3A, D-H (as *Eubaeocera*, characters, records)
Löbl, 1989a: 10, Fig. 1 (characters, records)
DISTRIBUTION. Algeria; Tunisia.

Baeocera karamea Löbl & Leschen

Baeocera karamea Löbl & Leschen, 2003b: 34, 48, Figs 67, 68, Map 7. Holotype male, NZAC; type locality: New Zealand: South Island, NN, Karamea Bluff.
DISTRIBUTION. New Zealand: South Island.

***Baeocera karen* Löbl**

Baeocera karen Löbl, 1990b: 528, Figs 28, 29. Holotype male, MHNG; type locality: Thailand: Kaeng Krachan Nat. Park, 450 m.

DISTRIBUTION. Thailand.

***Baeocera keralensis* Löbl**

Baeocera keralensis Löbl, 1979a: 96, Figs 11, 12. Holotype male, MHNG; type locality: India: Kerala, between Pambanar and Peermade, 950 m.

DISTRIBUTION. India: Kerala.

***Baeocera khasiana* Löbl**

Baeocera khasiana Löbl, 1984a: 77, Figs 26, 27. Holotype male, MHNG; type locality: India: Mawphlang, 1800 m, Khasi Hills, Meghalaya.
Löbl, 1990b: 530 (records)
Löbl, 1992a: 515 (records)

DISTRIBUTION. India: Meghalaya; Nepal; Thailand.

***Baeocera kinabalua* Löbl**

Baeocera kinabalua Löbl, 1987a: 86, Figs 1–3. Holotype male, MHNG; type locality: East Malaysia: Sabah, Kinabalu Nat. Park.

DISTRIBUTION. East Malaysia: Sabah.

***Baeocera koreana* Hoshina & Park**

Baeocera koreana Hoshina & Park, 2011: 50, Figs 6, 13–15. Holotype male, CNUIC; type locality: South Korea: Geumsubong, Gyesab-myeon, Yuseong-ri, Daejeon, Chungnam Prov.

DISTRIBUTION. South Korea.

***Baeocera kubani* Löbl**

Baeocera kubani Löbl, 1999: 729, Figs 56, 57. Holotype male, MHNG; type locality: China: Jizu Mts, 2800 m, Yunnan.
Löbl, 2003a: 62 (records)

DISTRIBUTION. China: Fujian, Jiangxi, Yunnan.

***Baeocera kurbatovi* Löbl**

Baeocera kurbatovi Löbl, 1993: 35, Figs 1, 2. Holotype male, MHNG; type locality: Russia: Kamenushka, East Ussuriysk.

DISTRIBUTION. Russia: Far East.

Baeocera kuscheli Löbl

Baeocera kuscheli Löbl, 1980c: 384, Figs 9, 10. Holotype male, NZAC; type locality: Fiji: Ndelaikoro, 500 m, Vanua Levu.

DISTRIBUTION. Fiji.

Baeocera kuscheliana Löbl

Baeocera kuscheliana Löbl, 1980c: 386, Figs 15, 16. Holotype male, NZAC; type locality: Fiji: Nandarivatu ridge, 1000 m, Viti Levu.

DISTRIBUTION. Fiji.

Baeocera laevis (Reitter)

Scaphisoma laeve Reitter, 1880a: 47. Lectotype male, MNHN; type locality: "Nordamerika".

Eubaeocera mitchelli Cornell, 1967: 15, Figs 2K, 3G, 4H. Holotype male, USNM; type locality: USA: Hobe Shelter, Game Refuge, NC.

Kirk, 1969: 30 (as *E. mitchelli*, records)

Löbl, 1987b: 316, Figs 3, 4 (lectotype designation and transfer of *S. laeve* to *Baeocera*, synonymy with *E. mitchelli*)

Löbl & Stephan, 1993: 708, Figs 41, 42, 131, 132 (characters, records, habitat)

DISTRIBUTION. United States: Alabama, Florida, Georgia, Louisiana, Mississippi, Missouri, New Jersey, New York, North Carolina, Oklahoma, South Carolina, Texas, West Virginia.

Baeocera laminula Löbl

Baeocera laminula Löbl, 1992a: 517, Figs 99, 100. Holotype male, MHNG; type locality: Nepal: west Bagarchap, 2200 m, Manang District.

DISTRIBUTION. Nepal.

Baeocera lasciva Löbl

Baeocera lasciva Löbl, 2003a: 66, Figs 4, 5. Holotype male, MHNG; type locality: China: Fujian: Wuyi Shan near Guadung, 1150 m.

DISTRIBUTION. China: Fujian.

Baeocera latens (Blackburn)

Sciatrophes latens Blackburn, 1903: 100. Lectotype female, NHML; type locality: Australia: Dividing Range, Victoria.

Scaphisoma bryophaga Elston, 1921: 144. Syntypes, ANIC/SAMA/NHML; type locality: Australia: Myponga, South Australia.

Löbl, 1977e: 6, Figs 1–4 (lectotype fixed by inference for *S. latens*, synonymy of *S. bryophaga* with *S. latens*, characters, records)

DISTRIBUTION. Australia: New South Wales, South Australia, Tasmania, Victoria, Western Australia.

***Baeocera leleupi* Löbl**

Baeocera leleupi Löbl, 1989b: 282, Figs 6, 7. Holotype male, MRAC; type locality: Democratic Republic of the Congo: Bitale, 1600 m, NE Kahusi, T. Kalehe, Kivu.

DISTRIBUTION. Democratic Republic of the Congo.

***Baeocera lenczyi* Löbl & Stephan**

Baeocera lenczyi Löbl & Stephan, 1993: 702, Figs 52, 111–113. Holotype male, USNM; type locality: USA: Arizona, Santa Rita Mts, Madera Canyon, Santa Cruz Co.

DISTRIBUTION. United States: Arizona.

***Baeocera lenta* (Löbl)**

Eubaeocera lenta Löbl, 1971c: 949, Figs 8, 9, 18. Holotype male, MHNG; type locality: Sri Lanka: Alut Oya.

Baeocera pseudolenta Löbl, 1979a: 95, Fig. 10, 19. Holotype male, MHNG; type locality: India: Kaikatty, Neliampathi hills, 900 m, Kerala.
Löbl, 1984a: 68 (as *B. pseudolenta*, records)
Löbl, 1992a: 511 (synonymy of *B. pseudolenta* with *B. lenta*, records)

DISTRIBUTION. India: Kerala, Meghalaya, Tamil Nadu; Nepal; Sri Lanka.

***Baeocera lindae* Löbl**

Baeocera lindae Löbl, 2012a: 125, Figs 20–22. Holotype male, MHNG; type locality: Taiwan: Tengchih, 1610 m, Kaohsiung Co.

DISTRIBUTION. Taiwan.

***Baeocera longicornis* (Löbl)**

Eubaeocera longicornis Löbl, 1971c: 955, 14, 15, 21. Holotype male, MHNG; type locality: Sri Lanka: Inginiyagala.
Löbl, 1990b: 527, Fig. 21 (characters, records)
Löbl, 1992a: 511 (records)
Rougemont, 1996: 12 (records)
Löbl, 1999: 724 (records)
Löbl, 2003a: 62 (records)
Löbl, 2012: 127 (records)

DISTRIBUTION. China: Hong Kong, Jiangxi, Yunnan; India: Assam, Meghalaya, Uttarakhand, West Bengal (Darjeeling District); Nepal; Sri Lanka; Thailand; Taiwan.

Baeocera louisi **Löbl**

Baeocera louisi Löbl, 2012d: 370, Figs 43–46. Holotype male, MHNG; type locality: Philippines: Banguio, 1500 m, near Crystal Caves, Luzon.

DISTRIBUTION. Philippines: Luzon.

Baeocera macrops **(Löbl)**

Eubaeocera macrops Löbl, 1973c: 173, Figs 24, 25. Holotype male, MHNG; type locality: Singapore: Bukit Timah Nat. Reserve.

DISTRIBUTION. Singapore.

Baeocera major **Matthews**

Baeocera major Matthews, 1888: 169, Figs 3, 4. Lectotype male, NHML; type locality: Guatemala: Zapote.

Löbl, 1992b: 382 (lectotype designation, characters)

DISTRIBUTION. Guatemala; Mexico.

Baeocera manasensis **Löbl**

Baeocera manasensis Löbl, 1984a: 68, Figs 10–12. Holotype male, MHNG; type locality: India: Assam, Manas Wilf Life Sanctuary, 200 m.

DISTRIBUTION. India: Assam.

Baeocera martensi **Löbl**

Baeocera martensi Löbl, 1992a: 520, Figs 109–111. Holotype male, MHNG; type locality: Nepal: Arun Valley, below Num, 1050 m, Sankhuwasabha District.

DISTRIBUTION. Nepal.

Baeocera matthewsi **(Löbl)**

Sciatrophes matthewsi Löbl, 1977e: 10, Figs 7, 8. Holotype male, SAMA; type locality: Australia: Dorrigo, New South Wales.

DISTRIBUTION. Australia: New South Wales, Queensland.

Baeocera mendax **Löbl**

Baeocera mendax Löbl, 1979a: 95, Figs 8, 9, 18. Holotype male, MHNG; type locality: India: Kerala, Thekkady, near Periyar, 900 m.

DISTRIBUTION. India: Kerala.

Baeocera mexicana **Reitter**

Baeocera mexicana Reitter, 1880a: 45. Lectotype female, MNHN; type locality: Mexico.

Löbl, 1992b: 379 (lectotype designation, characters)

DISTRIBUTION. Mexico.

Baeocera microps Löbl

Baeocera microps Löbl, 1984a: 76, Fig. 23. Holotype male, MHNG; type locality: India: West Bengal, Tiger Hill, 2500–2600 m.

DISTRIBUTION. India: West Bengal (Darjeeling District).

Baeocera microptera Löbl

Baeocera microptera Löbl, 1986c: 351 Figs 1–3. Holotype male, MHNG; type locality: Pakistan: near Murre, 2100 m, Punjab.
Löbl, 1988a: 373 (records)
Newton & Stephenson, 1990: 212 (slime mould hosts)
Löbl, 1992a: 514 (records)

DISTRIBUTION. India: Himachal Pradesh, Uttarakhand; Nepal; Pakistan.

Baeocera mindanaosa Löbl

Baeocera mindanaosa Löbl, 2012d: 371, Figs 47–50. Holotype male, SMNS; type locality: Philippines: Mindanao, Misamis Occ., 1700 m, Don Victoriano.

DISTRIBUTION. Philippines: Mindanao.

Baeocera monstrosa (Löbl)

Eubaeocera monstrosa Löbl, 1971c: 959, Figs 22–24. Holotype male, MHNG; type locality: Sri Lanka: Nedunleni.
Löbl, 1979a: 98 (records)
Löbl, 1986c: 345 (records)
Löbl, 1992a: 515 (records)

DISTRIBUTION. India: Himachal Pradesh, Kerala, Tamil Nadu, Uttarakhand (Kumaon); Sri Lanka.

Baeocera monstrosetibialis Löbl

Baeocera monstrosetibialis Löbl, 1984a: 80, Figs 29–31. Holotype male, MHNG; type locality: India: Meghalaya, Shillong Peak, 1850–1950 m, Khasi Hills.

DISTRIBUTION. India: Meghalaya.

Baeocera montana (Pic)

Scaphosoma montanum Pic, 1955: 51. Lectotype female, MRAC; type locality: Burundi: Nyamasumu, east Usumbura.
Löbl, 1987d: 845 (lectotype designation, transfer to *Baeocera*, characters)

DISTRIBUTION. Burundi, Rwanda.

Baeocera montanella **Löbl**

Baeocera montana Löbl, 1979: 92, Figs 5, 6, 16. Holotype male, MHNG; type locality: India: Tamil Nadu, Coonoor, 1600 m, Nilgiri Hills.
Baeocera montanella Löbl, 1992a: 504; replacement name for *Baeocera montana* Löbl, 1979 (nec *Baeocera montana* (Pic, 1955).
DISTRIBUTION. India: Tamil Nadu.

Baeocera monticola **Vinson**

Baeocera monticola Vinson, 1943: 198, Fig. 19. Holotype, NHML; type locality: Mauritius: Mt. Cocotte.
Löbl, 1977f: 40, Fig. 2 (characters)
DISTRIBUTION. Mauritius.

Baeocera murphyi **(Löbl)**

Eubaeocera murphyi Löbl, 1973c: 166, Figs 15, 16. Holotype male, MHNG; type locality: Singapore: Bukit Timah Nat. Reserve.
DISTRIBUTION. Singapore.

Baeocera mussardiana **Löbl**

Baeocera mussardiana Löbl, 1979a: 90, Figs 3, 4, 15. Holotype male, MHNG; type locality: India: above Aliyar Dam, 550 m, Anaimalai Hills.
DISTRIBUTION. India: Kerala, Tamil Nadu.

Baeocera mustangensis **Löbl**

Baeocera mustangensis Löbl, 1992a: 522, Figs 114, 115. Holotype male, MHNG; type locality: Nepal: Lete, 2550 m, Mustang District.
Löbl, 2005: 180 (records)
DISTRIBUTION. Nepal.

Baeocera mutata **Löbl**

Baeocera mutata Löbl, 2012a: 127, Figs 23–26. Holotype male, MHNG; type locality: Taiwan: Pingtung Co., Peitawushan trail, 1500 m.
DISTRIBUTION. Taiwan.

Baeocera myrmidon **(Achard)**

Scaphosoma myrmidon Achard, 1923: 116. Lectotype female, NMPC; type locality: Japan: Nagasaki.
Miwa & Mitono, 1943: 544 (as *Scaphosoma*, characters)
Nakane, 1955c: 57 (as *Scaphisoma*, characters)
Löbl, 1966a: 1, Figs 1–3 (lectotype fixed by inference, transfer to *Eubaeocera*, characters)

Löbl, 1969b: 322, Figs 12, 13 (as *Eubaeocera*, characters)
Löbl, 1980a: 94, Figs 7, 8 (characters, records)
Morimoto, 1985: 258 (characters)
Ogawa & Löbl, 2013: 309, Figs 3, 12, 25, 48, 50, 63 (characters, records)

DISTRIBUTION. Japan: Kyushu; Taiwan.

Baeocera nakanei (Löbl)

Eubaeocera nakanei Löbl, 1968a: 1, Fig. 1. Holotype female, SNMC; type locality: Japan: Kasuga.
Löbl, 1969b: 341 (as *Eubaeocera*, characters)
Löbl, 1984b: 188, Fig. 14 (characters)
Morimoto, 1985: 258 (characters)
Ogawa & Löbl, 2013: 320, Figs 1, 8, 17, 30, 47, 55, 60, 68, 70, 76, 77right, 81, 83 (characters, records)

DISTRIBUTION. Japan: Honshu, Kyushu, Shikoku.

Baeocera nana Casey

Baeocera nana Casey, 1893: 521. Lectotype, USNM; type locality: USA: Rhode Island.

Baeocera rubriventris Casey, 1900: 58. Lectotype, USNM; type locality: USA: Rhode Island.
Cornell, 1967: 15, Figs 2I, 3F, 4Q (lectotype fixed by inference for *B. nana*, transfer to *Eubaeocera*, synonymy of *B. nana* with *B. rubriventris*, characters, records, fungal hosts)
Kirk, 1969: 30 (records)
Lawrence & Newton, 1980: 137 (slime mould hosts)
Löbl & Stephan, 1993: 693, Figs 11–13, 88, 89 (lectotype designation of *Baeocera rubriventris*, characters, records, habitat, fungal hosts)
Majka et al., 2011: 88 (distribution)

DISTRIBUTION. Canada: Ontario, Quebec; United States: Alabama, Arkansas, Connecticut, District of Columbia, Florida, Georgia, Illinois, Indiana, Iowa, Kansas, Kentucky, Maryland, Massachusetts, Maine, Michigan, Mississippi, Missouri, New Hampshire, New Jersey, North Carolina, North Dakota, Ohio, Oklahoma, Rhode Island, South Carolina, Tennessee, Texas, Virginia, West Virginia.

Baeocera nanula Löbl

Baeocera nanula Löbl, 1980a: 96, Figs 9, 10. Holotype male, MHNG; type locality: Taiwan: Akau.
Löbl, 2012a: 128 (records)

DISTRIBUTION. Taiwan.

Baeocera nitida (Löbl)

Eubaeocera nitida Löbl, 1975b: 273, Figs 3, 4. Holotype male, MHNG; type locality: West Malaysia: Fraser's Hill, 4200 ft., Selangor.

DISTRIBUTION. West Malaysia.

Baeocera nobilis nobilis Reitter

Baeocera nobilis Reitter, 1884: 371. Lectotype male, MCSN (Dodero coll.); type locality: Italy: Sardinia, Santadi.

Baeocera devillei Reitter, 1899: 157. Syntypes, females, HNHM; type locality: France: Corsica, Vizzavona.

Scaphosoma laeve Guillebeau, 1893: cccxxvii (nec *Scaphosoma laeve* Reitter, 1880). Holotype female, MNHN; type locality: Algeria: Philippeville.

Scaphosoma reitteri Csiki, 1904: 85; replacement name for *Scaphosoma laeve* Guillebeau, 1893.

Reitter, 1885b: 362 (characters)
Reitter, 1887a: 5 (characters)
Normand, 1934: 45 (records)
Löbl, 1965d: 335 (*S. laeve* as possible synonym of *B. devillei*)
Löbl, 1969b: 333, Figs 14a, 15, 29 (as *Eubaeocera*, lectotype fixed by inference, synonymy of *B. nobilis* with *B. devillei* and *S. laeve*, characters, records)
Tamanini, 1969b: 356, Fig. 3B (as *Eubaeocera*, characters, records)
Tamanini, 1969c: Fig. 46 (as *Eubaeocera*, characters)
Tamanini, 1970: 8, Figs 4A, B (as *Eubaeocera*, characters, records)
Löbl, 1989a: 10 (records)

DISTRIBUTION. Algeria; France: Corsica; Italy; Tunisia.

Baeocera nobilis besucheti (Löbl)

Eubaeocera nobilis besucheti Löbl, 1969b: 334, Fig. 14b. Holotype male, MHNG; type locality: Turkey: Bolu, between Elmalık and Bakacak, 850 m.

DISTRIBUTION. Turkey.

Baeocera nonguensis Löbl

Baeocera nonguensis Löbl, 1983b: 165, Figs 9, 10. Holotype male, MHNG; type locality: Chile: Estero Nonguen, Concepción.

Achard, 1921b: 88 (as *Toxidium chilense* Pic, characters)

DISTRIBUTION. Argentina; Chile.

Baeocera obesa Löbl & Stephan

Baeocera obesa Löbl & Stephan, 1993: 686, Figs 68. 69. Holotype male, CNCI; type locality: USA: Texas Guadeloupe Mt. Nat. Park, Mc. Kittrick Can., Culberson Co.

DISTRIBUTION. United States: Texas.

Baeocera obliqua (Löbl)

Eubaeocera obliqua Löbl, 1973c: 165, Figs 13, 14. Holotype male, MHNG; type localiy: Malaysia: Sabah, Quoin Hill, Tawau.

DISTRIBUTION. East Malaysia: Sabah.

Baeocera ornata (Löbl)

Eubaeocera ornata Löbl, 1973c: 162, Figs 9, 10. Holotype male, MHNG; type locality: Singapore: Bukit Timah Nat. Reserve.

DISTRIBUTION. Singapore.

Baeocera ovalis Löbl

Baeocera ovalis Löbl, 1980c: 382, Figs 7, 8. Holotype male, NZNC; type locality: Fiji: Nandarivatu ridge, 1000 m, Viti Levu.

DISTRIBUTIONi Fiji.

Baeocera ovicula Löbl

Baeocera ovicula Löbl, 2002a: 9, Figs 16–18. Holotype male, MHNG; type locality: Papua New Guinea: Morobe District, Mt. Kaindi, 1350 m.

DISTRIBUTION. Papua New Guinea.

Baeocera pacifica Löbl

Baeocera pacifica Löbl, 1981b: 72, Figs 1, 2. Holotype male, BPBM; type locality: Micronesia: Caroline Islands, Dugoy, Yap.

DISTRIBUTION. Micronesia: Caroline Islands.

Baeocera palawana (Löbl)

Eubaeocera palawana Löbl, 1971a: 247. Holotype female, ZMUC; type locality: Philippines: Palawan, Uring Uring.

Löbl, 2012d: 355, Figs 6–9 (characters, records)

DISTRIBUTION. Philippines: Palawan.

Baeocera pallida Casey

Baeocera pallida Casey, 1900: 58. Lectotype, USNM; type locality: USA: near Philadelphia, Pennsylvania.

Cornell, 1967: 7, Figs 2H, 3M, 4M (lectotype fixed by inference, transfer to *Eubaeocera*, characters, records)

Löbl & Stephan, 1993: 685, Figs 3, 4, 60, 61 (characters, records, habitat)

DISTRIBUTION. Canada: Ontario, Quebec; United States: Illinois, Missouri, Nebraska, New Hampshire, Ohio, Oklahoma, Tennessee, Texas, Vermont.

Baeocera palmi Löbl

Baeocera palmi Löbl, 1987d: 848, Figs 12, 13. Holotype male, MHNG; type locality: Kenya: Mt. Elgon, 2000 m.

DISTRIBUTION. Kenya.

Baeocera papua (Löbl)

Eubaeocera papua Löbl, 1975a: 376, Figs 9, 10. Holotype male, HNHM; type locality: Papua New Guinea: Friedrich Wilhelms-Hafen [Madang], Astrolabe Bay.

DISTRIBUTION. Papua New Guinea.

Baeocera paradoxa (Löbl)

Eubaeocera paradoxa Löbl, 1971c: 961, Figs 25–27. Holotype male, MHNG; type locality: Sri Lanka: Nedunleni.

DISTRIBUTION. Sri Lanka.

Baeocera parallela Löbl

Baeocera parallela Löbl, 1980c: 381, Figs 1, 2. Holotype male, NZAC; type locality: Fiji: Nandrau, 750 m, Viti Levu.

DISTRIBUTION. Fiji.

Baeocera pecki Löbl & Stephan

Baeocera pecki Löbl & Stephan, 1993: 687, Figs 60, 51, 70, 71. Holotype male, CNCI; type locality: USA: Florida Long Key, Layton, Monroe Co.

DISTRIBUTION. United States: Florida.

Baeocera picea Casey

Baeocera picea Casey, 1893: 520. Lectotype, USNM; type locality: USA: Penssylvania.

Blatchley, 1910: 494 (characters, records)
Cornell, 1967: 8, Figs 2L, 4A, J (lectotype fixed by inference, transfer to *Eubaeocera*, characters, records)
Kirk, 1969: 30 (records)
Lawrence & Newton, 1980: 137 (slime mould hosts)
Newton, 1984: 319 (slime mould hosts)
Newton, 1991: 338 (larvae, fungal hosts)
Löbl & Leschen, 1993: 688, Figs 7, 8, 72–74 (characters, records, habitat, fungal hosts)
Betz et al., 2003: 203, Fig. 13 (characters)
Majka et al., 2011: 88 (distribution)

DISTRIBUTION. Canada: Ontario, Quebec; Mexico; United States: Arkansas, Florida, Georgia, Illinois, Indiana, Kansas, Kentuky, Maine, Maryland, Massachusett, Michigan, Mississipi, Missouri, New Hampshire, New Yersey, New York, North Carolina, Ohio, Oklahoma, Pennsylvania, Rhode Island, Texas, Virginia, West Virginia.

Baeocera pigra (Löbl)

Eubaeocera pigra Löbl, 1971c: 953, Figs 12, 13, 19. Holotype male, MHNG; type locality: Sri Lanka: Dialuma Falls.

Eubaeocera decipiens Löbl, 1973c: 160, Figs 5, 6. Holotype male, NHMW; type locality: Thailand: forest above Kachon Reserve Station near Trang.

Löbl, 1979a: 93 (records)

Löbl, 1984a: 67 (characters, records)

Löbl, 1986c: 345 (records)

Löbl, 1990b: 525, Fig. 20 (synonymy of *B. decipiens* with *B. pigra*, characters, records)

Löbl, 1992a: 512 (records)

Löbl, 1999: 724 (records)

DISTRIBUTION. China: Yunnan; India: Assam, Kerala, Meghalaya, Tamil Nadu, Uttarakhand (Kumaon); Nepal; Sri Lanka; Thailand.

Baeocera plana Löbl

Baeocera plana Löbl, 1980c: 382, Figs 5, 6. Holotype male, NZAC; type locality: Fiji: Nandarinvatu, 1000 m, Viti Levu.

DISTRIBUTION. Fiji.

Baeocera praedicta Löbl

Baeocera praedicta Löbl, 2002a: 11, Figs 19–21. Holotype male, MHNG; type locality: Papua New Guinea: Morobe District, Biaru Road, Mt. Kolorong, 2200 m.

DISTRIBUTION. Papua New Guinea.

Baeocera praesignis Löbl

Baeocera praesignis Löbl, 2002a: 14, Figs 25–27. Holotype male, MHNG; type locality: Papua New Guinea: Morobe District, Biaru Rd., Mt. Kolorong, 1900 m.

DISTRIBUTION. Papua New Guinea.

Baeocera problematica Löbl

Baeocera problematica Löbl, 1987d: 842, Figs 1–4. Holotype male, MHNG; type locality: Kenya: near Limuru, 2300 m, Kiambu District.

DISTRIBUTION. Kenya.

Baeocera procerula **Löbl**

Baeocera procerula Löbl, 2012d: 373, Figs 51–54. Holotype male, MHNG; type locality: Philippines: Mt. Makiling, 400 m, Luzon.

DISTRIBUTION. Philippines: Luzon.

Baeocera prodroma **Löbl**

Baeocera prodroma Löbl, 2002a: 8, Figs 10–12. Holotype male, MHNG; type locality: Papua New Guinea: Bulolo Road at Wau.

DISTRIBUTION. Papua New Guinea.

Baeocera producta **(Pic)**

Scaphosoma productum Pic, 1926e: 143. Lectotype male, MNHN; type locality: Vietnam: Tonkin, Hoa Binh.

Löbl, 1975b: 270, Fig. 2 (lectotype fixed by inference, transfer to *Baeocera*, characters)

DISTRIBUTION. Vietnam.

Baeocera profana **Löbl**

Baeocera profana Löbl, 2012d: 357, Figs 10–13. Holotype male, MHNG; type locality: Philippines: Mountain Prov., Sagada, Luzon.

DISTRIBUTION. Philippines: Luzon.

Baeocera prolixa **Löbl**

Baeocera prolixa Löbl, 2002a: 4, Figs 4–6. Holotype male, HNHM; type locality: Papua New Guinea: Kiunga.

DISTRIBUTION. Papua New Guinea.

Baeocera promelas **(Löbl)**

Sciatrophes promelas Löbl, 1977e: 14, Figs 13, 14. Holotype male, SAMA; type locality: Australia: Ord River.

DISTRIBUTION. Western Australia.

Baeocera proseminata **Löbl**

Baeocera proseminata Löbl, 2003a: 66, Figs 6, 7. Holotype male, ZIBC; type locality: China: Wuyi Shan, road to Guadun, 1300 m, Fujian.

DISTRIBUTION. China: Fujian.

Baeocera prospecta **Löbl**

Baeocera prospecta Löbl, 2002a: 12, Figs 22–24. Holotype male, MHNG; type locality: Papua New Guinea: Morobe District, Biaru Road, Mt. Kolorong, 2200 m.

DISTRIBUTION. Papua New Guinea.

Baeocera provida Löbl

Baeocera provida Löbl, 2002a: 4, Figs 1–3. Holotype male, MHNG; type locality: Papua New Guinea: Bulolo Road at Wau.

DISTRIBUTION. Papua New Guinea.

Baeocera pseudincisa Löbl

Baeocera pseudincisa Löbl, 1984a: 68, Figs 7–9. Holotype male, MHNG; type locality: India: 10–12 km NW Dawki, 500–900 m, Khasi Hills, Meghalaya.
Löbl, 2004a: 345 (records)

DISTRIBUTION. India: Madhya Pradesh, Meghalaya.

Baeocera pseudinculta Löbl

Baeocera pseudinculta Löbl, 1990b: 527, Figs 22, 23. Holotype male, MHNG; type locality: Thailand: Khao Yai Nat. Park, 750–850 m.
Löbl, 1999: 724 (records)

DISTRIBUTION. China: Yunnan; Thailand.

Baeocera pseudovilis Löbl

Baeocera pseudovilis Löbl, 1986c: 353, Figs 4, 5. Holotype male, MHNG; type locality: India: Uttarakhand, Bhim Tal, 1500 m, Kumaon.
Löbl, 1992a: 513 (records)

DISTRIBUTION. India: Uttarakhand (Kumaon), West Bengal (Darjeeling District); Nepal.

Baeocera pubiventris Löbl

Baeocera pubiventris Löbl, 1990b: 534, Figs 38–41. Holotype male, MHNG; type locality: Thailand: 1 km below Mae Nang Kaeo, 54 km NE Chiang Mai.
Löbl, 1992a: 516 (records)
Löbl, 1999: 725 (characters)

DISTRIBUTION. China: Yunnan; India: Himachal Pradesh; Nepal; Thailand.

Baeocera pulchella Fierros-López

Baeocera pulchella Fierros-López, 2010: 202, Figs 1–6. Holotype male, CZUG; type locality: Mexico: El Floripondio, 2300 m, km 4 RMO Las Viboras, Municipality San Gabriel, Jalisco.
Fierros-López, 2010: 204 (fungal hosts)

DISTRIBUTION. Mexico: Jalisco.

Baeocera punctata **(Löbl)**

Eubaeocera punctata Löbl, 1975a: 378, Figs 11, 12. Holotype male, HNHM; type locality: Papua New Guinea: Sattelberg [Sattelburg], Huon Golf.

DISTRIBUTION. Papua New Guinea.

Baeocera punctatissima **Löbl & Leschen**

Baeocera punctatissima Löbl & Leschen, 2003b: 34, 48, Figs 13, 43–45, Map 8. Holotype male, NZAC; type locality: New Zealand: South Island, MC, Mt. Somers, Petrifying Creek, 610 m.

DISTRIBUTION. New Zealand: South Island.

Baeocera puncticollis **Löbl**

Baeocera puncticollis Löbl, 1977a: 253, Figs 3, 4. Holotype male, NHMB; type locality: India: West Bengal, Tiger Hill, 2150 m, Darjeeling District.

Löbl, 1984a: 73, Fig. 20 (characters, records)

DISTRIBUTION. India: Meghalaya, Sikkim, West Bengal (Darjeeling District).

Baeocera punctipennides **Newton**

Baeocera punctipennis Matthews, 1888: 170. Lectotype male, NHML; type locality: Nicaragua: Chontales.

Baeocera punctipennides Newton, 2017: 11; replacement name for *Baeocera punctipennis* Matthews, 1888 (nec *Baeocera punctipennis* (Macleay, 1871).

Champion, 1913: 70 (doubtfully in *Scaphisoma*, characters)

Löbl, 1992b: 380, Figs 1, 2 (lectotype designation, characters)

DISTRIBUTION. Nicaragua.

Baeocera punctipennis **(Macleay)**

Scaphisoma punctipenne Macleay, 1871: 156. Syntypes, AMSC; type locality: Australia: Queensland, Gayndah.

Löbl, 1977e: 3 (nomen dubium in *Scaphisoma*, possibly member of "*Sciatrophes*")

Newton, 2017: 11 (implicitly transferred to *Baeocera*)

DISTRIBUTION. Australia: Queensland.

Baeocera reducta **Löbl**

Baeocera reducta Löbl, 1980c: 388. Holotype female, NZAC; type locality: Fiji: Nandarivatu Ridge, 1000 m, Viti Levu.

DISTRIBUTION. Fiji.

***Baeocera reductula* Löbl, 1997**

Baeocera reducta Löbl, 1992a: 518, Figs 101, 102. Holotype male, SMNS; type locality: Nepal: above Dumbus, 2100 m, Kaski District.

Baeocera reductula Löbl, 1997: xi; replacement name for *Baeocera reducta* Löbl, 1992 (nec *Baeocera reducta* Löbl, 1980).

DISTRIBUTION. Nepal.

***Baeocera robertiana* Löbl**

Baeocera roberti Löbl, 1986b: 87, Figs 1–3. Holotype male, MHNG; type locality: Indonesia: Sumatra, Lake Toba, Pradot.

Baeocera robertiana Löbl, 1990b: 536; replacement name for *Baeocera roberti* Löbl, 1986 (nec *Baeocera mussardi roberti* Löbl, 1979, now in *Kasibaeocera*).

DISTRIBUTION. Indonesia: Sumatra.

***Baeocera robustula* Casey**

Baeocera robustula Casey, 1893: 519. Lectotype, USNM; type locality: USA: Texas.

Eubaeocera kingsolveri Cornell, 1967: 6, Figs 2D, 3J, 4O. Holotype male, USNM; type locality: USA: near Aberdeen, North Carolina.

Löbl & Stephan, 1993: 707, Figs 43, 127–128 (lectotype designation of *B. robustula*, synonymy of *B. robustula* with *B. kingsolveri*, characters, records, habitat)

DISTRIBUTION. United States: Arkansas, Florida, Mississippi, New Jersey, North Carolina, Ohio, Oklahoma, Tennessee, Texas.

***Baeocera rubripennis* Reitter**

Baeocera rubripennis Reitter, 1880a: 44. Syntypes, MNHN; type locality: Colombia: La Vega.

DISTRIBUTION. Colombia.

***Baeocera rufoguttata* (Fairmaire)**

Scaphisoma rufoguttatum Fairmaire, 1898a: 223. Lectotype female, MNHN; type locality: Madagascar: Suberbieville [in Maevatanana District].

Achard, 1924b: 28 (transfer to *Cyparella*, characters)

Löbl, 1987d: 855 (lectotype designation, transfer from *Cyparella* to *Baeocera*, characters)

DISTRIBUTION. Madagascar.

Baeocera rufula **(Löbl)**

Eubaeocera rufula Löbl, 1973c: 164, Figs 11, 12. Holotype male, MHNG; type locality: Singapore: Bukit Timah Nat. Reserve.

DISTRIBUTION. Singapore.

Baeocera sarawakensis **Löbl**

Baeocera sarawakensis Löbl, 1987a: 88, Figs 4, 5. Holotype male, NHML; type locality: Malyasia: Sarawak, Gunung Mulu Nat. Park, near Camp 1, 150–200 m.

DISTRIBUTION. East Malyasia: Sarawak.

Baeocera satana **Nakane**

Baeocera satana Nakane, 1963a: 22. Holotype male, HUSJ; type locality: Japan: Sata, Ohsumi.

Nakane, 1963b: 79 (characters)
Löbl, 1968a: 2, Fig. 2 (as *Eubaeocera*, characters)
Löbl, 1969b: 340, Fig. 26 (as *Eubaeocera*, characters)
Löbl, 1984b: 188, Figs 12, 13 (characters, records)
Morimoto, 1985: 258 (characters)
Löbl, 1999: 726 (records)
Ogawa & Löbl, 2013: 322, Figs 10, 19, 32, 56, 61, 72 (characters, records)

DISTRIBUTION. China: Guangxi; Japan: Honshu, Kyushu, Shikoku.

Baeocera sauteri **Löbl**

Baeocera sauteri Löbl, 1980a: 93, Figs 3, 4. Holotype male, HNHM; type locality: Taiwan: Pilam.

DISTRIBUTION. Taiwan.

Baeocera schawalleri **Löbl**

Baeocera schawalleri Löbl, 1992a: 521, Figs 112, 113. Holotype male, SMNS; type locality: Nepal: Hellok, Tamur Valley, 2000 m, Taplejung District.

DISTRIBUTION. Nepal.

Baeocera schirmeri **Reitter**

Baeocera schirmeri Reitter, 1880d: 45. Lectotype male, HNHM; type locality: Croatia: Dalmatia.

Baeocera schirmeri Reitter, 1880c: 221 [subsequent description]; type locality: Croatia: Narenta.

Kuthy, 1897: 86 (records)
Ganglbauer, 1899: 345 (characters, records)

Novak, 1952: 74 (records)
Löbl, 1969b: 331, Figs 10, 11 (as *Eubaeocera*) (lectotype fixed by inference, characters, records)
Tamanini, 1969b: 356, Fig. 3C (as *Eubaeocera*) (characters, records)
Tamanini, 1970: 8, Figs 4C, D (as *Eubaeocera*) (characters, records, habitat)
Löbl, 1982g: 47 (records)
Löbl & Ogawa, 2016c: 38 (note)

DISTRIBUTION. Azerbaijan; Croatia; Bosnia and Herzegovina; Hungary; France: Corsica; Iran; Israel, Italy; Montenegro; Romania; Serbia.

Baeocera schreyeri **Löbl**

Baeocera schreyeri Löbl, 1990b: 523, Figs 16, 17. Holotype male, MHNG; type locality: Thailand: Doi Inthanon, near Forestry Department, 1250 m.

DISTRIBUTION. Thailand.

Baeocera scylla **(Cornell)**

Eubaeocera scylla Cornell, 1967: 8, Figs 2C, 4D, G. Holotype male, USNM; type locality: USA: Whispering Pines, Moor Co., North Carolina.
Löbl & Stephan, 1993: 692, Figs 78, 79 (characters, records, habitat)

DISTRIBUTION. United States: Alabama, Mississippi, New Jersey, North Carolina, Oklahoma, Texas, Virginia.

Baeocera securiforma **(Cornell)**

Eubaeocera securiforma Cornell, 1967: 13, Figs 3A, D, 4X. Holotype male, USNM; type locality: USA: Marietta, Ohio.
Löbl & Stephan, 1993: 703, Figs 116, 117 (characters, records)
Webster et al., 2012: 244 (records, habitat)

DISTRIBUTION. Canada: Manitoba, New Brunswick, Ontario, Quebec; United States: Florida, Massachusetts, New York, Ohio, Oklahoma, Virginia.

Baeocera semiglobosa **(Achard)**

Scaphosoma semiglobosum Achard, 1921b: 87. Lectotype male, NBCL; type locality: Taiwan: Koroton.
Miwa & Mitono, 1943: 542 (as *Scaphosoma*, characters)
Löbl, 1980a: 94, Figs 5, 6 (lectotype fixed by inference, transfer to *Baeocera*, characters, records)
Löbl, 2012: 129 (records)

DISTRIBUTION. Taiwan.

Baeocera senilis Löbl

Baeocera senilis Löbl, 1984a: 65, Fig. 3. Holotype male, MHNG; type locality: India: Meghalaya, Mawphlang, 1800 m, Khasi Hills.

DISTRIBUTION. India: Meghalaya.

Baeocera serendibensis (Löbl)

Eubaeocera serendibensis Löbl, 1971c: 946, Figs 4, 5. Holotype male, MHNG; type locality: Sri Lanka: Mululla, above village 750 m.

Löbl, 1979a: 898 (records)
Löbl, 1984a: 65 (characters, records)
Löbl, 1986c: 344 (records)
Löbl, 1990b: 517 (characters, records)
Löbl, 1992a: 510 (records)
Rougemont, 1996: 12 (records)
Löbl, 1999: 724 (records)
Löbl, 2004 (records)

DISTRIBUTION. China: Hong Kong; India: Assam, Kerala, Mahdya Pradesh, Meghalaya, Uttarakhand, West Bengal (Darjeeling District), Nepal; Pakistan; Sri Lanka; Thailand.

Baeocera signata Löbl

Baeocera signata Löbl, 1984a: 70, Figs 14, 15. Holotype male, MHNG; type locality: India: West Bengal, Algarah, 1800 m, Darjeeling District.

Löbl, 1992a: 513 (records)

DISTRIBUTION. India: West Bengal (Darjeeling District); Nepal.

Baeocera similaris Löbl & Stephan

Baeocera similaris Löbl & Stephan, 1993: 692, Figs 80, 81. Holotype male, FSCA; type locality: USA: Latimer Co., Oklahoma.

DISTRIBUTION. United States: Alabama, Arkansas, Mississippi, Oklahoma.

Baeocera simoni (Pic)

Scaphosoma simoni Pic, 1920b: 5. Lectotype female, MNHN; type locality: Philippines: Antipolo.

Löbl, 1971a: 248 (lectotype fixed by inference, transfer to *Eubaeocera*, characters)
Löbl, 1972b: 102 (invalid lectotype designation, characters)
Löbl, 2012d: 358, fifs 14–16 (characters, records)
Note. Pic (l.c.) credited the species epithet to Portevin.

DISTRIBUTION. Philippines: Luzon.

Baeocera socotrana Löbl

Baeocera socotrana Löbl, 2012e: 142, Figs 1–8. Holotype male, NMPC; type locality: Yemen: Al Haghier Mts., Scan Mt., 1450 m, Socotra.

DISTRIBUTION. Yemen: Socotra.

Baeocera solida Löbl & Stephan

Baeocera solida Löbl & Stephan, 1993: 704, Figs 29, 30, 118, 119. Holotype male, FSCA; type locality: USA: Arizona, Madera Cyn., Santa Rita Mts.

DISTRIBUTION. United States: Arizona.

Baeocera sordida Löbl

Baeocera sordida Löbl, 1984b: 183, Figs 1–3. Holotype male, MHNG; type locality: Japan: Omogo, Mt. Ishizuchi, 700 m, Ehime.

Ogawa & Löbl, 2013: 316, Figs 4, 13, 26, 51, 64, 79, 80 (characters, records)

DISTRIBUTION. Japan: Honshu, Kyushu, Shikoku.

Baeocera sordidoides Löbl

Baeocera sordidoides Löbl, 1992a: 516, Figs 97, 98. Holotype male, MHNG; type locality: Nepal: Phulcoki, 2700 m, Patan District.

Löbl, 1999: 724 (records)

DISTRIBUTION. China: Sichuan; India: Uttarakhand (Garhwal); Nepal.

Baeocera speculifer Casey

Baeocera speculifer Casey, 1893: 518. Lectotype male, USNM; type locality: USA: Keokuk, Iowa.

Cornell, 1967: 11 (as synonym of *Eubaeocera congener*)

Kirk, 1969: 30 (records)

Löbl & Stephan, 1993: 704, Figs 120, 121 (lectotype designation, characters, records)

DISTRIBUTION. United States: Iowa.

Note: Record from South Carolina in Kirk, 1969: 30 unverified.

Baeocera sternalis Broun

Baeocera sternalis Broun, 1914: 173. Lectotype, NHML; type locality: New Zealand: South Island, Mc Clennans [McLennan's Bush]', near Methven.

Kuschel, 1990: 43 (as *Scaphisoma*, habitat)

Newton, 1996: 159 (transfer to *Baeocera*)

Löbl & Leschen, 2003b: 35, 48, Figs 15, 49–51, Map 9 (lectotype designation, characters)

DISTRIBUTION. New Zealand: South Island.

Baeocera sticta Löbl & Stephan

Baeocera sticta Löbl & Stephan, 1993: 686, Figs 5, 6, 64, 65. Holotype male, CNCI; type locality: 7 mi westeast Turkey Crk., Chiricuhua Mts.

DISTRIBUTION. United States: Arizona.

Baeocera subaenea (Fauvel)

Scaphosoma subaeneum Fauvel, 1903: 292. Lectotype female, ISNB; type locality: New Caledonia: Boulari.

Löbl, 1969a: 1, Figs 1, 2 (lectotype fixed by inference, transfer to *Eubaeocera*, characters)

Löbl, 1973b: 309, Figs 1, 2 (characters, records)

Löbl, 1974b: 407 (records)

Löbl, 1977i: 817 (records)

Löbl, 1981a: 350, Fig. 3 (characters, records)

DISTRIBUTION. New Caledonia.

Baeocera subferruginea (Reitter)

Scaphosoma subferrugineum Reitter, 1908: 32. Holotype female, NHMW; type locality: Tanzania: Amani.

Löbl, 1989b: 281 (transfer to *Baeocera*, characters)

DISTRIBUTION. Tanzania.

Baeocera sumatrensis (Löbl)

Eubaeocera sumatrensis Löbl, 1973c: 169, Figs 19, 20. Holotype male, MHNG; type locality: Indonesia: Sumatra, Balighe.

DISTRIBUTION. Indonesia: Sumatra.

Baeocera suthepensis Löbl

Baeocera suthepensis Löbl, 1990b: 527, Figs 24, 25. Holotype male, MHNG; type locality: Thailand: Doi Suthep, north slope, 1550 m, Chiang Mai.

DISTRIBUTION. Thailand.

Baeocera takizawai Löbl

Baeocera takizawai Löbl, 1984b: 190, Figs 15–17. Holotype male, HUSJ; type locality: Japan: Kabira, Isigaki.

Löbl, 2003a: 62 (records)

Löbl, 2012a: 129 (characters, records)

Ogawa & Löbl, 2013: 321, Figs 9, 18, 31, 69, 71 (characters, records)

DISTRIBUTION. China: Jiangxi, Japan: Ryukyus; Taiwan.

Baeocera tamil Löbl

Baeocera tamil Löbl, 1979a: 96, Figs 13, 14. Holotype male, MHNG; type locality: India: Tamil Nadu, 16 km east Kodaikanal, 1400 m, Palni Hills.

DISTRIBUTION. India: Tamil Nadu.

Baeocera taylori (Löbl)

Eubaeocera taylori Löbl, 1973c: 168, Figs 17, 18. Holotype male, MHNG; type locality: Malaysia: Sarawak, Santubong near Kuching, 600 ft.

DISTRIBUTION. East Malaysia: Sarawak.

Baeocera tekootii Löbl & Leschen

Baeocera tekootii Löbl & Leschen, 2003b: 36, 48, Figs 14, 46–48, Map 10. Holotype male, NZAC; type locality: New Zealand: GB, Taikaeakawa.

DISTRIBUTION. New Zealand: North and South Islands.

Baeocera tensingi Löbl & Leschen

Baeocera tensingi Löbl & Leschen, 2003b: 37, 49, Figs 57–59, Map 11. Holotype male, NZAC; type locality: New Zealand: South Island, BR, Tawhai SF, Big Red Road 3 km south of Reefton.

DISTRIBUTION. New Zealand: South Island.

Baeocera tenuis Löbl & Leschen

Baeocera tenuis Löbl & Leschen, 2003b: 37, 49, Figs 16, 52, 53, Map 12. Holotype male, NZAC; type locality: New Zealand: North Island, TO, Ohakune Mountain Rd, near Mangowhero Lodge.

DISTRIBUTION. New Zealand: North Island.

Baeocera texana Casey

Baeocera texana Casey, 1893: 520. Lectotype female, USNM; type locality: USA: Columbus, Texas.

Eubaeocera dybasi Cornell, 1967: 6, Figs 2B, 3K, 4I. Holotype male, USNM; type locality: USA: San Diego, Texas.

Löbl & Stephan, 1993: 708, Figs 45–47, 129, 130 (lectotype designation for *B. texana*, synonymy of *B. texana* with *B. dybasi,* characters, records)

DISTRIBUTION. United States: Florida, Texas.

Baeocera thoracica Löbl

Baeocera thoracica Löbl, 1992a: 523, Figs 117–119. Holotype male, MHNG; type locality: Nepal: Lete, 2550 m, Mustang District.

DISTRIBUTION. India: Himachal Pradesh; Nepal.

Baeocera tibialis **Löbl**

Baeocera tibialis Löbl, 1977f: 39, Figs 1, 3, 4. Holotype male, MHNG; type locality: La Réunion: forest trail to Bélouve, ca 1400 m.
Lecoq, 2015: 211 (records)
DISTRIBUTION. La Réunion.

Baeocera tuberculosa **Löbl**

Baeocera tuberculosa Löbl, 1992a: 523, Figs 120–122. Holotype male, MHNG; type locality: Nepal: Ghoropani Pass, 2850 m, Parbat District.
DISTRIBUTION. Nepal.

Baeocera umtalica **Löbl**

Baeocera umtalica Löbl, 1987d: 851, Figs 14, 15. Holotype male, MHNG; type locality: Zimbabwe: Melsetter, Umtali, 1700 m.
DISTRIBUTION. Zimbabwe.

Baeocera vafra **Löbl**

Baeocera vafra Löbl, 2015b: 168, Figs 6–10. Holotype male, SMNS; type locality: Indonesia: Maluku Islands, Halmahera, Tobelo, 850 m.
DISTRIBUTION. Indonesia: Maluku Islands: Halmahera.

Baeocera vagans **Löbl**

Baeocera vagans Löbl, 1987d: 850, Figs 14, 15. Holotype male, MHNG; type locality: Ivory Coast: Agboville, forest near Yapo Gare.
DISTRIBUTION. Ivory Coast.

Baeocera valdiviana **Löbl**

Toxidium chilense Pic, 1915e: 5. Lectotype female, MNHN; type locality: Chile: Valdivia.
Baeocera valdiviana Löbl, 1983b: 167; replacement name for *Toxidium chilense* Pic, 1915 (nec *Baeocera chilensis* Reitter, 1880).
Löbl, 1983b: 167 (lectotype designation for *T. chilense*, characters)
DISTRIBUTION. Chile.

Baeocera valida **(Löbl)**

Sciatrophes valida Löbl, 1976c: 209, Figs 2, 5, 9, 10. Holotype male, CNCI; type locality: USA: Arizona, Wet Cn., 6100 ft, Pinaleno Mts, Graham Co.
Löbl & Stephan, 1993: 698, Figs 31, 32, 100, 101 (characters, records, habitat)
DISTRIBUTION. United States: Arizona, Colorado, New Mexico.

Baeocera vanuana Löbl

Baeocera vanuana Löbl, 1980c: 381, Figs 3, 4. Holotype male, NZAC; type locality: Fiji: Ndelaikoro, 500 m, Vanua Levu.

DISTRIBUTION. Fiji.

Baeocera variata Löbl

Baeocera variata Löbl, 2015b: 168, Figs 11–15. Holotype male, SMNS; type locality: Indonesia: Maluku Islands, Halmahera, Tobelo.

DISTRIBUTION. Indonesia: Maluku Islands: Halmahera.

Baeocera ventralis (Löbl)

Eubaeocera ventralis Löbl, 1973c: 157, Figs 1, 2. Holotype male, NHMW; type locality: Thailand: Kaopun near Trang.

Baeocera bhutanensis Löbl, 1977a: 251, Figs 1, 2. Holotype male, NHMB; type locality: Bhutan: Samchi.

Löbl, 1984a: 76, Figs 24, 25 (synonymy of *B. bhutanensis* with *B. ventralis*, characters, records)
Löbl, 1986c: 344 (records)
Löbl, 1990b: 522 (records)
Löbl, 1992a: 514 (records)
Löbl, 2004a: 346 (records)

DISTRIBUTION. Bhutan; India: Assam, Himachal Pradesh, Mahdya Pradesh, Meghalaya, Uttarakhand, West Bengal (Darjeeling District); Nepal; Pakistan; Thailand.

Baeocera vesiculata Löbl

Baeocera vesiculata Löbl, 1979a: 93, Figs 7, 17. Holotype male, MHNG; type locality: India: Thekkady near Periyar, 900 m, Kerala.

Löbl, 1984a: 67 (records)
Löbl, 1986c: 344 (records)
Löbl, 1992a: 512 (records)

DISTRIBUTION. India: Kerala, Meghalaya; Nepal.

Baeocera vicina Löbl

Baeocera vicina Löbl, 2015b: 166, Figs 1–5. Holotype male, SNBS; type locality: Indonesia: Maluku Islands, Halmahera, 28 km south Tobelo, Togoliua, 200 m.

DISTRIBUTION. Indonesia: Maluku Islands: Halmahera, Morotai.

Baeocera vidua **Löbl**

Baeocera vidua Löbl, 1990b: 525, Figs 18, 19. Holotype male, MHNG; type locality: Thailand: Doi Inthanon, near Forestry Department, 1250 m.
Löbl, 1999: 724 (records)

DISTRIBUTION. China: Hubei, Guanxi, Sichuan, Yunnan; Thailand.

Baeocera vilis **Löbl**

Baeocera vilis Löbl, 1984a: 71, Figs 16, 17. Holotype male, MHNG; type locality: India: between Ghoom and Lopchu, 2000 m, Darjeeling District.
Löbl, 1986c: 345 (records)
Löbl, 1992a: 513 (records)

DISTRIBUTION. Bhutan; India: Sikkim, Uttarakhand (Kumaon), West Bengal (Darjeeling District); Nepal.

Baeocera wheeleri **Löbl**

Baeocera wheeleri Löbl, 1992b: 382, Figs 5, 6. Holotype male, CUIC; type locality: Mexico: 7 mi east Cuernavaca, Morelos.

DISTRIBUTION. Mexico.

Baeocera wittmeri **Löbl**

Baeocera wittmeri Löbl, 1977a: 254, Figs 5, 6. Holotype male, NHMB; type locality: India: Lebong, 1600–1800 m, Darjeeling District.
Löbl, 1984a: 67, Fig. 6 (characters, records)
Löbl, 1986c: 345 (records)
Löbl, 1992: 512 (records)

DISTRIBUTION. India: Uttarakhand (Kumaon), West Bengal (Darjeeling District); Nepal.

Baeocera wolfgangi **Löbl**

Baeocera wolfgangi Löbl, 2012d: 374, Figs 55–58. Holotype male, SMNS; type locality: Philippines: Visca north Baybay, 200–500 m, Leyte.

DISTRIBUTION. Philippines: Leyte.

Baeocera xichangana **Löbl**

Baeocera xichangana Löbl, 1999: 726, Figs 51, 52. Holotype male, MHNG; type locality: China: Xichang, 1600 m, Sichuan.

DISTRIBUTION. China: Sichuan.

Baeocera yamdena **Löbl, 2014**

Baeocera yandena Löbl, 2014a: 51, Figs 4–7. Holotype male, MHNG; type locality: Indonesia: Maluku Islands, Yamdena 22 km north Saumlaki.

DISTRIBUTION. Indonesia: Maluku Islands.

Baeocera youngi **(Cornell)**

Eubaeocera youngi Cornell, 1967: 11, Figs 2M, 3L, 4W. Holotype male, USNM; type locality: USA: Glenwood, Colorado.

Kirk, 1969: 30 (records)

Löbl, 1987b: 315 (as *Baeocera falsata* Achard, characters)

Löbl & Stephan, 1993: 705, Figs 122–124 (characters, records)

Majka et al., 2011: 88 (distribution)

Webster et al., 2012: 244 (records, habitat)

DISTRIBUTION. Canada: Manitoba, New Brunswick, Newfoundland, Nova Scotia, Ontario, Quebec, Saskatchewan; United States: Louisiana, Massachussets, Maine, Minnesota, North Carolina, Oklahoma, Texas.

Baeocera yunnanensis **Löbl**

Baeocera yunnanensis Löbl, 1999: 731, Fogs 60–62. Holotype male, MHNG; type locality: China: Yunnan, Mengyang Nat. Res., 500 m.

DISTRIBUTION. China: Yunnan.

Baeoceridium Reitter

Baeoceridium Reitter, 1889: 6. Type species: *Baeoceridium depressipes* Reitter, 1889; by monotypy. Gender: neuter.

[*Baeceridium* Pic, 1925a: 195, misspelling].

Leschen & Löbl, 2005: 36, Figs 52, 54 (characters), 44 (termitophily)

Ogawa, 2015: 76 (characters)

Baeoceridium celebense **Löbl**

Baeoceridium celebense Löbl, 1982c: 29, Figs 1–5. Holotype male, MZBI; type locality: Indonesia: Sulawesi, Keti Kesu near Rantepao, Tana Toraja.

Ogawa, 2015: 77, Figs 4-12d-g (characters, records)

Ogawa & Maeto, 2015: 303, Figs 2D-G (characters, records)

DISTRIBUTION. Indonesia: Sulawesi.

Baeoceridium conuroides **Pic**

Baeoceridium conuroides Pic, 1920b: 3. Syntypes, MNHN; type locality: Sierra Leone.

DISTRIBUTION. Sierra Leone.

Baeoceridium depressipes **Reitter**

Baeoceridium depressipes Reitter, 1889: 6. Lectotype, NBCL; type locality: Angola: Humpata.

Scaphisoma pallipes Kraatz, 1895: 154. Holotype, DEIC; type locality: Togo.

Reitter, 1908: 35 (possible synonymy of *S. pallipes* with *B. depressipes*)

Achard, 1921b: 88 (lectotype fixed by inference for *B. depressipes*, characters)
Paulian & Villers, 1940: 75, Fig. 2 (*S. pallipes* in synonymy with *B. depressipes*, characters, records)

DISTRIBUTION. Angola; ?Cameroon; Togo.

Baeoceridium glabrum **Löbl**

Baeoceridium glabrum Löbl, 1979b: 323, Figs 3–11. Holotype male, MZBI; type locality: Indonesia: Sumatra, Lankat Reserve near Bukit Lawang, ex fungus garden of *Odontotermes takensis* Ahmad.

DISTRIBUTION. Indonesia: Sumatra.

Baeoceridium piceonotatum **(Pic)**

Toxidium piceonotatum Pic, 1954b: 39. Syntypes, MRAC/MNHN; type locality: Democratic Republic of the Congo, Kapanga, Lulua.
Löbl & Leschen, 2010: 91, Figs 23, 25 (transfer from *Toxidium*, characters)

DISTRIBUTION. Democratic Republic of the Congo.

Baeoceroxidium **Ogawa & Löbl**

Baeoceroxidium Ogawa & Löbl, 2013: 323. Type species *Scaphosoma micros* Achard, 1923 by original designation. Gender: neuter.

Baeoceroxidium micros **(Achard)**

Scaphosoma micros Achard, 1923: 116. Lectotype male, NHML; type locality: Japan: Kurigahara.
Miwa & Mitono, 1943: 544 (as *Scaphosoma*, characters)
Nakane, 1955c: 56 (as *Scaphosoma*, characters)
Löbl, 1969b: 335, Figs 18, 19, 28 (as *Eubaeocera*, lectotype fixed by inference, characters)
Löbl, 1984b: 184, Figs 4, 5 (as *Baeocera*, characters)
Ogawa & Löbl, 2013: 324, Figs 22, 23, 33, 41–46, 57, 74, 75 (transfer to *Baeoceroxidium*, characters, records)

DISTRIBUTION. Japan: Hokkaido, Honshu, Kyushu, Shikoku.

Baeoceroxidium piliferum **(Löbl)**

Baeocera pilifera Löbl, 1984a: 66, Figs 4, 5. Holotype male, MHNG; type locality: India: West Bengal, 13 km north Ghoom, 1500 m.
Ogawa & Löbl, 2013: 324 (transfer to *Baeoceroxidium*)

DISTRIBUTION. India: West Bengal (Darjeeling District).

Baeoceroxidium pyricola **(Löbl)**

Baeocera pyricola Löbl, 1990b: 519, Figs 7–9. Holotype male, MHNG; type locality: Thailand: Tham Lok Forest Park, 8 km north Sop Pong, Mae Hong Son.
Ogawa & Löbl, 2013: 324 (transfer to *Baeoceroxidium*)

DISTRIBUTION. Thailand.

Baeoceroxidium schwendingeri **(Löbl)**

Baeocera schwendingeri Löbl, 1990b: 520, Figs 10–12. Holotype male, MHNG; type locality: Thailand: Sai Yok Nat. Park, 100 m, Kanchanaburi.
Ogawa & Löbl, 2013: 325 (transfer to *Baeoceroxidium*)

DISTRIBUTION. Thailand.

Baeoceroxidium uncatum **(Löbl)**

Baeocera uncata Löbl, 1990b: 522, Figs 13–15. Holotype male, MHNG; type locality: Thailand: mDoi Suthep, 1550 m, Chiang Mai.
Ogawa & Löbl, 2013: 325 (transfer to *Baeoceroxidium*)

DISTRIBUTION. Thailand.

Bertiscapha Leschen & Löbl

Bertiscapha Leschen & Löbl, 2005: 20; type species *Bertiscapha compacta* Leschen & Löbl, 2005; by original designation. Gender: feminine.

Bertiscapha burlischi **Leschen & Löbl**

Bertiscapha burlischi Leschen & Löbl, 2005: 21, Figs 10, 11. Holotype male, MHNG; type locality: Madagascar: Vohitrosa forest 2 km from peak, 1400–1670 m, 30 km SSE of Betroka.

DISTRIBUTION. Madagascar.

Bertiscapha compacta **Leschen & Löbl**

Bertiscapha compacta Leschen & Löbl, 2005: 22, Figs 2–9. Holotype male, MHNG; type locality: Madagascar: N. Andringitra, Vohindray rdg., 3–4 km SSE Amboarafibe, 1600–1700 m.

DISTRIBUTION. Madagascar.

Bertiscapha striata **Leschen & Löbl**

Bertiscapha striata Leschen & Löbl, 2005: 23, Figs 12, 13. Holotype male, MHNG; type locality: Madagascar: Mt. Ambondrombe, 1500–1600 m.

DISTRIBUTION. Madagascar.

Birocera Löbl

Birocera Löbl, 1970b: 130. Type species: *Amalocera punctatissima* Reitter, 1880; by original designation. Gender: feminine.
Löbl, 2011a: 302 (key to species)
Ogawa, 2015: 100 (characters)

Birocera basicollis Löbl

Birocera basicollis Löbl, 2011a: 302, Figs 1–5. Holotype male, MHNG; type locality: Philippines: Luzon, Mt. Banahaw above Kinabuhayan, Cristalino Tail.
DISTRIBUTION. Philippines: Leyte, Luzon, Mindanao.

Birocera derougemonti Löbl

Birocera derougemonti Löbl, 1983a: 292, Figs 11, 12. Holotype male, MHNG; type locality: Indonesia: Sulawesi, Makale.
Ogawa, 2015: 100, Figs 4-17a, b (characters, records)
DISTRIBUTION. Indonesia: Sulawesi.

Birocera punctatissima (Reitter)

Omalocera [misspelling for *Amalocera*] *punctatissima* Reitter, 1880a: 43. Lectotype female, MNHN; type locality: Indonesia: Sulawesi.
Löbl, 1970b: 130 (lectotype designation, transfer from *Amalocera*, characters)
Löbl, 2011a: 304, Figs 6–8 (characters, records)
Ogawa, 2015: 100, Figs 4-17c, d (characters, records)
DISTRIBUTION. Philippines: Leyte, Luzon, Mindanao; Indonesia: Sulawesi.

Bironium Csiki

Bironium Csiki, 1909: 341. Type species: *Bironium longipes* Csiki, 1909 [=*Heteroscapha basicollis* Pic, 1956]; by monotypy. Gender: neuter.
Heteroscapha Achard, 1914: 394. Type species: *Heteroscapha feai* Achard, 1914; by oiginal designation. Gender: feminine.
Scutotoxidium Pic, 1915b: 30. Type species: *Scutotoxidium nigrolineatum* Pic, 1915; by monotypy. Gender: neuter.
Arachnoscaphula Heller, 1917: 48. Type species: *Arachnoscaphula trisulcata* Heller, 1917; by monotypy. Gender: feminine.
Pic, 1925a: 194 (characters)
Champion, 1927: 272 (as *Heteroscapha*, characters)
Löbl, 1971c: 1002 (synonymy of *Bironium* with *Heteroscapha, Scutotoxidium* and *Arachnoscaphula*, characters)
Löbl, 1975a: 418 (key to New Guinean species)

Löbl, 1987a: 106 (key to Bornean species)
Löbl, 1989c: 374 (key to New Guinean species)
Löbl, 1990b: 614 (characters)
Löbl, 1992a: 576 (characters)
Löbl & Leschen, 2005: 36, Fig. 49
Löbl, 2011a: 306 (key to Philippine species)
Ogawa, 2015: 101 (characters)

***Bironium amicale* Löbl**

Bironium amicale Löbl, 2012b: 174, Figs 1–4. Holotype male, NMPC; type locality: Malaysia: Cameron Highlands, Tanah Rata village, near Gunung Jasat, 1470–1705 m, Pahang.

DISTRIBUTION. West Malaysia.

***Bironium atripenne* (Pic)**

Scutotoxidium atripenne Pic, 1921c: 5. Syntypes, MNHN; type locality: Indonesia: Sumatra.
Pic, 1925a: 194 (as *Heteroscapha*, characters)
Pic, 1947b: 7 (as *Heteroscapha*, characters)

DISTRIBUTION. Indonesia: Sumatra.

***Bironium basicolle* (Pic)**

Heteroscapha basicollis Pic, 1956b: 72. Holotype female, HNHM; type locality: Papua New Guinea: Sattelberg [Sattelburg].

Bironium longipes Csiki, 1909: 341 (secondary junior homonym with *Scaphicoma longipes* Reitter, 1880). Lectotype male, HNHM; type locality: Papua New Guinea: Sattelberg [Sattelburg].
Löbl, 1975a: 418, Figs 79, 80 (lectotype fixed by inference for *B. longipes*, characters)
Löbl, 1989c: 368 (characters)

DISTRIBUTION. Papua New Guinea.

***Bironium bidens* Löbl**

Bironium bidens Löbl, 1990b: 617, Figs 176, 181. Holotype male, MHNG; type locality: Thailand: Doi Suthep, north slope, 1400 m, Chiang Mai.
Löbl, 1999: 733 (records)

DISTRIBUTION. China: Yunnan; Thailand.

Bironium biplagatum (Achard)

Heteroscapha biplagatum Achard, 1920l: 6. Syntypes, MNHN/NMPC; type localities: Indonesia: Sumatra: Palembang, Mentawei: Sipora.
Achard, 1920d: 136 (records)
DISTRIBUTION. Indonesia: Mentawei: Sipura, Sumatra.

Bironium biroi (Pic)

Heteroscapha biroi Pic, 1956b: 71. Holotype male, HNHM; type locality: Papua New Guinea: Stephansort, Astrolabe Bay.
Löbl, 1975a: 419, Figs 81, 82 (characters, records)
Löbl, 1989c: 368 (characters)
DISTRIBUTION. Papua New Guinea.

Bironium bisulcatum Löbl

Bironium bisulcatum Löbl, 1972b: 97, Figs 47, 48. Holotype male, ZMUB; type locality: Philippines: Palawan, Binaluan.
Löbl, 2011a: 307 (records)
DISTRIBUTION. Philippines: Palawan.

Bironium borneense Löbl

Bironium borneense Löbl, 1987a: 106, Fig. 34. Holotype male, NHML; type locality: Malaysia: Sarawak, Gunung Mulu Nat. Park, near Base Camp, 50–100 m.
DISTRIBUTION. East Malaysia: Sarawak.

Bironium coomani (Pic)

Heteroscapha coomani Pic, 1940a: 2. Syntypes, MNHN; type locality: Vietnam: Tonkin.
Löbl, 1990b: 616 (type material, characters)
DISTRIBUTION. Vietnam.

Bironium distinctum (Achard)

Heteroscapha distincta Achard, 1920j: 265. Syntypes, MCSN/NMPC; type locality: Myanmar: Carin Chebà, 900–1100 m [approx. 19°13'N 96°35'E].
Achard, 1920l: 8 (as new species, subsequent description)
Achard, 1920d: 135 (*Heteroscapha*, records)
Pic, 1920f: 24 (*Heteroscapha*, characters)
Pic, 1921a: 163 (*Heteroscapha*, characters)

Champion, 1927: 272 (*Heteroscapha*, characters, records)
Löbl, 1984a: 106 (characters, records)
Löbl, 1990b: 616, Figs 174, 177 (characters, records)
DISTRIBUTION. India: Assam, Meghalaya, West Bengal (Darjeeling District); Myanmar; Thailand.

Bironium elegans **Löbl**
Bironium elegans Löbl, 1977d: 60, Fig. 2. Holotype male, MHNG; type locality: Singapore: Bukit Timah Nat. Res.
DISTRIBUTION. Singapore.

Bironium feai **(Achard)**
Heteroscapha feai Achard, 1914: 395, Fig. 1. Syntypes, NMPC; type locality: Myanmar: Carin Chebà, 900–1100 m [approx. 19°13'N 96°35'E].
Achard, 1920d: 135 (as *Heteroscapha*, characters, records)
Pic, 1921a: 163 (as *Heteroscapha*, records)
Löbl, 1990b: 617 (characters, records)
DISTRIBUTION. Myanmar; Thailand.

Bironium flavapex **Löbl**
Bironium flavapex Löbl, 2015b: 170, Figs 16–19. Holotype male, SMNS; type locality: Indonesia: Maluku Islands, Morotai, west Daruba Raja, ca 250 m.
DISTRIBUTION. Indonesia: Maluku Islands: Halmahera, Morotai.

Bironium glabrum **Löbl**
Bironium glabrum Löbl, 1989c: 370, Figs 4–6. Holotype male, MHNG; type locality: Papua New Guinea: Wau.
DISTRIBUTION. Papua New Guinea; Indonesia: Western New Guinea.

Bironium grouvellei **(Achard)**
Heteroscapha grouvellei Achard, 1920j: 5. Syntypes, MNHN; type locality: Indonesia: Sumatra [Palembang].
DISTRIBUTION. Indonesia: Sumatra.

Bironium loksai **Löbl**
Bironium loksai Löbl, 1989c: 368, Figs 1–3. Holotype male, HNHM; type locality: Papua New Guinea: Wau.
DISTRIBUTION. Papua New Guinea.

***Bironium longipes* (Reitter)**

Scaphicoma longipes Reitter, 1880a: 49. Syntypes, MNHN; type locality: Indonesia: Mysol.

Achard, 1924b: 31 (transfer to *Heteroscapha*)

Löbl, 1989c: 373, Figs 9–11 (as cf. *Bironium longipes*, characters, records)

DISTRIBUTION. Indonesia: Misool; Papua New Guinea.

***Bironium maculatum* Löbl**

Bironium maculatum Löbl, 1989c: 371, Figs 7, 8. Holotype male, MHNG; type locality: Papua New Guinea: Popondetta.

DISTRIBUTION. Papua New Guinea.

***Bironium minutum* (Achard)**

Heteroscapha minuta Achard, 1920l: 7. Lectotype male, MNHN; type locality: Indonesia: Sumatra.

Heteroscapha jacobsoni Pic, 1947b: 7. Lectotype female, MNHN; type locality: Indonesia: Sumatra, Fort de Kock [Bukittinggi].

Löbl, 1977d: 59, Figs 1–2 (lectotype fixed by inference for *H. minutum*, lectotype designation for *B. jacobsoni*, synonymy of *H. jacobsoni* with *H. minutum*, characters, records)

DISTRIBUTION. Indonesia: Sumatra; East Malaysia: Sarawak.

***Bironium nepalense* Löbl**

Bironium nepalense Löbl, 1992a: 576, Figs 174–176. Holotype male, MHNG; type locality: Nepal: Arun Valley, Chichila, 2200 m, Sankhuwasabha District.

DISTRIBUTION. Nepal.

***Bironium nigrolineatum* (Pic)**

Scutotoxidium nigrolineatum Pic, 1915b: 31. Syntypes, MNHN; type locality: Sri Lanka.

Löbl, 1971c: 1004, Figs 80–83 (transfer to *Bironium*, records)

DISTRIBUTION. Sri Lanka.

***Bironium pustulatum* Löbl**

Bironium pustulatum Löbl, 2011a: 306, Figs 9, 10. Holotype male, SMNS; type locality: Philippines: Mindanao, 25 km west of New Batan, 1200 m.

DISTRIBUTION. Philippines: Mindanao.

***Bironium quadrimaculatum* Löbl**

Bironium quadrimaculatum Löbl, 1979a: 124, Figs 60, 61. Holotype male, MHNG; type locality: India: Mundakayam, 100 m, Kerala.

DISTRIBUTION. India: Kerala, Tamil Nadu.

Bironium rufescens Löbl

Bironium rufescens Löbl, 1972b: 99, Figs 51, 52. Holotype male, ZMUB; type locality: Philippines: Mindanao, Surigao.
Löbl, 2011a: 307 (records)

DISTRIBUTION. Philippines: Mindanao.

Bironium rufonotatum (Pic)

Heteroscapha rufonotata Pic, 1940a: 2. Syntypes, MNHN; type locality: Vietnam: Tonkin.

DISTRIBUTION. Vietnam.

Bironium sumatranum (Achard)

Heteroscapha sumatrana Achard, 1920l: 6. Syntypes, MNHN; type locality: Indonesia: Sumatra.
Löbl, 1977d: 61, Fig. 4 (characters)

DISTRIBUTION. Indonesia: Sumatra.

Bironium testaceum (Pic)

Heteroscapha testacea Pic, 1931a: 3. Syntypes, MNHN; type locality: Indonesia: Java [Pengarengan, west Java].

DISTRIBUTION. Indonesia: Java.

Bironium tonkineum (Pic)

Heteroscapha tonkinea Pic, 1922b: 1. Syntypes, MNHN; type locality: Vietnam: Tonkin [Chapa].
Löbl, 1990b: 616 (characters)

DISTRIBUTION. Vietnam.

Bironium trisulcatum (Heller)

Arachnoscaphula trisulcata Heller, 1917: 49, Fig. 1. Syntypes, SMTD; type locality: Philippines: Luzon, Mt. Makiling.
Löbl, 1972b: 99, Figs 49, 50 (characters)
Löbl, 2011a: 308 (records)

DISTRIBUTION. Philippines: Luzon.

Bironium troglophilum Löbl

Bironium troglophilum Löbl, 1990b: 616, Figs 178–180. Holotype male, MHNG; type locality: Thailand: Sop Pong, Tam "Plaa", Mae Hong Son.

DISTRIBUTION. Thailand.

Brachynoposoma Löbl

Brachynoposoma Löbl, 1973b: 329. Type species: *Brachynoposoma punctatum* Löbl, 1973; by original designation. Gender: neuter.
Löbl, 1981a: 353 (characters, key)

Brachynoposoma major Löbl

Brachynoposoma major Löbl, 1973b: 334. Holotype female, NHMW; type locality: New Caledonia: Rivière Bleu.
Löbl, 1981a: 353 (characters, records)
DISTRIBUTION. New Caledonia.

Brachynoposoma punctatum Löbl

Brachynoposoma punctatum Löbl, 1973b: 332, Figs 33–41. Holotype male, NHMW; type locality: New Caledonia: Table d'Union.
Löbl, 1981a: 353 (characters, records)
DISTRIBUTION. New Caledonia.

Brachynopus Broun

Brachynopus Broun, 1881a: 664. Type species: *Brachynopus latus* Broun, 1881; by monotypy. Gender: masculine.
Löbl & Leschen, 2003b: 18 (characters, key)

Brachynopus latus Broun

Brachynopus latus Broun, 1881a: 664. Holotype, NHML; type locality: New Zealand: Parua.
Baeocera fulvicollis Broun, 1881b: 891. Lectotype male, NHML; type locality: New Zealand: Tiritiri Island.
Kuschel, 1990: 43 (habitat)
Löbl & Leschen, 2003b: 21, 49, Figs 4, 20–22, 84, 87–95, 97–102, 106, 116, Map 14 (lectotype designations, synonymy of *B. fulvicollis* with *B. latus*, characters, records, fungal and slime mould hosts)
Löbl & Leschen, 2005: 37, Figs 55–58, 60, 62 (characters)
DISTRIBUTION. New Zealand: Three Kings Islands, North Island.

Brachynopus scutellaris (Redtenbacher)

Scaphisoma scutellare Redtenbacher, 1868: 32. Lectotype, NHMW; type locality: New Zealand.
Scaphisoma tenellum Pascoe, 1876: 48. Holotype male, NHML; type locality: New Zealand: Tairoa, Auckland.
Baeocera rufipes Broun, 1886: 833. Lectotype mal, e, NHML; type locality: New Zealand: Otago.

Broun, 1880: 159 (*Scaphisoma*, characters)
Reitter, 1880a: 44 (*Baeocera*, synonymy of *B. scutellare* with *S. tenellum*, characters)
Hudson, 1934: 49, Pl. VI (*Scaphisoma*, habitat, characters)
Kuschel, 1990: 43, Figs 23, 122 (habitat, fungal hosts)
Löbl & Leschen, 2003b: 23, 51, Figs 5, 23–25, Map 16 (lectotype designations for *B. rufipes*, *S. scutellare* and *S. tenellum*, synonymy of *B. rufipes* with *B. scutellaris*, transfer to *Brachynopus*, characters, records, fungal and slime mould hosts)
Leschen et al., 2008: 7 (populations, genetics)
DISTRIBUTION. New Zealand.

Collartium Pic

Collartium Pic, 1928b: 39. Type species: *Collartium irregulare* Pic, 1928; by monotypy. Gender: neuter.

Collartium irregulare Pic

Collartium irregulare Pic, 1928b: 39. Syntypes, MRAC/MNHN; type locality: Democratic Republic of the Congo, Tchiobo, N'Goy.
DISTRIBUTION. Democratic Republic of the Congo.

Curtoscaphosoma Pic

Curtoscaphosoma Pic, 1951c: 1099. Type species: *Curtoscaphosoma nigrum* Pic, 1951; by original designation. Gender: neuter.

Curtoscaphosoma nigrum Pic

Curtoscaphosoma nigrum Pic, 1951c: 1099. Holotype, IFAN; type locality: Guinea: northeast Mont Nimba.
DISTRIBUTION. Guinea.

Irianscapha Löbl

Irianscapha Löbl, 2012f: 310. Type species *Irianscapha dimorpha* Löbl, 2012, by original designation. Gender: feminine.

Irianscapha dimorpha Löbl

Irianscapha dimorpha Löbl, 2012f: 311, Figs 1–10. Holotype male, NMEC; type locality: Indonesia: Western New Guinea, Nabire area, road Nabire-Ilaga, km 54, 750 m.
DISTRIBUTION. Indonesia: Western New Guinea.

Kasibaeocera Leschen & Löbl

Kasibaeocera Leschen & Löbl, 2005: 23. Type species: *Eubaeocera mussardi* Löbl, 1971; by original designation. Gender: feminine.

Kasibaeocera mussardi (Löbl)

Eubaeocera mussardi Löbl, 1971c: 944, Figs 2, 3. Holotype male, MHNG; type locality: Sri Lanka: Alut Oya.

Baeocera mussardi roberti Löbl, 1979a: 89. Holotype male, MHNG; type locality: India: Palni Hills, 39 km east Kodaikanal, 650 m, Tamil Nadu.

Löbl, 1984a: 64 (as *Baeocera*, records)

Löbl, 1990b: 517 (as *Baeocera*, records)

Löbl, 1992a: 511 (as *Baeocera*, records)

Rougemont, 1996: 12 (as *Baeocera*, records)

Löbl, 1999: 724 (as *Baeocera*, records)

Leschen & Löbl, 2005: 24, Figs 14–18 (transfer from *Baeocera*, synonymy of *B. mussardi roberti*)

DISTRIBUTION. Bhutan; China: Hong Kong, Zhejiang, Yunnan; India: Kerala, Tamil Nadu, Uttar Pradesh (Garhwal), West Bengal (Darjeeling District); Nepal; Sri Lanka; Thailand.

Kathetopodion Löbl

Kathetopodion Löbl, 1982d: 792. Type species: *Amalocera borneensis* Pic, 1935; by original designation. Gender: neuter.

Kathetopodion borneense (Pic)

Amalocera borneensis Pic, 1935: 471, Figs 8–11. Lectotype femjale, NHML; type locality: East Malaysia: Sabah, Bettotan near Sandakan.

Löbl, 1982d: 794 (lectotype designation, transfer to *Kathetopodion*, characters)

DISTRIBUTION. East Malaysia: Sabah.

Kathetopodion kiunganum Leschen & Löbl

Kathetopodion kiunganum Leschen & Löbl, 2005: 62, Figs 38–41. Holotype male, HNHM; type locality: Papua New Guinea: Kiunga.

DISTRIBUTION. Papua New Guinea.

Mordelloscaphium Pic

Mordelloscaphium Pic, 1915c: 35. Type species: *Mordelloscaphium testaceimembre* Pic; by monotypy. Gender: neuter.

***Mordelloscaphium testaceimembre* Pic**

Mordelloscaphium testaceimembris [sic] Pic, 1915c: 35. Syntypes, MNHN; type locality: Indonesia: Kalimantan, Martapoera [Riam Kalan].

DISTRIBUTION. Indonesia: Kalimantan.

Nesoscapha Vinson

Nesoscapha Vinson, 1943: 204. Type species: *Nesoscapha hughscotti* Vinson, 1943; by original designation. Gender: feminine.

***Nesoscapha gomyi* Löbl**

Nesoscapha gomyi Löbl, 1977f: 40, Figs 5, 6. Holotype male, MHNG; type locality: La Réunion: forest trail to Plaine d'Affouche, 1400 m.

Lecoq, 2015: 212 (records)

DISTRIBUTION. La Réunion.

***Nesoscapha hughscotti* Vinson**

Nesoscapha hughscotti Vinson, 1943: 206, Fig. 23. Holotype female, NHML; type locality: Mauritius: Mt. Cocotte.

Löbl, 1977f: 42 (characters)

DISTRIBUTION. Mauritius.

Notonewtonia Löbl & Leschen

Notonewtonia Löbl & Leschen, 2003b: 24. Type species *Notonewtonia thayerae* Löbl & Leschen, 2003; by original designation. Gender: feminine.

***Notonewtonia thayerae* Löbl & Leschen**

Notonewtonia thayerae Löbl & Leschen, 2003b: 25, Figs 8, 30, 31, 77–79, Map 19. Holotype male, NZAC; type locality: New Zealand: BR, 1.8 km north Punakaiki, 50 m.

DISTRIBUTION. New Zealand: North and South Islands.

***Notonewtonia watti* Löbl & Leschen**

Notonewtonia watti Löbl & Leschen, 2003b: 26, Figs 9, 32, 33, 80–82, Map 20. Holotype male, NZAC; type locality: New Zealand: WO, Hapua Kohe Ra. 3 km SW of Kaihere.

DISTRIBUTION. New Zealand: North Island.

Paratoxidium Vinson

Paratoxidium Vinson, 1943: 203. Type species: *Paratoxidium pollicis* Vinson, 1943; by original designation. Gender: neuter.

***Paratoxidium pollicis* Vinson**

Paratoxidium pollicis Vinson, 1943: 204, Fig. 22. Holotype, NHML; type locality: Mauritius: The Pouce.

Distribution. Mauritius.

***Pseudobironiella* Löbl**

Pseudobironiella Löbl, 1973d: 19. Type species: *Scaphisoma madeccasa* Brancsik, 1893; by original designation. Gender: feminine.

***Pseudobironiella madeccasa* (Brancsik)**

Scaphisoma madecassa Brancsik, 1891: 221. Lectotype male, FMNH; type locality: Madagascar: Nossibé.

Löbl, 1973d: 20, Figs 1–5 (lectotype fixed by inference, transfer, characters)

Distribution. Madagascar.

***Pseudobironiella maxima* (Pic)**

Scaphosoma maximum Pic, 1920b: 5. Lectotype female, MNHN; type locality: Madagascar [Diego Suarez].

Scaphosoma maximum var. *diversicolor* Pic, 1920b: 6. Lectotype male, MNHN; type locality: Madagascar [Diego Suarez].

Löbl, 1974c: 315, Figs 1, 2 (lectotype fixations by inference for *S. maximum* and var. *diversicolor*, synonymy of *diversicolor* with *S. maximum*, transfer to *Pseudobironiella*, characters)

Distribution. Madagascar.

***Pseudobironiella pilifera* Löbl**

Pseudobironiella pilifera Löbl, 1974c: 317, Figs 3, 4. Holotype male, MNHN; type locality: Madagascar, Antongil Bay.

Distribution. Madagascar.

Pseudobironium Pic

Pseudobironium Pic, 1920d: 15; type species: *Pseudobironium subovatum* Pic, 1920; by monotypy. Gender: neuter.

Morphoscapha Achard, 1920d: 131; type species: *Morphoscapha grossum* Achard, 1920; by original designation. Gender: feminine.

Achard, 1923: 117 (characters)

Löbl, 1969b: 322 (characters, key to Palaearctic species)

Löbl, 1982e: 159 (key to Indian species)

Iablokoff-Khnzorian, 1985: 138 (characters)

Löbl, 1990b: 513 (key to Thai species)

Löbl, 1992a: 500 (species groups, characters, key to Nepalese species)
Löbl, 1999: 721 (key to Chinese species)
Löbl, 2011b: 205 (key to Taiwanese species)
Löbl & Tang, 2013 (review, species groups, key, characters)

***Pseudobironium achardi* (Pic)**

Amalocera achardi Pic, 1920h: 242. Lectotype female, NMPC; type locality: Indonesia: Sumatra, Palembang.
Löbl, 1975b: 269 (transfer from to *Pseudobironium*)
Löbl & Tang, 2013: 730 (lectotype designation, characters)
DISTRIBUTION. Indonesia: Sumatra.

***Pseudobironium almoranum* Champion**

Pseudobironium almoranum Champion, 1927: 273. Lectotype male, NHML; type locality: India: W. Almora, Kumaon.
Löbl, 1969b: 324 (records)
Löbl, 1984: 62 (records)
Löbl, 1986c: 343 (records)
Löbl, 1992a: 502, Fig. 89 (lectotype designation, characters, records)
Löbl, 2005: 180 (records)
Löbl & Tang, 2013: 694 (characters, records)
DISTRIBUTION. India: Arunachal Pradesh. Himachal Pradesh, Uttarakhand; Nepal.

***Pseudobironium antennatum* Löbl & Tang**

Pseudobironium antennatum Löbl & Tang, 2013: 716. Holotype male, NHMB; type locality: West Malaysia: Benom Mts., 15 km east Kampong Dong, 700 m.
DISTRIBUTION. West Malaysia.

***Pseudobironium augur* Löbl & Tang**

Pseudobironium augur Löbl & Tang, 2013: 689. Holotype male, NHML; type locality: Japan: Miyanoshita.
DISTRIBUTION. Japan: Shikoku, Kyushu.

***Pseudobironium banosense* (Pic)**

Scaphosoma banosense Pic, 1931a: 3. Lectotype female, MNHN; type locality: Luzon: Los Banos.
Scaphosoma obscuricolle Pic, 1947c: 2. Lectotype female, MNHN; type locality: Luzon, Banguio.

Löbl, 1970b: 126 (misspelled *banonense*, lectotype fixations by inference for *S. banosense* and *S. obscuricolle*, transfer to *Pseudobironium*)

Löbl, 2011a: 308, Figs 11–13 (synonymy of *S. obscuricolle*, misspelled *banosense*, records)

Löbl & Tang, 2013: 694 (misspelled *banonense*, characters, records)

DISTRIBUTION. Philippines: Luzon.

***Pseudobironium bicolor* Löbl**

Pseudobironium bicolor Löbl, 1992a: 501, 585, Figs 25, 88. Holotype male, MHNG; type locality: Nepal: Arun Valley below Num, 1050 m, Sankhuwasabha District.

Löbl, 1982e: 160 (as *Pseudobironium* sp., characters)

Löbl, 1984: 62 (as *Pseudobironium* sp., characters)

Löbl, 1990b: 514 (as *Pseudobironium* sp., characters)

Löbl & Tang, 2013: 696 (characters, records)

DISTRIBUTION. India: Meghalaya, Nepal; Thailand.

***Pseudobironium bilobum* Löbl & Tang**

Pseudobironium bilobum Löbl & Tang, 2013: 696. Holotype male, NHMW; type locality: West Malaysia: Pahang, Cameron Highlands, Tanah Rata, Gn. Jasar, 1400–1500 m.

DISTRIBUTION. West Malaysia.

***Pseudobironium brancuccii* Löbl & Tang**

Pseudobironium brancuccii Löbl & Tang, 2013: 690. Holotype male, NMHB; type locality: Nepal: Arun Valley, Hedangan-Num, 800 m.

DISTRIBUTION. India: Meghalaya; Nepal.

***Pseudobironium carinense* (Achard)**

Morphoscapha carinense Achard, 1920d: 134. Lectotype female, MNHN; type locality: Myanmar: Carin Chebà, 900–1100 m [approx. 19°13'N 96°35'E].

Pseudobironium castaneum Pic, 1923b: 17. Lectotype male, MNHN; type locality: Vietnam: Tonkin [Lac Tho].

Pic, 1921a: 163 (records)

Löbl, 1990b: 514 (records)

Löbl, 1992a: 504, Figs 95, 96 (lectotype designation for *P. castaneum*, characters)

Löbl & Tang, 2013: 718 (lectotype designation for *M. carinense*, synonymy of *P. castaneum* with *carinense*, characters, records)

DISTRIBUTION. China: Yunnan; Laos; Myanmar; Thailand; Vietnam.

***Pseudobironium confusum* Löbl & Tang**

Pseudobironium confusum Löbl & Tang, 2013: 697. Holotype male, SNUC; type locality: China: Ledong County, Hainan, Jianfengling, 1000 m.

DISTRIBUTION. China: Hainan; Vietnam.

***Pseudobironium conspectum* Löbl & Tang**

Pseudobironium conspectum Löbl & Tang, 2013: 720. Holotype male, MHNG; type locality: Nepal: Arun Valley below Num, 1050 m.

DISTRIBUTION. China: Yunnan; India: Kerala, Tamil Nadu, West Bengal; Laos; Myanmar; Nepal; Thailand.

***Pseudobironium convexum* Löbl & Tang**

Pseudobironium convexum Löbl & Tang, 2013: 722. Holotype male, MHNG; type locality: Indonesia: Sumatra, Mt. Leuser Nat. Park, Ketambe 800 m.

DISTRIBUTION. China: Hainan; Indonesia: Kalimantan, Mentawei, Sumatra; West Malaysia; Vietnam.

***Pseudobironium fasciatum* Löbl**

Pseudobironium fasciatum Löbl, 1982e: 157, Figs 1–3. Holotype male, EUMJ; type locality: India: Cinchona, 3500 ft., Anaimalai Hills, Tamil Nadu.

Löbl, 1979a: 86 (as *Pseudobironium* sp.)

Löbl & Tang, 2013: 692 (characters, records)

DISTRIBUTION. India: Kerala, Tamil Nadu.

***Pseudobironium feai* Pic**

Pseudobironium feai Pic, 1920f: 24. Lectotype male, MCSN; type locality: Myanmar: Carin Chebà, 900–1100 m [approx. 19°13'N 96°35'E].

Pic, 1921a: 163 (characters)

Löbl & Tang, 2013: 699 (lectotype designation, characters, records)

DISTRIBUTION. China: Guangxi, Hainan, Xizang; India: Meghalaya; Laos; Myanmar.

***Pseudobironium flagellatum* Löbl & Tang**

Pseudobironium flagellatum Löbl & Tang, 2013: 692. Holotype male, NHMB; type locality: Laos: Kham Mouan Prov., Ban Khoum Ngeum, about 200 m.

DISTRIBUTION. Laos.

***Pseudobironium fujianum* Löbl & Tang**

Pseudobironium fujianum Löbl & Tang, 2013: 682. Holotype male, SNUC; type locality: China: Fujian, Wuyishan City, Guadun, 1200–1300 m.

DISTRIBUTION. China: Fujian.

***Pseudobironium grossum* (Achard)**
Morphoscapha grossum Achard, 1920d: 132. Holotype male, NMPC; type locality: Laos.
Löbl & Tang, 2013: 724 (characters)
DISTRIBUTION. Laos.

***Pseudobironium hisamatsui* Löbl**
Pseudobironium hisamatsui Löbl, 2011b: 203, Figs 7–9. Holotype male, MHNG; type locality: Taiwan: Tienhsiang, Hualien Co.
Löbl & Tang, 2013: 724 (characters, records)
DISTRIBUTION. Taiwan.

***Pseudobironium horaki* Löbl & Tang**
Pseudobironium horaki Löbl & Tang, 2013: 701. Holotype male, MHNG; type locality: East Malaysia: Sarawak, Rumah Ugap village, Sut river.
DISTRIBUTION. Indonesia: Java; East Malaysia: Sabah, Sarawak.

***Pseudobironium impressipenne* Löbl**
Pseudobironium impressipenne Löbl, 1973a: 149, Figs 1, 2. Holotype male, MCSN; type locality: Indonesia: Sumatra, Si-Rambé [near Balige].
Löbl & Tang, 2013: 693 (characters)
DISTRIBUTION. Indonesia: Sumatra.

***Pseudobironium incisum* Löbl & Tang**
Pseudobironium incisum Löbl & Tang, 2013: 703. Holotype male, MHNG; type locality: Vietnam: Tuyen Quang Prov., Na Hang Reserve, 360 m.
DISTRIBUTION. Vietnam.

***Pseudobironium ineptum* Löbl**
Pseudobironium ineptum Löbl, 1992a: 503, Figs 90, 91. Holotype male, MHNG; type locality: Nepal: Arun Valley below Num, Sankhuwasabha District.
Löbl & Tang, 2013: 693 (characters)
DISTRIBUTION. Nepal.

***Pseudobironium javanum* Löbl**
Pseudobironium javanum Löbl & Tang, 2013: 679. Holotype male, MZBI; type locality: Indonesia: Java, Tapos, Mt. Ojedo, 700 m.
DISTRIBUTION. Indonesia: Java.

Pseudobironium languei **(Achard)**

Morphoscapha languei Achard, 1920d: 133. Syntypes, MNHN; type locality: Vietnam: Tonkin, Laokai [Hoa Binh].

Cyparium monticola Miwa & Mitono, 1943: 536. Lectotype male, TARI; type locality: Taiwan: Kuaru.

Pseudobironium tonkineum Pic, 1923d: 195. Lectotype female, MNHN; type locality: Vietnam: Tonkin, Hoa Binh.

Löbl, 2011b: 203, Figs 1–6 (lectotype designation of *C. monticola*, synonymy with *P. languei*, characters)

Löbl, 2012b: 183 (records)

Löbl & Tang, 2013: 705 (lectotype designation of *P. tonkineum* and synonymy with *P. languei*, characters, records)

DISTRIBUTION. Laos; Taiwan; Thailand; Vietnam; West Malayasia.

Pseudobironium lewisi **Achard**

Pseudobironium lewisi Achard, 1923: 118. Lectotype male, NHML; type locality: Japan: Nagasaki.

Achard, 1923: 118 (records)

Nakane, 1955b: 53 (characters)

Nakane, 1963b: 79 (characters)

Löbl, 1969b: 324, Figs 2, 4, 5 (lectotype designation, characters, records)

Hisamatsu, 1977: 193 (records)

Morimoto, 1985: Pl. 45, Fig. 28 (characters)

Löbl & Tang, 2013: 705 (characters, records)

DISTRIBUTION. Japan: Honshu, Kyushu, Shikoku.

Pseudobironium merkli **Löbl & Tang**

Pseudobironium merkli Löbl & Tang, 2013: 684. Holotype male, HNHM; type locality: Vietnam: Cuc Phuong, Minh Binh.

DISTRIBUTION. China: Yunnan; Vietnam.

Pseudobironium montanum **Löbl & Tang**

Pseudobironium montanum Löbl & Tang, 2013: 685. Holotype male, PCAP; type locality: China: Yunnan, Gaoligong Shan, side valley 19 km NW Liuku, 2730 m.

DISTRIBUTION. China: Yunnan; Vietnam.

Pseudobironium parabicolor **Löbl & Tang**

Pseudobironium parabicolor Löbl & Tang, 2013: 708. Holotype male, MHNG; type locality: China: Yunnan, Mengyang Nat. Res., ca 500 m.

DISTRIBUTION. China: Yunnan.

***Pseudobironium plagifer* Löbl**

Pseudobironium plagifer Löbl, 1980a: 92, Figs 1, 2. Holotype male, MHNG; type locality: Taiwan.
Löbl, 2011b: 203 (records)
Löbl & Tang, 2013: 680 (characters)
DISTRIBUTION. Taiwan.

***Pseudobironium pseudobicolor* Löbl & Tang**

Pseudobironium pseudobicolor Löbl & Tang, 2013: 709. Holotype male, MHNG; type locality: Taiwan: Sandimen township, Rd 29, Saijia, 1053 m.
DISTRIBUTION. Taiwan.

***Pseudobironium pubiventer* Löbl & Tang**

Pseudobironium pubiventer Löbl & Tang, 2013: 710. Holotype male, NHMB; type locality: East Malaysia: Mt. Kinabalu Nat. Park, 1700 m.
DISTRIBUTION. East Malaysia: Sabah.

***Pseudobironium rufitarse* Löbl**

Pseudobironium rufitarse Löbl, 1992a: 502, Figs 92–94. Holotype male, SMNS; type locality: Nepal: Yamputhin, 1650–1800 m, Taplejung District.
Löbl & Tang, 2013: 694 (characters, records)
DISTRIBUTION. India: Sikkim; Nepal.

***Pseudobironium schuhi* Löbl & Tang**

Pseudobironium schuhi Löbl & Tang, 2013: 687. Holotype male, MHNG; type locality: Indonesia: Java, Mt. Gede, 1400–1600 m.
DISTRIBUTION. Indonesia: Java.

***Pseudobironium sinicum* Pic**

Pseudobironium sinicum Pic, 1954c: 58. Lectotype female, NHRM; type locality: China: Fujian, Kuatun.
Löbl, 1999: 721 (lectotype designation)
Löbl & Tang, 2013: 731 (characters)
DISTRIBUTION. China: Fujian.

***Pseudobironium sparsepunctatum* (Pic)**

Amalocera sparsepunctata Pic, 1915b: 31. Lectotype female, MNHN; type locality: East Malaysia: Banguey [Banggi Is.].
Morphoscapha bangueyi Achard, 1920d: 134. Lectotype male, NMPC; type locality: East Malaysia: Banguey [Banggi Is.].

Löbl, 1982d: 790, Figs 1–4 (lectotype designations for *A. sprasepunctata* and *M. bangueyi*, transfer of *A. sprasepunctata* to *Pseudobironium*, synonymy, characters, records)
Löbl, 1990b: 514 (records)
Löbl & Tang, 2013: 712 (characters, records, comments)

DISTRIBUTION. East Malaysia: Banggi, Sabah, Sarawak; Indonesia: Java; Philippines: Palawan.

Pseudobironium spinipes **Löbl**

Pseudobironium spinipes Löbl & Tang, 2013: 680. Holotype male, SNUC; type locality: China: Zhejian, Longquan city, Fengyang shan 1430 m.

DISTRIBUTION. China: Zhejiang.

Pseudobironium stewarti **Löbl & Tang**

Pseudobironium stewarti Löbl & Tang, 2013: 713. Holotype male, MHNG; type locality: Vietnam: Tuyen Quang Prov., Na Hang Reserve, 300 m.

DISTRIBUTION. Vietnam.

Pseudobironium subglabrum **Löbl**

Pseudobironium subglabrum Löbl, 1990b: 514, Figs 3, 4. Holotype male, MHNG; type locality: Thailand: Doi Inthanon, Chiang Mai, 1650 m.

DISTRIBUTION. Thailand.

Pseudobironium subovatum **Pic**

Pseudobironium subovatum Pic, 1920d: 15. Lectotype male, MNHN; type locality: Indonesia: Sumatra, Palembang.
Löbl & Tang, 2013: 726 (lectotype designation, characters, records)

DISTRIBUTION. Indonesia: Kalimantan, Sumatra; Laos; East Malaysia: Sabah; West Malaysia; Thailand.

Pseudobironium ussuricum **Löbl**

Pseudobironium ussuricum Löbl, 1969b: 325, Figs 6, 7. Holotype male, ZMAS; type locality: Russia: Vinogradovka, Ussuri.
Löbl, 1984c: 993 (records)
Hoshina & Ahn, 2005: 522 (records)
Löbl & Tang, 2013: 715 (characters, records)
Li, 2015: 39, Fig. 5 (characters, records)

DISTRIBUTION. China: Anhui, Heilongjiang, Yunnan; Russia: Far East; South Korea.

***Pseudobironium vitalisi* (Achard)**

Morphoscapha vitalisi Achard, 1920d: 133. Lectotype male, MNHN; type locality: Indonesia: "Sumatra est".

Morphoscapha dohertyi Achard, 1920d: 135. Lectotype female, NMPC; type locality: Indonesia: Kalimantan, Martapura.

Löbl & Tang, 2013: 728 (lectotype designations for *M. vitalisi* and *M. dohertyi*, synonymy of *M. dohertyi*, characters, records)

DISTRIBUTION. Indonesia: Java, Sumatra, Kalimantan; East Malaysia: Sarawak, Sabah.

***Sapitia* Achard**

Sapitia Achard, 1920g: 207. Type species: *Sapitia lombokiana* Achard, 1920; by monotypy. Gender: feminine.

Baeoceridiolum (sg. of *Baeoceridium*) Pic, 1922a: 2. Type species: *Baeoceridium* (*Baeoceridiolum*) *sericeum* Pic, 1922 [=*Sapitia lombokiana* Achard, 1920]; by monotypy. Gender: neuter.

[*Baeceridiolum* Pic, 1925a: 195, misspelling].

Löbl, 1978a: 53. 57 (synonymy of *Baeoceridiolum* with *Sapitia*, characters, key)

***Sapitia lombokiana* Achard**

Sapitia lombokiana Achard, 1920g: 208 Lectotype female, MNHN; type locality: Indonesia: Lombok, Sapit.

Baeoceridium (*Baeoceridiolum*) *sericeum* Pic, 1922a: 2. Lectotype female, MNHN; type locality: Vietnam: Hoa Binh.

Löbl, 1971a: 251 (doubtfull record)

Löbl, 1978a: 55, Figs 1, 2 (lectotype fixations by inference for *S. lombokiana* and *B. sericeum*, synonymy, characters, records)

Löbl, 1990b: 618 (records)

Löbl & Ogawa, 2016b: 1343 (records)

DISTRIBUTION. Indonesia: Lombok, Sumatra; Philippines: Palawan; Thailand; Vietnam.

***Sapitia sumatrana* Löbl**

Sapitia sumatrana Löbl, 1978a: 56, Figs 5, 6. Holotype male, MHNG; type locality: Indonesia: Sumatra.

DISTRIBUTION. Indonesia: Sumatra.

***Sapitia versicolor* (Pic)**

Baeoceridium versicolor Pic, 1920h: 242. Lectotype male, NMPC; type locality: Indonesia: Sumatra, Palembang.

Löbl, 1978a: 56, Figs 3, 4 (lectotype designation, transfer to *Sapitia*, characters, records)
Löbl, 1990b: 618 (records)

DISTRIBUTION. Indonesia: Sumatra; East Malaysia: Sabah; Thailand.

Scaphicoma Motschulsky

Scaphicoma Motschulsky, 1863: 435. Type species: *Scaphicoma flavovittata* Motschulsky, 1863; by monotypy. Gender: feminine.

Lepteroscapha Achard, 1921b: 88. Type species *Lepteroscapha pallens* Achard, 1921; designated subsequently. Gender: feminine.
Löbl, 1971c: 1000 (synonymy of *Lepteroscapha* with *Scaphicoma*, key to Sri Lanka species, characters)
Löbl, 1992a: 571, Figs 177, 187, 188 (characters)
Löbl & Leschen, 2010: 72 (characters)
Löbl, 2011a: 309 (key to Philippine speies)
Löbl & Leschen, 2005: 36, Figs 51, 53
Ogawa & Hoshina, 2012: 263 (characters)
Ogawa et al., 2014: 1 (fixation of type species for *Lepteroscapha*, keys to Sulawesi species)
Ogawa, 2015: 116 (characters, Sulawesi species)

Scaphicoma antennalis (Achard)

Lepteroscapha antennalis Achard, 1922c: 44. Lectotype male, NMPC; type locality: Equatorial Guinea: Bioko Island, Musola.

Toxidium collarti Pic, 1930a: 87. Lectotype male, MRAC; type locality: Democratic Republic of the Congo: Uluku, Buhunde.
Löbl, 1971: 1001 (lectotype designation for *L. antennalis*)
Löbl & Leschen, 2010: 73, Figs 1–3 (lectotype designation for *T. collarti*, synonymy of *T. collarti* with *S. antennalis*, characters)

DISTRIBUTION. Equatorial Guinea: Bioko Island [=Fernando Poo]; Democratic Republic of the Congo.

Scaphicoma apicalis (Pic)

Toxidium apicale Pic, 1923d: 195. Lectotype, MNHN; type locality: Vietnam: Tonkin, Hoa Binh.
Löbl, 1973a: 159 (lectotype designation, transfer to *Scaphicoma*)

DISTRIBUTION. Vietnam.

Scaphicoma arcuata (Champion)

Toxidium arcuatum Champion, 1927: 272. Holotype male, NHML; type locality: India: Haldwani District, Kumaon.

Löbl, 1984a: 105 (transfer to *Scaphicoma*, records)
Löbl, 1990b: 614 (records)
Löbl, 1992a: 571, Figs 177, 187, 188 (characters, records)
Löbl, 1999: 734 (records)

DISTRIBUTION. China: Yunnan; India: Assam, Uttarakhand (Kumaon), West Bengal (Darjeeling District); Nepal; Thailand.

Scaphicoma bidentia Ogawa & Löbl

Scaphicoma bidentia Ogawa & Löbl, 2014: 6, Figs 1b, 3. Holotype male, MZBI; type locality: Indonesia: Sulawesi, Mt. Tilongkabila, ca 500–800 m.
Ogawa, 2015: 119, Figs 4-23b, 4-25a-e (characters, records)

DISTRIBUTION. Indonesia: Sulawesi

Scaphicoma cincta (Pic)

Toxidium cinctum Pic, 1920f: 24. Lectotype male, MNHN; type locality: Indonesia: Sumatra, Si Rambé.
Pic, 1920e: 97 (characters)
Löbl, 1973a: 159, Figs 16. 17 (lectotype designation, transfer to *Scaphicoma*, characters)

DISTRIBUTION. Indonesia: Sumatra.

Scaphicoma dohertyi (Pic)

Toxidium dohertyi Pic, 1915c: 35. Syntypes, MNHN; type locality: Indonesia: Kalimantan, Martapoera.
Löbl, 1984a: 96 (transfer from *Toxidium*)

DISTRIBUTION. Indonesia: Kalimantan.

Scaphicoma flavovittata Motschulsky

Scaphicoma flavovittata Motschulsky, 1863: 436. Lectotype female, ZMUM; type locality: Sri Lanka: Nura-Ellia [Nuwara Eliya].
Lepteroscapha filiformis Achard, 1921b: 90. Lectotype male, NMPC; type locality: Sri Lanka: Nuwara Eliya.
Löbl, 1971c: 1001, Fig. 78 (lectotype fixations by inference for *S. flavovittata* and *L. filiformis*, synonymy, characters, records)

DISTRIBUTION. Sri Lanka.

Scaphicoma gracilis Löbl

Scaphicoma gracilis Löbl, 1971a: 251, Fig. 6. Holotype male, ZMUC; type locality: Papua New Guinea: New Ireland, Bismark Is., Danu, Kalili Bay.

DISTRIBUTION. Papua New Guinea: New Ireland.

Scaphicoma hiranoi (Hoshina)

Scaphoxium hiranoi Hoshina, 2008b: 57, Figs 1–7. Holotype male, MNHA; type locality: Japan: Omoto, Ishigaki Is., Yaeyama Group, Ryukyus.
Ogawa & Hoshina, 2012: 265, Figs 1a-c, 2a-f (transfer to *Scaphicoma*, characters, records)

DISTRIBUTION. Japan: Ryukyus: Ishigaki Is.

Scaphicoma kejvali Löbl

Scaphicoma kejvali Löbl, 2003d: 96, Figs 3–5. Holotype male, MHNG; type locality: India: Cardamom Hills, 300 m, ca 50 km NW Pathanamthitta, Kerala.

DISTRIBUTION. India: Kerala.

Scaphicoma monteithi Löbl & Leschen

Scaphicoma monteithi Löbl & Leschen, 2010: 74. Holotype female, QSCI; type locality: Australia: Millaa Millaa Falls, Queensland.

DISTRIBUTION. Australia: Northen Queensland.

Scaphicoma nigrovittata (Achard)

Lepteroscapha nigrovittata Achard, 1921b: 90. Lectotype female, NHML; type locality: Sri Lanka: Dikoya.
Löbl, 1971c: 1002, Fig. 79 (lectotype designation, characters)

DISTRIBUTION. Sri Lanka.

Scaphicoma ophthalmica (Achard)

Toxidium ophthalmicum Achard, 1920d: 136. Syntypes, MNHN; type locality: Indonesia: Java, Buitenzorg [Bogor].
Löbl, 1984a: 96 (transfer from *Toxidium*)

DISTRIBUTION. Indonesia: Java.

Scaphicoma pallens (Achard)

Lepteroscapha pallens Achard, 1921b: 89. Syntypes, NBCL; type locality: Indonesia: Java, Goenoeng, Oengaran.
Löbl, 1971a: 252, Fib. 8 (characters)

DISTRIBUTION. Indonesia: Java.

Scaphicoma pulex (Heller)

Toxidium pulex Heller, 1917: 50. Lectotype male, SMTD; type locality: Philippines: Luzon, Mt. Makiling.

Löbl, 1971a: 252, Fig. 7 (lectotype fixed by inference, transfer to *Scaphicoma*, characters)
Löbl, 2011a: 309, Fig. 15 (characters, records)
DISTRIBUTION. Philippines: Luzon.

Scaphicoma quadrifasciata Ogawa & Löbl
Scaphicoma quadrifaciata Ogawa & Löbl, 2014: 8, Figs 1c, 4. Holotype male, MZBI; type locality: Indonesia: Sulawesi, Mt. Tilongkabila, ca 1300 m.
Ogawa, 2015: 121, Figs 4-23c, 4-26a-e (characters, records)
DISTRIBUTION. Indonesia: Sulawesi.

Scaphicoma rufa (Pic)
Toxidium rufum Pic, 1923d: 195. Syntypes, MNHM; type locality: Indonesia: Sumatra, Sibolangit.
Löbl & Leschen, 2010: 77 (characters)
DISTRIBUTION. Indonesia: Sumatra.

Scaphicoma subflava Ogawa & Löbl
Scaphicoma subflava Ogawa & Löbl, 2014: 3, Figs 1a, 2, 5. Holotype male, MZBI; type locality: Indonesia: Sulawesi, Mt. Tilongkabila, ca 800 m.
Ogawa, 2015: 117, Figs 4-23a, 4-24a-e, 4-27a-h (characters, records)
DISTRIBUTION. Indonesia: Sulawesi.

Scaphicoma yapo Löbl & Leschen
Scaphicoma yapo Löbl & Leschen, 2010: 74, Figs 4–6. Holotype male, MHNG; type locality: Ivory Coast, forest near Yapo Gare.
DISTRIBUTION. Ivory Coast.

Scaphischema Reitter

Scaphischema Reitter, 1880a: 44. Type species: *Scaphisoma poupillieri* Reiche, 1864, by monotypy. Gender: neuter.
[*Scaphischema* Reitter, 1880d: 43 (subsequent description as new genus)].
Scaphoschema Reitter, 1885b: 361 (unjustified emendation).
Reitter, 1880d: 43 (characters)
Reitter, 1887a: 5 (characters)
Löbl, 1969b: 327 (characters)

Scaphischema poupillieri (Reiche)
Scaphisoma poupillieri Reiche, 1864: 238. Syntypes, MNHN not traced; type locality: Algeria.

Reitter, 1885b: 361 (characters)
Reitter, 1886: 5 (characters)
Kocher, 1958: 88 (records)
Löbl, 1969b: 327, Figs 1, 8, 9 (characters, records)
Tamanini, 1969b: 355, Figs 2D-G (misspelled as *S. poupilleri*, characters, records)
Löbl, 1974a: 61 (records)
Löbl, 1989a: 9 (records)

DISTRIBUTION. Algeria; Morocco; South Spain.

Scaphisoma Leach

Scaphisoma Leach, 1815: 89. Type species: *Silpha agaricina* Linnaeus, 1758; by monotypy. Gender: neuter.

Scaphosoma Agazzi, 1846: 332 [injustified emendation]. Gender: neuter.

Caryoscapha Ganglbauer, 1899: 343 (as subgenus of *Scaphisoma*). Type species: *Scaphisoma limbatum* Erichson, 1945; by monotypy. Gender: feminine.

Scaphiomicrus Cassey, 1900: 58. Type species: *Scaphisoma pusilla* LeConte, 1860; by original designation. Gender: masculine.

Pseudoscaphosoma Pic, 1915b: 31. Type species: *Pseudoscaphosoma testaceo-maculatum* Pic, 1915; by original designation. Gender: neuter.

Scutoscaphosoma Pic, 1916b: 3 (as subgenus of *Scaphosoma*). Type species: *Scaphisoma* (*Scutoscaphosoma*) *rouyeri* Pic, 1916; by monotypy. Gender: neuter.

Scaphella Achard, 1924b: 29. Type species: *Scaphisoma antennatum* Achard, 1920; by original designation. Gender: feminine.

Macrobaeocera Pic, 1925a: 195. Type species: *Scaphisoma phungi* Pic, 1922, by monotypy. Gender: feminine.

Macroscaphosoma Pic, 1928b: 33. Type species: *Macroscaphosoma collarti* Pic, 1928, by monotypy. Gender: neuter. Note: established as a subgenus not assigned to any generic name; deemed unavailable and credited to Löbl, 1970a: 128 by Löbl, 1997: xi (see ICZN, 1999, Art. 5.1, 6.1; deemed available under ICZN, 1999, Art. 43.1.).

Mimoscaphosoma Pic, 1928d: 49 (as subgenus of *Scaphisoma*). Type species: *Scaphisoma bruchi* Pic, 1928; by original designation. Gender: neuter.

Metalloscapha Löbl, 1975a: 384. Type species: *Metalloscapha papua* Löbl, 1975; by original designation. Gender: feminine.

[*Macroscaphosoma* Löbl, 1997: xi. Type species *Macroscaphosoma collarti* Pic, 1928 (credited to Löbl, 1970). Gender: neuter.]

Stephens, 1830: 3 (characters)
Westwood, 1838: 11 (characters)
Heer, 1841: 372 (characters)

Erichson, 1845: 8 (characters)
Redtenbacher, 1847: 147 (characters)
Lacordaire, 1854: 240 (characters)
Fairmaire & Laboulbène, 1855: 341 (characters)
Jacquelin du Val, 1858: 123 (characters)
Gutfleisch, 1859: 222 (characters)
Thomson, 1862: 127 (characters)
Reitter, 1880a: 46 (key to species)
Matthews, 1888: 170 (characters, key to Central American species)
Casey, 1893: 523 (characters, key to North American species)
Ganglbauer, 1899: 344 (characters, key to European species)
Everts, 1903: 446 (characters)
Reitter, 1909: 276 (*Caryoscapha* raised to genus rank, characters, key)
Fall, 1910: 119 (synonymy of *Scaphiomicrus* with *Scaphisoma*)
Reitter, 1909: 276 (key to Central european species)
Blatchley, 1910: 495 (key to Indiana species)
Achard, 1915c 292 (as subgenus of *Scaphosoma*)
Achard, 1923: 110 (characters)
Miwa & Mitono, 1943: 538 (key to Japanese and Taiwanese species)
Vinson, 1943: 186 (key to Mauricius species)
Lundblad, 1952: 28 (key to Swedish species)
Tamanini, 1954: 85 (characters)
Löbl, 1964b: 2 (key to European species)
Palm, 1966: 44 (key to European species of the *S. subalpinum* group)
Kasule, 1966: 266 (immature stages)
Kasule, 1968: 117 (immature stages)
Kofler, 1968: 39 (key to Tirolean species)
Tamanini, 1969c: Figs 25, 29, 35–37, 39, 40 (characters)
Tamanini, 1970: 10 (key to Italian species)
Löbl, 1970a: 10 (characters, key to Polish species, characters)
Löbl, 1970c: 730 (synonymy of *Scaphella* with *Scaphisoma*, species groups, key to Palaearctic species, characters)
Löbl, 1971c: 962 (species groups, key to Sri Lankan species)
Freude, 1971: 345 (characters, key to Central European species)
Löbl, 1975a: 386 (key to New Guinean species)
Löbl, 1975b: 270 (synonymy of *Pseudoscaphosoma* and *Macrobaeocera* with *Scaphisoma*, species groups)
Löbl, 1977e: 15, 17 (species groups, key to Australian species)
Löbl, 1977f: 50 (key to Réunion species)
Löbl, 1977i: 827 (key to Fijian species)

Löbl, 1979a: 99 (key to South Indian species, species groups)
Löbl, 1980a: 99 (key to Taiwanese species)
Löbl, 1980c: 389 (key to Fijian species)
Löbl, 1981a: 358 (species groups, key to New Caledonian species)
Löbl, 1981b: 73 (key to Micronesian species)
Löbl, 1981d: 156, 165 (synonymy of *Scutoscaphosoma* with *Scaphisoma*, key to species of the *S. rouyeri* group)
Ashe, 1984: 361 (immature stages, natural history)
Newton, 1984: 318 (fungal hosts)
Iablokoff-Khnzorian, 1985: 139 (key to species of the former Soviet Union, characters)
Löbl, 1986a: 138 (key to Northeast Indian and Bhutanese species)
Löbl, 1986e: 389 (key to Pakistani species)
Löbl, 1987c: 390 (key to species of *Caryoscapha*)
Leschen, 1988a: 226 (immature stages: larvae, natural history)
Löbl, 1990a: 121 (key to species of Himachal Pradesh, India)
Leschen et al., 1990: 278 (key to Ozark Highlands (USA) species, characters)
Löbl, 1990b: 538 (key to Thai species)
Newton, 1991: 338 (immature stages: larvae)
Löbl, 1992a: 525 (key to Himalayan species, characters)
Löbl, 1993: 39 (key to Far East Russian species)
Leschen, 1993: 73 (larval mandible)
Leschen, 1994: 3 (natural history)
Márquez Luna & Navarrete Heredia, 1995: 35 (association with debris of *Atta mexicana* (F. Smith)
Downie & Arnett, 1996: 365 (key to northeast American species)
Hanley, 1996: 37 (immature stages, natural history)
Newton et al., 2000: 375 (notes)
Löbl, 2000: 603 (key to Chinese species)
Löbl & Leschen, 2003b: 38 (characters, key to New Zealand species)
Löbl, 2003c: 182 (synonymy of *Metalloscapha* with *Scaphisoma*)
Fierros-López, 2005: 3 (immature stages: larva, pupa)
Fierros-López, 2006a: 67 (key to Mexican species with maculate elytra, fungal hosts)
Leschen & Löbl, 2005: 24 (synonymy of *Caryoscapha* and *Macroscaphosoma* with *Scaphisoma*)
Löbl, 2012g: 203 (characters, key to Central European species)
Löbl, 2014a: 59 (key to Maluku Islands species)
Löbl, 2015a: 93 (key to Lesser Sunda species)
Löbl, 2015c: 143 (key to Kalimantan species)

Ogawa, 2015: 80 (characters, key to Sulawesi species)
Harrison, 2016: 63 (key to British species)
Löbl & Ogawa, 2016b: 1343 (species groups, key to Philippine species)

***Scaphisoma ablutum* Löbl**
Scaphisoma ablutum Löbl, 2015a: 92, Figs 60–62. Holotype male, MHNG; type locality: Indonesia: Lombok, G. Rinjani, 1950 m.
DISTRIBUTION. Indonesia: Lombok.

***Scaphisoma absurdum* Löbl**
Scaphisoma absurdum Löbl, 1986a: 177, Figs 46–51. Holotype male, MHNG; type locality: India: Algarah, 1800 m, Darjeeling District.
Löbl, 1992a: 538 (records)
DISTRIBUTION. India: West Bengal (Darjeeling District); Nepal.

***Scaphisoma acclivum* Löbl**
Scaphisoma acclivum Löbl, 2000: 626, Figs 29, 30. Holotype male, ZMUB; type locality: China: Sichuan, Shimian Co., Xiaoxiang Ling, side valley above Nanya Cun near Caluo, 11 km south Shimian, ca 1200 m.
DISTRIBUTION. China: Sichuan.

***Scaphisoma achardianum* Scott**
Scaphosoma achardianum Scott, 1922: 225, Pl. 21, Fig. 22. Syntypes, NHML/CUMZ; type localities: Seychelles: Silhouette: Mare aux Cochons; Mahé: high forest of Morne Blanc.
DISTRIBUTION. Seychelles.

***Scaphisoma aciculare* Löbl**
Scaphisoma aciculare Löbl, 2000: 617, Figs 9–11. Holotype male, MHNG; type locality: China: Yunnan, Gaoligong Mts, 2200–2500 m.
DISTRIBUTION. China: Yunnan.

***Scaphisoma activum* Löbl**
Scaphisoma activum Löbl, 2015a: 98, Figs 72, 73. Holotype male, SMNS; type locality: Indonesia: W. Sumbawa, Batudulang, 30 km south of Besar, 1000 m.
DISTRIBUTION. Indonesia: Sumbawa.

***Scaphisoma acutatum* Löbl**
Scaphisoma activum Löbl, 2015a: 99, Figs 74, 75. Holotype male, SMNS; type locality: Indonesia: Lombok, Sapit – Sembalun Bumbung, 900–1500 m.
DISTRIBUTION. Indonesia: Lombok.

***Scaphisoma acuticauda* Fairmaire**

Scaphisoma acuticauda Fairmaire, 1897c: 465. Syntypes, MNHN; type locality: Madagascar: Suberbieville [in Maevatanana District].

DISTRIBUTION. Madagascar.

***Scaphisoma acutulum* Löbl & Ogawa**

Scaphisoma acutulum Löbl & Ogawa, 2012b: 1416, Figs 180, 181. Holotype male, SMNS; type locality: Philippines: Leyte, Visca, north Baybay, 200–500 m.

DISTRIBUTION. Philippines: Leyte.

***Scaphisoma acutum* Löbl**

Scaphisoma acutum Löbl, 2015a: 103, Figs 82–84. Holotype male, MHNG; type locality: Indonesia: Lombok, Batu Koq, north of Rinjani, 500 m.

DISTRIBUTION. Indonesia: Lombok.

***Scaphisoma adivasis* Löbl**

Scaphisoma adivasis Löbl, 2012c: 88, Figs 6–10. Holotype male, MHNG; type locality: India: South Andaman Island, Chiriyatapu.

DISTRIBUTION. India: Andaman Islands.

***Scaphisoma adjacens* Löbl**

Scaphisoma adjacens Löbl, 1992a: 546, Fig. 134. Holotype male, MHNG; type locality: Nepal: Chichila, 2200 m, Sankhuwasabha District.

DISTRIBUTION. Nepal.

***Scaphisoma adjunctum* Löbl**

Scaphisoma adjuctum Löbl, 2015a: 94, Figs 63–65. Holotype male, SMNS; type locality: Indonesia: Lombok, Senaro, north slope of Rinjani, 1100 m.

DISTRIBUTION. Indonesia: Lombok.

***Scaphisoma adnexum* Löbl**

Scaphisoma adnexum Löbl, 1972a: 117, Figs 3, 4. Holotype male, UZMH; type locality: Japan: Tateshima, Nagano.

DISTRIBUTION. Japan: Honshu.

***Scaphisoma adscitum* Löbl**

Scaphisoma adscitum Löbl, 2015a: 97, Figs 70, 71. Holotype male, SMNS; type locality: Indonesia: Lombok, Senaro, north slope of Rinjani, 1100 m.

DISTRIBUTION. Indonesia: Lombok.

Scaphisoma adustum **Löbl**

Scaphisoma adustum Löbl, 1980a: 112, Figs 29, 30. Holotype male, HNHM; type locality: Taiwan: Pilam.

DISTRIBUTION. Taiwan.

Scaphisoma aemulum **Löbl**

Scaphisoma aemulum Löbl, 1973b: 312, Figs 5, 6. Holotype male, NHMW; type locality: New Caledonia: Table d'Union near Col d'Amieu.
Löbl, 1981a: 362 (characters, records)

DISTRIBUTION. New Caledonia.

Scaphisoma aequatum **Löbl**

Scaphisoma aequatum Löbl, 1977i: 819, Figs 3, 4. Holotype male, BPBM; type locality: Fiji: Belt Road, 42–44 mi west of Suva, Viti Levu.
Löbl, 1980c: 390, Figs 23, 24 (characters, records)

DISTRIBUTION. Fiji.

Scaphisoma aequum **Löbl**

Scaphisoma aequum Löbl, 2015a: 95, Figs 66, 67. Holotype male, SMNS; type locality: Indonesia: W. Sumbawa, Batudulang, 30 km south of Besar, 1000 m.

DISTRIBUTION. Indonesia: Lombok.

Scaphisoma aereum **Löbl**

Scaphisoma aereum Löbl, 2015a: 97, Figs 68, 69. Holotype male, MHNG; type locality: Indonesia: W. Sumbawa, Batudulang, 30 km south of Sumb. Besar, 1000

DISTRIBUTION. Indonesia: Sumbawa.

Scaphisoma affabile **Löbl**

Scaphisoma affabile Löbl, 2015a: 99, Figs 76, 77. Holotype male, MHNG; type locality: Indonesia: Timor, 16 km north Soe.

DISTRIBUTION. Indonesia: Bali.

Scaphisoma affectum **Löbl**

Scaphisoma affectum Löbl, 2015a: 107, Figs 91–94. Holotype male, MHNG; type locality: Indonesia: Bali, Mt. Batukari, 500–700 m.

DISTRIBUTION. Indonesia: Timor.

Scaphisoma agaricinum **(Linnaeus)**

Silpha agaricina Linnaeus, 1758: 360. Syntype, LCLS; type locality: "habitat in Agaricis".

Scaphosoma agaricinum var. *major* Sahlberg, 1889: 146. Lectotype male, UZMH; type locality: Finland: Karislojo [=Karjalohja].

Scaphosoma agaricinum var. *punctipenne* Petz, 1905: 100 (primary junior homonym with *Scaphisoma punctipenne* Macleay, 1871). Syntypes, not traced; type locality: Austria: Damberg bei Steyr.

Scaphosoma agaricinum var. *petzi* Csiki, 1908: 162; replacement name for *Scaphosoma agaricinum* var. *punctipenne* Petz, 1905.

Scaphosoma podoces Lundblad, 1952: 28, Fig. 2E. Holotype male, NHRS; type locality: Sweden: Uppland, Tursbo, Kirchspiel Vassunda.

Gmelin, 1790: 1619 (*Silpha*, characters)
Olivier, 1790: 20: 5, pl. 1 (*Scaphidium*, characters)
Fabricius, 1792: 510 (*Scaphidium*, characters)
Panzer, 1792: 2 (characters)
Panzer, 1793: 15 (*Scaphidium*, characters)
Herbst, 1793: 134, pl. 49 (*Scaphidium*, characters)
Paykull, 1800: 239 (*Scaphidium*, characters)
Fabricius, 1801: 576 (*Scaphidium*, characters)
Latreille, 1804: 248 (*Scaphidium*, characters)
Leach, 1815: 89 (in *Scaphisoma*)
Duftschmid, 1825: 71 (*Scaphidium*, characters)
Stephens, 1830: 3 (characters)
Laporte, 1840: 19 (characters)
Heer, 1841: 373 (characters)
Erichson, 1845: 9 (characters)
Redtenbacher, 1847: 147 (characters)
Jacquelin du Val, 1858: 123, Pl. 34, Fig. 166 (characters)
Gutfleisch, 1859: 222 (characters)
Thomson, 1862: 127 (as *S. agaricina*, characters)
Solsky, 1871: 351 (records)
Redtenbacher, 1872: 335 (characters)
Perris, 1876: 269, Figs 1–8 (immature stages, natural history)
Fowler, 1889: 347 (characters)
Ganglbauer, 1899: 345 (characters)
Sahlberg, 1889: 80 (records)
Johnson & Halbert, 1902: 698 (records)
Reitter, 1909: 277, pl. 65 (characters)
Jakobson, 1910: 637 (records)
Saalas, 1917: 395 (records, fungal hosts)
Rüschkamp, 1929: 38 (characters, *punctipenne* as possible variety of *S. assimile*)
Scheerpeltz & Höfler, 1948: 149 (fungal hosts, habitat)

Palm, 1951: 145 (habitat)
Novak, 1952: 74 (records)
Benick, 1952: 46 (fungal hosts)
Palm, 1953: 172 (synonymy of *S. podoces* with *S. agaricinum*, characters)
Kocher, 1958: 88 (record based in misidentification)
Palm, 1959: 223 (habitat)
Koch & Lucht, 1962: 30 (records)
Löbl, 1965d: 336, Fig. 2 (characters)
Löbl, 1965g: 733 (records)
Kasule, 1966: 266, Fig. 18 (immature stages)
Löbl, 1967d: 34, Figs 1a, b (characters)
Kasule, 1968: 117, Fig. 7 (immature stages)
Kofler, 1968: 41, Figs 1D, 2D (characters, records)
Tamanini, 1969a: 487, Figs 9, 11 (characters, records)
Tamanini, 1969b: 367, Figs 6A, B, 7C, F (characters, records)
Tamanini, 1969c: Fig. 43 (characters)
Tamanini, 1970: 22, Figs FG, 6A, B, 8F, 8C, F, 10M (characters, fungal hosts)
Kofler, 1970: 56 (records)
Löbl, 1970a: 12, Figs 8, 18, 25 (characters, records)
Löbl, 1970c: 738, Figs 1, 3, 4 (invalid neotype designation, synonymy of *major* and *punctipenne* with *S. agaricinum*, characters, records)
Franz, 1970: 267 (records)
Freude, 1971: 347, Fig. 3: 1 (characters)
Nuss, 1975: 110 (fungal hosts)
Iablokoff-Khnzorian, 1985: 139 (records)
Klimaszewski & Peck, 1987: 543 (fungal hosts)
Burakowski et al., 1978: 232 (records, habitat, fungal hosts)
Schawaller, 1990: 235 (records)
Merkl, 1996: 261 (records)
Nikitsky et al., 1996: 24 (fungal hosts)
Krasutskij, 1996a: 276 (fungal hosts)
Merkl, 1987: 113 (records)
Krasutskij, 1996a: 96 (records, fungal hosts)
Krasutskij, 1997a: 307 (habitat, fungal hosts)
Kofler, 1998: 645–648 (fungal hosts)
Krasutskij, 2000: 80 (fungal hosts)
Nikitsky & Schigel, 2004: 8, 9, 13 (fungal hosts)
Schigel et al., 2004: 41 (fungal hosts)
Mateleshko, 2005: 128 (fungal hosts)
Krasutskij, 2005: 37–42 (fungal hosts)
Tronquet, 2006: 79 (records)

Byk et al., 2006: 337 (habitat, records)
Borowski, 2006: 52–58 (records, fungal hosts)
Nikitsky et al., 2008: 120 (records, habitat, fungal hosts)
Guèrguev & Ljubomirov, 2009: 247 (records)
Lott, 2009: 21 (misleading illustration of aedeagus, characters, records)
Tsurikov, 2009: 107 (records)
Krasutskij, 2010: 373 (fungal hosts)
Vinogradov et al., 2010: 16 (records, fungal hosts)
Zamotajlov & Nikitsky, 2010: 105 (fungal hosts)
Schigel, 2011: 329, 330, 331, 332, 334, 337, 339, 341 (fungal hosts)
Diéguez Fernández, 2014: 241 (records, fungal hosts)
Nikitsky et al., 2016: 138 (records, fungal hosts)
Semenov, 2017: 202 (records)
Vlasov & Nikitsky, 2017: 5 (records, fungal hosts)

DISTRIBUTION. Europe: all countries of West, North and Central Europe, including Ireland and Baltic states, Belarus, Ukraine, France; Russia including Siberia, Transbaikal, Far East; Mongolia; Southwest Europe: France, northern Spain; northeastern Italy; Southeast Europe including Albania, Bosnia and Herzegovina, Bulgaria, Croatia, Greece, Moldova; Romania, Serbia, Slovenia, European Turkey. Appears to be absent from Portugal, Swiss south of the Alps and most of Italy.

Note. Boheman, 1851: 558 reported the species from "terra Natalensi".

***Scaphisoma agile* Löbl**

Scaphisoma agile Löbl, 1990b: 550; Figs 56, 57. Holotype male, MHNG; type locality: Thailand: Doi Suthep, 1180 m, Chiang Mai.
Prokofiev, 2003: 3, Figs 1, 7 (characters, records)

DISTRIBUTION. Thailand; Vietnam.

***Scaphisoma alacre* Löbl**

Scaphisoma alacre Löbl, 1992a: 556, Fig. 153. Holotype male, MHNG; type locality: Nepal: Malemchi, 2800 m, Sindhupalcok District.
Löbl, 2005: 180 (records)

DISTRIBUTION. Nepal.

***Scaphisoma albertisi* Reitter**

Scaphisoma albertisi Reitter, 1881b: 141. Lectotype male, MNHN; type locality: Australia: Queensland, Somerset.
Löbl, 1977e: 58, Figs 88, 89 (lectotype designation, characters)

DISTRIBUTION. Australia: Queensland.

Scaphisoma alienum Löbl

Scaphisoma alienum Löbl, 1977i: 825, Figs 13, 14. Holotype male, BPBM; type locality: Fiji: Nandarivatu, 3700 ft, Viti Levu.
Löbl, 1980c: 394, Figs 33, 34 (characters, records)
DISTRIBUTION. Fiji.

Scaphisoma alluaudi Achard

Scaphosoma alluaudi Achard, 1920i: 241. Syntypes, NMPC; type locality: Madagascar: Tanala forest.
DISTRIBUTION. Madagascar.

Scaphisoma alternans Löbl

Scaphisoma alternans Löbl, 1973b: 319, Figs 15, 16. Holotype male, NHMW; type locality: New Caledonia: Col de Hau.
Löbl, 1977i: 818 (records)
Löbl, 1981a: 374 (characters, records)
DISTRIBUTION. New Caledonia.

Scaphisoma alticola Löbl & Ogawa

Scaphisoma alticola Löbl & Ogawa, 2016b: 1385, Figs 116, 117. Holotype male, EUMJ; type locality: Philippines: Luzon, Benguet Prov., Paoya near Sayangan, 2400 m.
DISTRIBUTION. Philippines: Luzon.

Scaphisoma alutaceum Achard

Scaphosoma alutaceum Achard, 1920i: 241. Syntypes, NMPC; type locality: Madagascar: Suberbieville [Maevatanana District].
DISTRIBUTION. Madagascar.

Scaphisoma amabile Löbl

Scaphisoma amabile Löbl, 1984c: 998, Figs 4, 5. Holotype male, MHNG; type locality: Myanmar: Taunggy, Shan State.
Löbl, 1990b: 566 (records)
DISTRIBUTION. Myanmar; Thailand.

Scaphisoma americanum (Löbl)

Caryoscapha americanum Löbl, 1987c: 387, Figs 1–6. Holotype male, CNCI; type locality: USA: Latimer Co., Oklahoma.
Ashe, 1984: 361, Figs 1–17 (as *Scaphisoma terminata*, immature stages, fungal hosts)

Leschen, 1988a: 228, Figs 6, 10 (immature stages, fungal hosts)
Leschen & Löbl, 2005: 25 (transfer to *Scaphisoma*)
DISTRIBUTION. United States: Arkansas, Florida, Illinois, Oklahoma.

***Scaphisoma anale* Motschulsky**
Scaphisoma anale Motschulsky, 1863: 434. Lectotype male, ZMUM; type locality: Sri Lanka: Nura-Ellie [Nuwara Eliya].
Löbl, 1971c: 979, Figs 46, 47 (lectotype fixed by inference, characters, records)
DISTRIBUTION. Sri Lanka.

***Scaphisoma anderssoni* Löbl**
Scaphisoma anderssoni Löbl, 1971c: 980, Figs 50, 51. Holotype male, MHNG; type locality: Sri Lanka: Kandy, hill south of catchment lake, ca 700 m.
DISTRIBUTION. Sri Lanka.

***Scaphisoma angulare* Löbl**
Scaphisoma angulare Löbl, 2015a: 105, Figs 87–90. Holotype male, SMNS; type locality: Indonesia: Sumbawa, Batudulang, 30 km south of Besar, 1000 m.
DISTRIBUTION. Indonesia: Sumbawa.

***Scaphisoma angulatum* Löbl**
Scaphisoma angulatum Löbl, 1975b: 284, Figs 20–22. Holotype male, MHNG; type locality: East Malayasia: Sabah, Mt. Kinabalu, 5800 ft.
DISTRIBUTION. East Malaysia: Sabah.

***Scaphisoma angulosum* Löbl & Ogawa**
Scaphisoma angulosum Löbl & Ogawa, 2016b: 1397, Figs 140, 141. Holotype male, SMNS; type locality: Philippines: Leyte, Visca north Baybay, 200–500 m.
DISTRIBUTION. Philippines: Leyte.

***Scaphisoma animatum* Löbl**
Scaphisoma animatum Löbl, 2015a: 101, Figs 80, 81. Holotype male, SMNS; type locality: Indonesia: Lombok, Sapit – Sembalun Bumbung, 900–1500 m.
DISTRIBUTION. Indonesia: Lombok.

***Scaphisoma annamitum* Pic**
Scaphosoma annamitum Pic, 1923b: 17. Syntypes, MNHN, not traced; type locality: Vietnam: Tonkin.
DISTRIBUTION. Vietnam.

Scaphisoma anomalum **Löbl**

Scaphisoma anomalum Löbl, 1972b: 91, Figs 29, 30. Holotype male, ZMUB; type locality: Philippines: Mindanao, Surigao.
Löbl & Ogawa, 2016b: 1389, Fig. 1 (characters, records)
DISTRIBUTION. Philippines: Mindanao.

Scaphisoma antennarum **Löbl**

Scaphisoma antennarum Löbl, 2015a: 101, Figs 78, 79. Holotype male, SMNS; type locality: Indonesia: Lombok, Sapit – Sembalun Bumbung, 900–1500 m.
DISTRIBUTION. Indonesia: Lombok.

Scaphisoma antennatum **Achard**

Scaphosoma antennatum Achard, 1920f: 362. Lectotype male, NMPC; type locality: India: West Bengal, Kurseong.
Achard, 1924b: 29 (transfer from *Scaphisoma* to *Scaphella*)
Löbl, 1986a: 155, Figs 18, 19 (lectotype designation, transfer from *Scaphella* to *Scaphisoma*, characters)
DISTRIBUTION. India: West Bengal (Darjeeling District).

Scaphisoma antongiliense **Achard**

Scaphosoma antongiliense Achard, 1920i: 241. Syntypes, NMPC/MNHN; type locality: Madagascar: Antongil Bay.
DISTRIBUTION. Madagascar.

Scaphisoma apertum **Löbl**

Scaphisoma apertum Löbl, 2000: 624, Figs 21–23. Holotype male, MHNG; type locality: China: Yunnan, Jizu Shan, 2500–2700.
DISTRIBUTION. China: Yunnan.

Scaphisoma apicale **Horn**

Scaphisoma apicale Horn, 1893: 363. Syntypes, MCZC; type locality: Mexico: La Chuparosa, Baja California.
Fall, 1910: 119 (characters)
DISTRIBUTION. Mexico: Baja California.

Scaphisoma apicefasciatum **Reitter**

Scaphosoma apicefasciatum Reitter, 1908: 32. Syntypes, NHMW; type locality: Tanzania: Mt. Bomole, Amani.
DISTRIBUTION. Tanzania.

***Scaphisoma apicenigrum* Pic**

Scaphosoma apicenigrum Pic, 1928b: 42. Syntypes, MRAC/MNHN; type localities: Democratic Republic of the Congo: Haut Uélé, Moto; Watsa.

DISTRIBUTION. Democratic Republic of the Congo.

***Scaphisoma apicerubrum* Oberthür**

Scaphisoma apicerubrum Oberthür, 1883: 14. Syntypes, MNHN; type locality: Ethiopia [Abyssinie].

DISTRIBUTION. Ethiopia.

***Scaphisoma apomontanum* Löbl & Ogawa**

Scaphisoma apomontanum Löbl & Ogawa, 2016b: 1417, Figs 182, 183. Holotype male, FMNH; type locality: Philippines: Mindanao, Davao Prov., east slope Mt. McKinley, 5200 ft.

DISTRIBUTION. Philippines: Mindanao.

***Scaphisoma apomontium* Löbl & Ogawa**

Scaphisoma apomontium Löbl & Ogawa, 2016b: 1363, Figs 70, 71. Holotype male, FMNH; type locality: Philippines: Mindanao, Mainit Riv., Mt Apo.

DISTRIBUTION. Philippines: Mindanao.

***Scaphisoma apparatum* Löbl**

Scaphisoma apparatum Löbl, 2015c: 130. Holotype female, NMPC; type locality: Indonesia: Kalimantan, ca 55 km west of Balikpapan, 100 m, PT Fajar Surya Swadaya area.

DISTRIBUTION. Indonesia: Kalimantan.

***Scaphisoma approximatum* Löbl**

Scaphisoma approximatum Löbl, 2015a: 105, Figs 85, 86. Holotype male, MHNG; type locality: Indonesia: Lombok, Mt. Rinjani above Senaro, 900–1100 m.

DISTRIBUTION. Indonesia: Lombok.

***Scaphisoma apterum* Löbl**

Scaphisoma apterum Löbl, 2014a: 55, Figs 10–13. Holotype male, MHNG; Indonesia: Maluku Islands, Kai Besar, Turkan, 300 m.

DISTRIBUTION. Indonesia: Maluku Islands.

***Scaphisoma arambourgi* Pic**

Scaphosoma arambourgi Pic, 1946: 84. Syntypes, MNHN; type locality: Kenya: Mt. Elgon.

DISTRIBUTION. Kenya.

Scaphisoma argutum **Löbl**

Scaphisoma argutum Löbl, 1986a: 182, Figs 58, 59. Holotype male, MHNG; type locality: India: Meghalaya, Tura Peak, 900 m, Garo Hills.

DISTRIBUTION. India: Meghalaya.

Scaphisoma armatum **Löbl**

Scaphisoam armatum Löbl, 1986a: 206, Figs 105–108. Holotype male, MHNG; type locality: India: Mahanadi near Kurseong, 1200 m, Darjeeling District.
Löbl, 1986c: 349 (records)
Löbl, 1990b: 582 (records)
Löbl, 1992a: 157 (records)

DISTRIBUTION. India: Himachal Pradesh, Sikkim, Uttarakhand, West Bengal (Darjeeling District); Nepal; Thailand.

Scaphisoma aspectum **Löbl**

Scaphisoma aspectum Löbl, 2015a: 108, Figs 95–98. Holotype male, SMNS; type locality: Indonesia: Bali, Danau Buyan, 1300 m.

DISTRIBUTION. Indonesia: Bali, Java.

Scaphisoma asper **Löbl**

Scaphisoma asper Löbl, 1980a: 113, Fig. 33. Holotype male, MHNG; type locality: Taiwan: Fenchihu, 1400 m.

DISTRIBUTION. Taiwan.

Scaphisoma assimile assimile **Erichson**

Scaphisoma assimile Erichson, 1845: 10. Lectotype male, ZMUB; type locality: Germany, "Berol." [Berlin].
Redtenbacher, 1847: 147 (characters)
Gutfleisch, 1859: 222 (characters)
Redtenbacher, 1872: 335 (characters, records)
Reitter, 1880d: 44 (characters)
Reitter, 1885a: 83 (characters)
Sahlberg, 1889: 81 (characters, records)
Fowler, 1889: 348 (characters)
Kuthy, 1897: 86 (records)
Ganglbauer, 1899: 344 (characters)
Jakobson, 1910: 637 (records)
Roubal, 1930: 298 (records)
Palm, 1951: 145 (fungal hosts, habitat)
Lundblad, 1952: 31, Figs 1C, 2C, 3C (characters)
Benick, 1952: 47 (fungal hosts)

Palm, 1959: 223 (fungal hosts)
Löbl, 1963a: 704, Figs 1–3 (lectotype designation, characters)
Kofler, 1968: 40, Figs 1A, 2A (characters, records)
Tamanini, 1969b: 366, Figs 5A-C (characters, records)
Kofler, 1970: 58 (records)
Löbl, 1970a: 14, Figs 16, 22, 28 (characters, records)
Löbl, 1970c: 785, Figs 70, 71 (characters, records)
Tamanini, 1970: 18, Figs 5A-C, 10D (characters, records)
Freude, 1971: 346, Fig. 3: 4 (characters)
Holzschuh, 1977: 31 (records)
Burakowski et al., 1978: 234 (records, habitat)
Iablokoff-Khnzorian, 1985: 141(records)
Walter, 1990: 31 (records)
Merkl, 1996: 261 (records)
Nikitsky et al., 1996: 25 (fungal hosts)
Krasutskij, 1996a: 96 (records, fungal hosts)
Byk, 2001: 354 (records)
Silfverberg, 2004: 38 (records)
Tsinkevich, 2004: 19 (records, fungal hosts)
Nikitsky & Schigel, 2004: 13 (fungal hosts)
Mateleshko, 2005: 128 (fungal hosts)
Tronquet, 2006: 79 (records)
Byk et al., 2006: 360 (habitat)
Borowski, 2006: 54, 58 (records, fungal hosts)
Krivosheyev, 2009: e14 (records)
Lott, 2009: 21 (records)
Levey, 2010: 117 (records)
Vinogradov et al., 2010: 17 (records, fungal hosts)
Levey, 2010: 117 (records)
Hamed & Vancl, 2016: 79 (records)
Vlasov & Nikitsky, 2017: 6 (records, fungal hosts)

DISTRIBUTION. Albania; Austria; Belarus; Bosnia and Herzegovina; Bulgaria; Croatia; Czech Republic; Denmark; England; Estonia; Finland; France; Germany; Greece; Hungary; Italy; Latvia; Liechtenstein; Lithuania; Luxemburg; Netherlands, Norway; Poland; Romania; Russia: European; Slovakia; Sweden; Switzerland; Turkey: Anatolia; Ukraine.

***Scaphisoma assimile curvistria* Reitter**

Scaphosoma curvistria Reitter, 1891: 22. Lectotype female, HNHM; type locality: Kazakhstan: Kasalinks am Aralsee.

Scaphosoma tedjense Löbl, 1964c: 487, Fig. 1. Holotype female, NHMW; type locality: Turkmenistan: Tedjen.
Jakobson, 1910: 637 (records)
Löbl, 1963a: 704 (lectotype designation)
Löbl, 1970c: 788 (synonymy of *S. tedjense* with *S. assimile curvistria*, characters, records)
Iablokoff-Khnzorian, 1985: 142 (records)
Löbl, 1986c: 347 (records)

DISTRIBUTION. India: Kashmir; Kazakhstan; Kyrgyzstan; Pakistan; Turkmenistan; Uzbekistan.

Scaphisoma atomarium Fairmaire

Scaphisoma atomarium Fairmaire, 1898b: 394. Syntypes, MNHN; type locality: Madagascar: Suberbieville [in Maevatanaba District].

DISTRIBUTION. Madagascar.

Scaphisoma atrofasciatum atrofasciatum Pic

Scaphosoma atrofasciatum Pic, 1928b: 44. Syntypes, MRAC/MNHN; type locality: Democratic Republic of the Congo: Moto, Haut-Uélé.

DISTRIBUTION. Democratic Republic of the Congo.

Scaphisoma atrofasciatum motoense Pic

Scaphosoma atrofasciatum var. *motoense* Pic, 1928b: 44. Holotype, MRAC; type locality: Democratic Republic of the Congo, Moto, Haut-Uélé.

DISTRIBUTION. Democratic Republic of the Congo.

Scaphisoma atronotatum Pic

Scaphosoma atronotatum Pic, 1920f: 24. Lectotype male, MCSN; type locality: Myanmar: Carin Chebà, 900–1100 m [approx. 19°13'N 96°35'E].
Pic, 1921a: 165 (characters)
Löbl, 1973a: 152, Figs 5, 6 (lectotype fixed by inference, characters)
Löbl, 1990b: 556, Figs 67, 68 (characters, records)
Löbl, 1992a: 535 (records)
Löbl, 2000: 607 (records)

DISTRIBUTION. China: Yunnan; Myanmar; Nepal; Thailand.

Scaphisoma atrosignatum Pic

Scaphosoma atrosignatum Pic, 1955: 52. Syntypes, MRAC; type locality: Rwanda: Tshuruyaga, 2400 m, terr. Astrida.

DISTRIBUTION. Rwanda.

Scaphisoma atrox Löbl

Scaphisoma atrox Löbl, 1981d: 164, Figs 10, 11. Holotype male, MHNG; type locality: Vietnam: Hoa Binh.
Löbl, 1986a: 179, Figs 52, 53 (characters, records)
DISTRIBUTION. India: Meghalaya; Vietnam.

Scaphisoma audax Löbl

Scaphisoma audax Löbl, 1975a: 407, Figs 60, 61. Holotype male, HNHM; type locality: Papua New Guinea: Simbang, Huon Golf.
DISTRIBUTION. Papua New Guinea.

Scaphisoma aurorae Löbl

Scaphisoma aurorae Löbl, 1992a: 549, Figs 27, 137, 138. Holotype male, MHNG; type locality: Nepal: Pun Hill near Ghoropani Pass, 3050–3100 m, Parbat District.
Löbl, 2011d: 183 (records)
DISTRIBUTION. Bhutan; Nepal.

Scaphisoma aurun Löbl

Scaphisoma aurum Löbl, 1979a: 110, Figs 31, 32. Holotype male, MHNG; type locality: India: Suruli Falls, 550 m, Varushanad Hills, Tamil Nadu.
Löbl, 1986a: 191 (records)
Löbl, 1986c: 348 (records)
Löbl, 1990b; 571 (records)
Löbl, 1992a: 535 (records)
DISTRIBUTION. India: Meghalaya, Tamil Nadu, Uttarakhand (Kumaon), West Bengal (Darjeeling District); Nepal; Thailand.

Scaphisoma austerum Löbl

Scaphosoma austerum Löbl, 1965h: 1, Fig. 1. Holotype female, HNHM; type locality: Japan: Gunma Pref., Mt. Tanigawa.
Löbl, 1970c: 793 (characters)
Morimoto, 1985: Pl. 45, Fig. 32 (characters)
DISTRIBUTION. Japan: Honshu.

Scaphisoma australicum Löbl

Scaphisoma australicum Löbl, 1977e: 46, Figs 68, 69. Holotype male, SAMA; type locality: Australia: Queensland, Maryborough.
DISTRIBUTION. Australia: Queensland.

Scaphisoma bacchusi **Löbl**

Scaphisoma bacchusi Löbl, 1975a: 399, Figs 44, 45. Holotype male, NHML; type locality: Papua New Guinea: Budemu, ca 4000 ft., Finisterre Mts.

DISTRIBUTION. Papua New Guinea.

Scaphisoma balcanicum **Taminini**

Scaphosoma balcanicum Tamanini, 1954: 85, Figs 1–7. Holotype male, MCRC; type locality: Albania: Miloti.

Löbl, 1964a: 71, Fig. (characters, key to similar species)
Löbl, 1964d: 50 (records)
Löbl, 1965d: 338 (records)
Löbl, 1965g: 733 (records)
Löbl, 1966b: 131 (records)
Palm, 1966: 43, Fig. A (characters, records)
Löbl, 1967b: 110 (records)
Löbl, 1967e: 8 (characters, records)
Tamanini, 1969a: 486 (records)
Tamanini, 1969c: Fig. 44 (characters)
Löbl, 1970a: 13, Figs 11, 23 (characters, records)
Löbl, 1970c: 763, Figs 38, 39 (characters, records)
Tamanini, 1970: 16, Figs 4E, 10B (characters, fungal hosts)
Franz, 1970: 267 (records)
Freude, 1971: 346, Fig. 3: 7 (characters)
Löbl, 1974a: 61 (records)
Ruter, 1977: 35 (records)
Holzschuh, 1977: 31 (records)
Burakowski et al., 1978: 235 (records)
Iablokoff-Khnzorian, 1985: 142 (records)
Nikitsky et al., 1996: 25 (fungal hosts)
Kofler, 1998: 643 (fungal hosts)
Rogé, 2000: 159 (misspelled *balkanicum*, records)
Byk, 2001: 354 (records)
Nikitsky & Schigel, 2004: 12 (fungal hosts)
Byk et al., 2006: 337 (habitat, records)
Nikitsky et al., 2008: 121 (records, fungal hosts)
Tsurikov, 2009: 107 (records)
Vinogradov et al., 2010: 17 (records, fungal hosts)
Zamotajlov & Nikitsky, 2010: 105 (records, fungal hosts)
Schülke, 2013: 149 (records)
Roosileht, 2015: 17 (records)

Harrison, 2016: 63, Figs 1, 2 (characters, records)
Holzer, 2016: 62 (records)
Solodovnikov, 2016: 55 (records)
Nikitsky et al., 2016: 138 (records, fungal hosts)
Semenov, 2017: 202 (records)
Vlasov & Nikitsky, 2017: 6 (records, fungal hosts)

DISTRIBUTION. Albania; Armenia; Austria; Belarus, Belgium; Bosnia and Herzegovina; Croatia; Czech Republic; Estonia; Finland; France; Germany; Great Britain; Hungary; Italy; Luxemburg; Netherlands; Norway; Poland; Romania; Russia; Slovakia; Spain; Sweden; Switzerland; Tajikistan; Turkey; Ukraine.

***Scaphisoma baloghi* Löbl**

Scaphisoma baloghi Löbl, 1975a: 413, Figs 69–70. Holotype male, HNHM; type locality: Papua New Guinea: Brown River, 40 km north Port Moresby.

DISTRIBUTION. Papua New Guinea.

***Scaphisoma baloo* Löbl**

Scaphisoma baloo Löbl, 1992a: 553, Figs 28, 144, 145. Holotype male, MHNG; type locality: Nepal: Phulcoki, 2650 m, Patan District.

DISTRIBUTION. Nepal.

***Scaphisoma balteatum* Matthews**

Scaphisoma balteatum Matthews, 1888: 172, Pl. 4, Fig. 12. Syntypes, NHML; type locality: Mexico: Cerro de Plumas, Jalapa.
Fierros-López, 2006a: 66, Fig. 2e (characters, records, fungal hosts)

DISTRIBUTION. Costa Rica; Mexico.

***Scaphisoma bancoense* (Pic)**

?Baeocera bancoensis Pic, 1948b: 72. Holotype male, MNHN; type locality: Ivory Coast: Banco.
Löbl, 1987d: 857, Fig. 21 (fransfer to *Scaphisoma*, characters, records)

DISTRIBUTION. Ivory Coast.

***Scaphisoma banguiense* Löbl**

Scaphisoma banguiense Löbl, 1972b: 95, Figs 34, 35. Holotype male, ZMUB; type locality: Philippines: Luzon, Bangui.
Löbl, 1981d: 163, Fig. 12 (characters)
Löbl & Ogawa, 2016b: 1408, Figs 160, 161 (on p. 1408 erroneously as "161, 162", characters)

DISTRIBUTION. Philippines: Luzon.

***Scaphisoma basale* Löbl**

Scaphisoma basale Löbl, 1977b: 29, Figs 17, 40, 41. Holotype male, SAMA; type locality: Australia: Tambourine Mts., Queensland.

DISTRIBUTION. Australia: New South Wales, Queensland.

***Scaphisoma basilewskyi* Pic**

Scaphosoma basilewskyi Pic, 1955: 51. Syntypes, MRAC/MNHN; type localities: Burundi: Rubengera, 1900 m, terr. Kibuye; Ruanda; Bururi, 1800 m.

[*Scaphosoma basilewskyi* var. *bururiense* Pic, 1955: 52. Holotype, MRAC; type locality: Burundi, Bururi, 1800 m. – Based on color patter stated variable within the nominate form and deemed infrasubspecific].

DISTRIBUTION. Burundi; Rwanda.

***Scaphisoma basipenne* (Pic)**

Pseudoscaphosoma basipenne Pic, 1930b: 58. Syntypes, NMPC; type locality: Myanmar: Tenasserim.

DISTRIBUTION. Myanmar.

***Scaphisoma bayau* Löbl**

Scaphisoma bayau Löbl, 1979a: 101, Figs 20, 21. Holotype male, MHNG; type locality: India: Valara Falls, 46 km SW Munnar, 450–500 m.

DISTRIBUTION. India: Kerala.

***Scaphisoma beccarii* Löbl**

Scaphisoma beccarii Löbl, 1975a: 407, Figs 62, 63. Holotype male, MCSN; type locality: Indonesia: Western New Guinea, Hatam.

DISTRIBUTION. Indonesia: Western New Guinea.

***Scaphisoma bedeli* Achard**

Scaphosoma bedeli Achard, 1920i: 240. Lectotype female, NMPC; type locality: India: West Bengal, Kurseong.

Löbl, 1992a: 541 (lectotype designation, characters)

DISTRIBUTION. India: West Bengal (Dargeling District).

***Scaphisoma bellax* Löbl**

Scaphisoma bellax Löbl, 1980a: 109, Figs 24, 25. Holotype male, MHNG; type locality: Taiwan.

DISTRIBUTION. Taiwan.

Scaphisoma besucheti Löbl

Scaphisoma besucheti Löbl, 1971c: 971, Figs 38, 39. Holotype male, MHNG; type locality: Sri Lanka: Kandy, about 600 m.

Löbl, 1979a: 102 (records)
Löbl, 1986c: 346 (records)
Löbl, 1990a: 121 (records)
Löbl, 1992a: 533 (records)
Löbl, 2004a: 346 (records)

DISTRIBUTION. India: Kerala, Himachal Pradesh, Mahdya Pradesh, Tamil Nadu, Uttarakhand; Nepal; Sri Lanka.

Scaphisoma bhareko Löbl

Scaphisoma bhareko Löbl, 1992a: 560, Figs 30, 163–165. Holotype male, MHNG; type locality: Nepal: Phulcoki, 2500 m, Patan District.

Löbl, 2005: 180 (records)

DISTRIBUTION. Nepal.

Scaphisoma bicinctum Löbl

Scaphisoma bicinctum Löbl, 1977c: 225, Figs 5, 6. Holotype male, MHNG; type locality: Vietnam: Saigon.

DISTRIBUTION. Vietnam.

Scaphisoma bicintum Zayas

Scaphosoma bicinta Zayas, 1988: 27, Fig. 18. Syntypes, FZCH; type locality: Cuba.

DISTRIBUTION. Cuba.

Scaphisoma bicolor Löbl & Ogawa

Scaphisoma bicolor Löbl & Ogawa, 2016b: 1444, Figs 3, 257. Holotype female, FMNH; type locality: Philippines: Mindanao, Davao Prov., Meran, east slope of Mt. Apo, 6000 ft.

DISTRIBUTION. Philippines: Mindanao.

Scaphisoma bicoloratum Löbl

Scaphisoma bicoloratum Löbl, 1977e: 39, Figs 56, 57. Holotype male, QMBA; type locality: Australia: Carnarvon, Western Australia.

DISTRIBUTION. Australia: Western Australia, Queensland.

Scaphisoma bicoloripenne Löbl & Ogawa

Scaphisoma bicoloripenne Löbl & Ogawa, 2016b: 1398, Figs 4, 142, 143. Holotype male, MHNG; type locality: Philippines: Bohol, Bilar.

DISTRIBUTION. Philippines: Bohol.

Scaphisoma bicuspidatum **Löbl & Ogawa**

Scaphisoma bicuspidatum Löbl & Ogawa, 2016b: 1387, Figs 5, 118, 119. Holotype male, MHNG; type locality: Philippines: Luzon, Lagunas Prov., Mt. Makiling, summit road, 400 m.

DISTRIBUTION. Philippines: Luzon.

Scaphisoma bifasciatum **Reitter**

Scaphisoma bifasciatum Reitter, 1881b: 140. Lectotype female, MNHN; type locality: Australia: Queensland, Somerset.
Löbl, 1977e: 58 (lectotype designation)

DISTRIBUTION. Australia: Queensland.

Scaphisoma bilimeki **Reitter**

Scaphisoma bilimeki Reitter, 1880a: 48. Syntypes, MNHN; type locality: Mexico.
Matthews, 1888: 175 (characters, records)
Löbl, 1997: 90 (misspelled *bilineki*)

DISTRIBUTION. Mexico.

Scaphisoma biliranense **Löbl**

Scaphisoma biliranense Löbl, 1972b: 86, Figs 19, 20. Holotype male, ZMUB; type locality: Philippines: Leyte, Biliran.
Löbl & Ogawa, 2016b: 1389, Fig. 2 (characters, records)

DISTRIBUTION. Philippines: Leyte.

Scaphisoma bilobum **Löbl & Ogawa**

Scaphisoma bilobum Löbl & Ogawa, 2016b: 1429, Figs 216–218. Holotype male, SMNS; type locality: Philippines: Leyte, Lake Danao, 500 m.

DISTRIBUTION. Philippines: Leyte.

Scaphisoma binaluanum **Pic**

Scaphosoma binaluanum Pic, 1947c: 3. Lectotype male, MNHN; type locality: Philippines: Palawan, Binaluan.
Löbl, 1972b: 105, Figs 53, 54 (lectotype fixed by inference, characters)
Löbl & Ogawa, 2016b: 1364, Figs 72, 73 (characters, records)

DISTRIBUTION. Philippines: Luzon, Leyte, Negros, Palawan, Mindanao.

Scaphisoma binhanum **(Pic)**

Pseudoscaphosoma binhanum Pic, 1922a: 2. Lectotype male, MNHN; type locality: Vietnam: Hoa Binh, Tonkin.
Löbl, 1977c: 222 (lectotype designation, characters, records)

Löbl, 1979a: 108 (records)
Löbl, 1979b: 328 (records)
Löbl, 1981f: 106, Fig. 4 (characters, records)
Löbl, 1986a: 174 (records)
Löbl, 1986c: 347 (records)
Löbl, 1990b: 556 (records)
Löbl, 1992a: 535, 587, Fig. 36 (records)

DISTRIBUTION. India: Assam, Kerala, Meghalaya, Uttarakhand (Garhwal), West Bengal (Darjeeling District); Indonesia: Java; Nepal; Thailand; Vietnam.

Scaphisoma binotatum **Achard**

Scaphosoma binotatum Achard, 1915a: 560. Lectotype male, NMPC; type locality: India: Chambaganor [Shembaganur, Palmi Hills], Tamil Nadu.

Scaphosoma binotatum var. *discoidale* Achard, 1915a: 561. Holotype female, NMPC; type locality: India: Chambaganor [Shembaganur, Palmi Hills], Tamil Nadu.

Löbl, 1979a: 102, Figs 22, 23 (lectotype designations for *S. binotatum*, synonymy of *discoidale* with *S. binotatum*, characters, records)

DISTRIBUTION. India: Kerala, Tamil Nadu.

Scaphisoma biplagiatum **Heller**

Scaphosoma biplagiatum Heller, 1917: 47. Lectotype male, SMTD; type locality: Philippines: Luzon, Mt. Makiling.

Pseudoscaphosoma implagiatum Pic, 1926b: 2. Lectotype female, MNHN; type locality: Philippines: Luzon.

Löbl, 1970b: 126, Fig. 2 (characters)
Löbl, 1972b: 107 (lectotype fixed by inference for *S. biplagiatum* and *S. implagiatum*, their synonymy, characters)
Löbl & Ogawa, 2016b: 1399, Figs 8, 144, 145 (characters, records)

DISTRIBUTION. Philippines: Bucas Grande, Leyte, Luzon, Masbate, Mindanao, Palawan.

Scaphisoma birmanicum **(Pic)**

Amalocera birmanica Pic, 1921a: 164. Lectotype male, MCSN; type locality: Myanmar: Carin Asciuii Cheba [Ghecù], 1200–1300 m.

Löbl, 1973a: 150, Figs 3, 4 (lectotype fixed by inference, transfer from *Amalocera* to *Scaphisoma*, characters)

DISTRIBUTION. Myanmar.

***Scaphisoma biroi* Pic**

Scaphosoma fasciatum Pic, 1956b: 72 (primary junior homonym with *Scaphisoma montanellum* var. *fasciatum* Vinson, 1943). Lectotype female, HNHM; type locality: Papua New Guinea: Stephansort, Astrolabe Bay.

Scaphosoma fasciatum var. *biroi* Pic, 1956b: 73. Lectotype male, HNHM; type locality: Papua New Guinea: Sattelberg [Sattelburg], Huon Golf.

Löbl, 1975a: 388, Figs 26, 27 (lectotype designations for *S. fasciatum* and *biroi*, their synonymy, characters, records)

DISTRIBUTION. Papua New Guinea.

***Scaphisoma bispinosum* Löbl**

Scaphisomas bispinosum Löbl, 1990b: 584, Figs 120–122. Holotype male, MHNG; type locality: Thailand: Doi Suthep, south slope, 1450 m, Chiang Mai.

DISTRIBUTION. Thailand.

***Scaphisoma blandum* Löbl**

Scaphisoma blandum Löbl, 1975a: 401, Figs 50, 51. Holotype male, HNHM; type locality: Papua New Guinea: Mt. Kaindi, 2400 m.

DISTRIBUTION. Papua New Guinea.

***Scaphisoma blefusca* Löbl & Ogawa**

Scaphisoma blefusca Löbl & Ogawa, 2016b: 1357, Figs 55, 56. Holotype male, MHNG; type locality: Philippines: Luzon, Lagunas Prov., Mt. Makiling, summit road, 600 m.

DISTRIBUTION. Philippines: Luzon, Palawan.

***Scaphisoma boettcheri* Pic**

Scaphosoma boettcheri Pic, 1947c: 2. Lectotype male, MNHN; type locality: Philippines: Palawan, Binaluan.

Scaphosoma boettcheri Pic var. *semiobscurum* Pic, 1947c: 2. Lectotype male, MNHN; type locality: Philippines: Palawan, Binaluan.

Löbl, 1970b: 106, Fig. 46 (lectotype designations for *S. boettcheri* and *semiobscurum*, synonymy, characters, records)

Löbl & Ogawa, 2016b: 1400 (record)

DISTRIBUTION. Philippines: Palawan.

***Scaphisoma boleti boleti* (Panzer)**

Scaphidium boleti Panzer, 1793: 16. Syntypes, ZMUB, not traced; type locality: Germany: Göttingen.

Scaphosoma agaricinum var. *robustior* Pic, 1905b: 169. Lectotype female, MNHN; type locality: Turkey: Mont Taurus.

Panzer, 1793: 16 (characters)
Marsham, 1802: 233 (as *Scaphidium*, characters)
Gyllenhal, 1808: 187 (as *Scaphidium*, in synonymy with *S. agaricinum*)
Stephens, 1830: 4 (characters, records)
Aragona, 1830: 18 (as *Scaphidium*, characters)
Heer, 1841: 373 (characters, fungal hosts)
Erichson, 1845: 9 (characters)
Redtenbacher, 1847: 147 (characters)
Gutfleisch, 1859: 222 (characters)
Thomson, 1862: 127 (as *S. assimila*, characters)
Redtenbacher, 1872: 335 (characters)
Reitter, 1885a: 83 (characters)
Fowler, 1889: 347 (characters)
Sahlberg, 1889: 81 (records)
Kuthy, 1897: 86 (records)
Ganglbauer, 1899: 345 (characters)
Johnson & Halbert, 1902: 698 (records)
Jakobson, 1910: 637 (records)
Roubal, 1930: 298 (records)
Horion, 1949: 255 (records)
Benick, 1952: 47 (fungal hosts, habitat)
Palm, 1951: 145 (habitat, fungal hosts)
Lundblad, 1952: 31, Figs 1D, 2D, 3D (as *S. agaricinum*, characters)
Palm, 1953: 172 (missidentified as *S. agaricinum* in Lundblad)
Tamanini, 1954: 88 (as *S. agaricinum*, characters)
Palm, 1959: 224 (habitat, fungal hosts)
Löbl, 1965g: 733 (records)
Koch, 1968: 85 (records)
Kofler, 1968: 40 (characters, records)
Tamanini, 1969a: 487 (stated previous missidentification as *S. agaricinum*, characters)
Tamanini, 1969b: 366, Figs 5G, 6A, B (characters, records, fungal hosts)
Kofler, 1970: 58 (records)
Löbl, 1970a: 13, Figs 14, 19, 26 (characters, records)
Löbl, 1970c: 753, Figs 25, 26 (lectotype fixed by inference for *robustior*, its synonymy with *S. boleti*, characters, records)
Tamanini, 1970: 20, Figs 5D-F, 10F (characters, records, fungal hosts)
Franz, 1970: 267 (records)
Freude, 1971: 347, Fig. 3: 3 (characters)
Nuss, 1975: 111 (fungal hosts)
Burakowski et al., 1978: 235 (records, habitat)

Iablokoff-Khnzorian, 1985: 142 (records)
Angelini, 1986: 55 (records)
Klimaszewski & Peck, 1987: 543 (fungal hosts)
Merkl, 1987: 113 (records)
Schawaller, 1990: 235 (records)
Nikitsky et al., 1996: 24 (fungal hosts)
Merkl, 1996: 261 (records)
Kofler, 1998: 644 (fungal hosts)
Byk, 2001: 354 (records)
Nikitsky & Schigel, 2004: 9, 14 (fungal hosts)
Tsinkevich, 2004: 19 (records, fungal hosts)
Silfverberg, 2004: 38 (records)
Tronquet, 2006: 79 (records)
Borowski, 2006: 52–58 (records, fungal hosts)
Byk et al., 2006: 337 (habitat, records)
Nikitsky et al., 2008: 121 (records, habitat, fungal hosts)
Guèrguev & Ljubomirov, 2009: 247 (records)
Lott, 2009: 21 (records)
Tsurikov, 2009: 107 (records)
Zamotajlov & Nikitsky, 2010: 105 (fungal hosts)
Vlasov & Nikitsky, 2017: 6 (records, fungal hosts)

DISTRIBUTION. Almost all European coutries including Ireland, Spain, and Caucasus; Georgia; Turkey including Anatolia. Not known from Cyprus, Malta, Portugal.

Scaphisoma boleti dilutum **Reitter**

Scaphosoma dilutum Reitter, 1885a: 83. Lectotype female, HNHM; type locality: Azerbaijan: Lenkoran.
Jakobson, 1910: 637 (records)
Löbl, 1964b: 3 (characters)
Löbl, 1970c: 754 (lectotype fixed by inference, downgraded to subspecies, characters, records)
Iablokoff-Khnzorian, 1985: 142 (records)
Ghahari et al., 2009: 1954 (as *boleti*, doubtful record)
Löbl & Ogawa, 2016c: 38 (records)

DISTRIBUTION. Azerbaijan; Iran; Turkmenistan.

Scaphisoma bolmarum **Löbl & Ogawa**

Scaphisoma bolmarum Löbl & Ogawa, 2016b: 1365, Figs 74, 75. Holotype male, SMNS; type locality: Philippines: Mindanao, Mt. Apo, Ilomavis, 1400 m.

DISTRIBUTION. Philippines: Mindanao.

Scaphisoma bonariense **Achard**

Scaphosoma bonariense Achard, 1920e: 350. Syntypes, NMPC; type locality: Argentina: Buenos Aires.

[*Scaphosoma bonariense* var. *deficiens* Achard, 1920e: 351. Syntypes, NMPC; type locality: Argentina: Buenos Aires. – Deemed unavailable, due to sympatry of three varieties with the nominate taxon]

[*Scaphosoma bonariense* var. *intermedium* Achard, 1920e: 351. Syntypes, NMPC; type locality: Argentina: Buenos Aires. – Deemed unavailable, due to sympatry of three varieties with the nominate taxon]

[*Scaphosoma bonariense* var. *rubromaculatum* Achard, 1920e: 351. Syntypes, NMPC; type locality: Argentina: Buenos Aires. – Deemed unavailable, due to sympatry of three varieties with the nominate taxon]

DISTRIBUTION. Argentina.

Scaphisoma boreale **Lundblad**

Scaphosoma boreale Lundbad, 1952: 29, Figs 1B, 2B, 3B. Holotype male, NHRS; type locality: Sweden: Upland, Fiby Forest, Kirchspiel Vänge.

Palm, 1953: 171 (records)
Palm, 1959: 224 (records, fungal hosts, habitat)
Löbl, 1964d: 49, Figs (characters, records)
Löbl, 1965d: 339 (records)
Löbl, 1965g: 733 (records)
Löbl, 1966b: 132 (records)
Strand, 1969: 18 (records)
Löbl, 1970a: 14, Figs 15, 21, 27 (characters, records)
Löbl, 1970c: 774, Figs 56, 57 (characters, records)
Tamanini, 1970: 20, Figs 4G, H, 10G (characters)
Burakowski et al., 1978: 236 (records)
Iablokoff-Khnzorian, 1985: 142 (records)
Angelini, 1986: 55 (records)
Angelini, 1987: 22 (records)
Schawaller, 1990: 235 (records)
Walter, 1990: 31 (records)
Nikitsky et al., 1996: 25 (fungal hosts)
Merkl, 1996: 261 (records)
Byk, 2001: 354 (records)
Byk et al., 2006: 337, 360 (habitat, records)
Schülke, 2013: 149 (records)
Nikitsky & Schigel, 2004: 13 (fungal hosts)
Schigel et al., 2004: 41 (fungal hosts)

Schmidl et al., 2005: 22 (records)
Borowski et al., 2005: 44 (records)
Borowski, 2006: 54–58 (records, fungal hosts)
Tsurikov, 2009: 107 (records)
Guèrguev & Ljubomirov, 2009: 247 (records)
Vinogradov et al., 2010: 16 (records, fungal hosts)
Schigel, 2011: 330, 332, 333, 337, 338, 340 (fungal hosts)
Monsevičius, 2013: 27 (records)
Roosileht, 2015: 17 (records)
Holzer, 2016: 62 (records)
Solodovnikov, 2016: 55 (records)
Löbl & Ogawa, 2016c: 38 (records)
Nikitsky et al., 2016: 138 (records, fungal hosts)
Semenov, 2017: 202 (records)
Vlasov & Nikitsky, 2017: 6 (records, fungal hosts)

DISTRIBUTION. Albania; Austria; Belarus; Bosnia and Herzegovina; Bulgaria; Czech Republic; Estonia; Finland; France; Germany; Hungary; Iran; Italy; Latvia; Liechtenstein; Lithuania; Norway; Poland; Romania; Russia: European, Altai; Slovakia; Spain; Sweden; Ukraine.

***Scaphisoma borneense* Pic**

Scaphosoma borneense Pic, 1916d: 7. Lectotype male, MNHN; type locality: Indonesia: Kalimantan [Riam Kanan, Martapura].
Löbl, 2015c: 131, Figs 1–3 (lectotype designation, characters)

DISTRIBUTION. Indonesia: Kalimantan.

***Scaphisoma bourbonense* Löbl**

Scaphisoma bourbonense Löbl, 1977f: 45, Figs 9, 10. Holotype male, MHNG; type locality: La Réunion: Plaine de Chicots, 1700 m.
Lecoq, 2015: 209 (records)

DISTRIBUTION. La Réunion.

***Scaphisoma breuili* Pic**

Scaphosoma breuili Pic, 1946: 83. Syntypes, MNHN; type locality: Ethiopia: cave Diré Daoua.

DISTRIBUTION. Ethiopia.

***Scaphisoma breve* Matthews**

Scaphisoma breve Matthews, 1888: 176. Syntypes, NHML; type locality: Panama: Bugaba.

DISTRIBUTION. Panama.

Scaphisoma breviatum Löbl & Ogawa

Scaphisoma breviatum Löbl & Ogawa, 2016b: 1382, Figs 110, 111. Holotype male, EUMJ; type locality: Philippines: Luzon, Ifugao Prov., Mt. Pangao, 2350 m, near Data.

DISTRIBUTION. Philippines: Luzon.

Scaphisoma brevicorne Reitter

Scaphosoma brevicorne Reitter, 1908: 33. Holotype, NHMW; type locality: Tanzania: Mt. Bomole, Amani.

DISTRIBUTION. Tanzania.

Scaphisoma brittoni Löbl

Scaphisoma brittoni Löbl, 1977e: 47, Figs 22, 70, 71. Holotype male, ANIC; type locality: Australia: New South Wales, Wahroonga.

DISTRIBUTION. Australia: Queensland, New South Wales, South Australia.

Scaphisoma bruchi Pic

Scaphosoma (*Mimoscaphososoma*) *bruchi* Pic, 1928d: 49. Syntypes, MACN/ MNHN; type locality: Argentina: Sierras de Cordoba, Alta Gracie la Granja.

DISTRIBUTION. Argentina.

Scaphisoma brunneipenne Pic

Scaphosoma brunneipenne Pic, 1916c: 19. Syntypes, MNHN; type locality: Brazil: Blumenau [SE Brazil].

DISTRIBUTION. Brazil.

Scaphisoma brunneonotatum Pic

Scaphosoma brunneonotatum Pic, 1923b: 17. Syntypes, MNHN; type locality: Vietnam: Tonkin [Lac Tho].

Löbl, 1980a: 120, Figs 27, 28 (characters, records)
Löbl, 1982b: 105 (characters, records)
Löbl, 1990b: 549 (records)
Rougemont, 1996: 12 (records)
Löbl, 2000: 607 (records)

DISTRIBUTION. China: Guangxi, Hong Kong; Japan: Ryukyus; Taiwan; Thailand; Vietnam.

Scaphisoma budemuense Löbl

Scaphisoma budemuense Löbl, 1975a: 395, Figs 36, 37. Holotype male, NHML; type locality: Papua New Guinea: Budemu, ca 4000 ft., Finistere Mts.

DISTRIBUTION. Papua New Guinea.

Scaphisoma bugi Löbl

Scaphisoma bugi Löbl, 1983a: 288, Figs 5, 6. Holotype male, MHNG; type locality: Indonesia: Sulawesi, Makale.
Ogawa, 2015: 82, Figs 4-14a, b (characters, records)
DISTRIBUTION. Indonesia: Sulawesi.

Scaphisoma bugiamapi Prokofiev

Scaphisoma bugiamapi Prokofiev, 2003: 2, Figs 3, 8. Holotype male, SIEE; type locality: Vietnam: Bonh Phuoc Prov., Phuoc Long District, Bu Gia Map National Park.
DISTRIBUTION. Vietnam.

Scaphisoma canaliculatum Löbl

Scaphisoma canaliculatum Löbl, 1990b: 554, Figs 65, 66. Holotype male, MHNG; type locality: Thailand: Prahu Pah, 50 km north Lampang, 620 m.
DISTRIBUTION. Thailand.

Scaphisoma cantrelli Löbl

Scaphisoma cantrelli Löbl, 1977e: 51. Holotype male, QMBA; type locality: Australia: Queensland, Mt. Glorious.
DISTRIBUTION. Australia: Queensland.

Scaphisoma carolinae Casey

Scaphisoma carolinae Casey, 1893: 531. Lectotype male, USNM; type locality: USA: North Carolina [Asheeville].
Scaphisoma arkansana Casey, 1893: 532. Holotype female, USNM; type locality: USA: Arkansas.
Brimley, 1938: 149 (records)
Leschen et al., 1990: 279, Figs 3, 12, 13 (lectotype designation for *S. carolinae*, synonymy of *S. arkansanum* with *S. carolinae*, characters, records)
Majka et al., 2011: 89 (distribution)
DISTRIBUTION. United States: Arkansas, Florida, Illinois, Maine, New Hampshire, North Carolina, Oklahoma.

Scaphisoma casiguran Löbl & Ogawa

Scaphisoma casiguran Löbl & Ogawa, 2016b: 1366, Figs 8, 76, 77. Holotype male, MHNG; type locality: Philippines: Luzon, Lagunas Prov., Mt. Makiling, summit road, 600 m.
DISTRIBUTION. Philippines: Luzon.

Scaphisoma castaneipenne Reitter

Scaphisoma castaneipennis Reitter 1877: 370. Lectotype male, MNHN; type locality: Japan.
Achard, 1923: 113 (characters)
Miwa & Mitono, 1943: 541 (characters)
Nakane, 1955c: 56 (characters)
Chûjô, 1961: 5 (records)
Löbl, 1965a: 44, Figs 1, 1a, b (lectotype designation, characters)
Löbl, 1970c: 764, Figs 42, 43 (characters, records)
Hisamatsu, 1977: 194 (records)
Morimoto, 1985: Pl. 45, Fig. 29 (characters)
DISTRIBUTION. Japan: Honshu, Kyushu, Shikoku, Tsushima.

Scaphisoma caudatoides Löbl & Ogawa

Scaphisoma caudatoides Löbl & Ogawa, 2016b: 1362, Fig. 69. Holotype male, MHNG; type locality: Philippines: Palawan, 4 km north Port Barton.
DISTRIBUTION. Philippines: Palawan.

Scaphisoma caudatum Löbl

Scaphisoma caudatum Löbl, 1975b: 281, Figs 16, 17. Holotype male, MHNG; type locality: Singapore: Bukit Timah Nat. Reserve.
DISTRIBUTION. Singapore.

Scaphisoma cederholmi Löbl

Scaphisoma cederholmi Löbl, 1971c: 983, Figs 52, 53. Holotype male, MHNG; type locality: Sri Lanka: Peradeniya, near Agricultural Station, ca 550 m.
Löbl, 1986a: 193, Fig. 74 (characters, records)
Löbl, 1990b: 572 (records)
Rougemont, 1996: 12 (records)
Löbl, 2000: 607 (records)
DISTRIBUTION. China: Hong Kong; India: Assam; Sri Lanka; Thailand.

Scaphisoma centripunctulum Löbl & Ogawa

Scaphisoma centripunctulum Löbl & Ogawa, 2016b: 1390, Figs 123, 124. Holotype male, FMNH; type locality: Philippines: Mindanao, Cotabato Prov., Burungkot, Upi, 1500 ft.
DISTRIBUTION. Philippines: Mindanao.

Scaphisoma centronotatum **(Pic)**

Pseudoscaphosoma centronotatum Pic, 1926d: 46. Lectotype female, MNHN; type locality: Vietnam: Tonkin, Lac Tho.
Löbl, 1975b: 289 (lectotype designation)
Löbl, 1981f: 105, Fig. 3 (characters, records)

DISTRIBUTION. Vietnam.

Scaphisoma cernense **Vinson**

Scaphosoma cernense Vinson, 1943: 194, Figs 4, 14. Holotype, NHML; type locality: Mauritius: The Pouce.

DISTRIBUTION. Mauritius.

Scaphisoma championi **Löbl**

Scaphosoma cribripenne Champion, 1927: 277. Lectotype male, NHML; type locality: India: Uttarakhand, Haldwani District.

Scaphisoma championi Löbl, 1981d: 158; replacement name for *Scaphisoma cribripenne* Champion, 1927 (nec *Scaphisoma cribripenne* Pic, 1923).
Löbl, 1992a: 541, Fig. 150 (lectotype designation, characters)

DISTRIBUTION. India: Uttarakhand (Kumaon).

Scaphisoma chujoi **Löbl**

Scaphisoma chujoi Löbl, 1982f: 6, Figs 2–4. Holotype male, PCMS; type locality: East Malaysia: Sarawak, Batu Kawa.

DISTRIBUTION. East Malaysia: Sarawak.

Scaphisoma cippum **Löbl**

Scaphisoma cippum Löbl, 1980a: 104, Figs 18, 19. Holotype male, MHNG; type locality: Taiwan: Fenchihu, 1400 m.

DISTRIBUTION. Taiwan.

Scaphisoma clavigerum **Löbl**

Scaphisoma clavigerum Löbl, 1992a: 554, Figs 29, 146–148. Holotype male, SMNS; type locality: Nepal: Dhorpar Kharka, 2700 m, Panchthar District.

DISTRIBUTION. India: Sikkim; Nepal.

Scaphisoma coalitum **Löbl**

Scaphisoma coalitum Löbl, 1992a: 558, Figs 158–160. Holotype male, MHNG; type locality: between Hunse and Mai Pokhara, 1600–2000 m, Ilam District.

DISTRIBUTION. Nepal.

Scaphisoma coarctatum Löbl

Scaphisoma coarctatum Löbl, 1976a: 9; Figs 3, 4. Holotype male, MHNG; type locality: Indonesia: Maluku Islands, Buru, Station 9.
Löbl, 2014a: 55 (records)
Löbl, 2015a: 109 (records)
DISTRIBUTION. Indonesia: Lombok, Maluku Islands: Buru.

Scaphisoma coeruleum Löbl

Scaphisoma coeruleum Löbl, 2003c: 182, Figs 1–2. Holotype male, HNHM; type locality: Papua New Guinea: Mt. Wilhelm Field Station.
DISTRIBUTION. Papua New Guinea.

Scaphisoma colasi Löbl

Scaphosoma colasi Löbl, 1965a: 55, Figs 8, 8a, b. Holotype male, MNHN; type locality: Japan: Tokyo, Botanical Garden.
Löbl, 1967c: 132 (records)
Löbl, 1970c: 754, Figs 27, 28 (characters, records)
DISTRIBUTION. Japan: Hokkaido, Honshu.

Scaphisoma collarti (Pic)

Macroscaphosoma collarti Pic, 1928b: 33. Lectotype male, MNHN; type locality: Democratic Republic of the Congo: Tibo.
Löbl, 1970b: 129, Figs 6, 8 (as *Macroscaphosoma*, lectotype designation, characters)
Löbl, 1997: xi (as *Macroscaphosoma*, credited to Löbl, 1970)
Leschen & Löbl, 2005: 25 (credited to Löbl, 1970, transfer to *Scaphisoma*)
Note: Pic assigned the species *collarti* to his genus-group name *Macroscaphosoma*, deemed unavailable by Löbl, 1997. The species epithet is deemed available under the ICZN 1999, Art. 11.9.3.1.
DISTRIBUTION. Democratic Republic of the Congo.

Scaphisoma commune Löbl

Scaphidium castaneum Motschulsky, 1845: 361. Syntypes, ZMUM; type locality: USA: California, San José.
Scaphisoma commune Löbl, 1997: xii, replacement name for *Scaphidium castaneum* Motschulsky, 1845 (nec *Scaphidium castaneum* Perty, 1830).
LeConte, 1860: 323 (transfer from *Scaphidium* to *Scaphisoma*, characters)
Casey, 1893: 526 (as *S. castanea*, characters, records)
Leng, 1920: 134 (as *S. castaneum*, records)

Hatch, 1957: 282, Pl. 36, Fig. 9 (as *S. castaneum*, characters, records, fungal hosts)
Hanley, 1996: 37, Figs 2–18 (as *S. castaneum*, immature stages, habitat, fungal hosts)
Betz et al., 2003: 203, Fig. 13 (characters)
DISTRIBUTION. Canada: British Columbia; United States: Arizona, California, Idaho, Oregon, Utah, Washington.

***Scaphisoma compactum* Löbl & Ogawa**
Scaphisoma compactum Löbl & Ogawa, 2016b: 1367, Figs 78, 79. Holotype male, MHNG; type locality: Philippines: Palawan: Roxas, 80 m, Region of Matalangao.
DISTRIBUTION. Philippines: Palawan.

***Scaphisoma complicans* Löbl**
Scaphisoma complicans Löbl, 1982f: 10, Figs 8, 9. Holotype male, MHNG; type locality: East Malaysia: Sarawak, Kuala Bok.
DISTRIBUTION. East Malaysia: Sarawak.

***Scaphisoma concolor* Löbl**
Scaphisoma concolor Löbl, 1981a: 361, Figs 20, 21. Holotype male, NZAC; type locality: New Caledonia: Table d'Union, 800 m, near Col d'Amieu.
DISTRIBUTION. New Caledonia.

***Scaphisoma conflictum* Löbl & Ogawa**
Scaphisoma conflictum Löbl & Ogawa, 2016b: 1418, Figs 184–189. Holotype male, MHNG; type locality: Luzon, Laguna Prov., Mt. Makiling, 4 km SE Los Banos.
DISTRIBUTION. Philippines: Luzon.

***Scaphisoma conforme conforme* Löbl**
Scaphisoma conforme Löbl, 1980a: 116, Figs 36, 37. Holotype male, MHNG; type locality: Taiwan: Fenchihu, 1400 m.
DISTRIBUTION. Taiwan.

***Scaphisoma conforme okinawense* Löbl**
Scaphisoma conforme okinavense Löbl, 1982b: 104, Fig. 4. Holotype male, PCMS; type locality: Japan: Yona, Okinawa.
DISTRIBUTION. Japan: Ryukyus.

Scaphisoma confrater Löbl

Scaphisoma confrater Löbl, 1981a: 363, Figs 22, 23. Holotype male, NZAC; type locality: New Caledpnia: Table d'Union, 800 m, near Col d'Amieu.
Lecoq, 2015: 209 (records)

DISTRIBUTION. New Caledonia.

Scaphisoma confusum Löbl & Ogawa

Scaphisoma confusum Löbl & Ogawa, 2016b: 1419, Figs 190–193. Holotype male, MHNG; type locality: Philippines: Mindanao, Mainit Hot Spring.

DISTRIBUTION. Philippines: Mindanao.

Scaphisoma congoanum Pic

Scaphosoma congoanum Pic, 1928b: 44. Syntypes, MRAC/MNHN; type localities: Democratic Republic of the Congo: Seke Banza, Mayumbé.
Pic, 1955: 52 (records)

DISTRIBUTION. Democratic Republic of the Congo; Rwanda.

Scaphisoma consimile Löbl

Scaphisoma consimile Löbl, 1973b: 326, Figs 25, 26. Holotype male, NHMW; type locality: New Caledonia: Mt. Koghis.
Löbl, 1981a: 373 (characters, records)

DISTRIBUTION. New Caledonia.

Scaphisoma conspicuum Löbl

Scaphisoma conspicuum Löbl, 1982h: 19, Figs 1, 2. Holotype male, EUMJ; type locality: Japan: Mt. Nanatsudake, Fukue, Gotô Islands.

DISTRIBUTION. Japan: Gotô Islands, Honshu, Shikoku.

Scaphisoma convexum Say

Scaphisoma convexa Say, 1825: 183. Type material probably destroyed; type locality: USA.
LeConte, 1860: 323 (characters, records)
Casey, 1893: 525 (characters, records)
Blatchley, 1910: 496 (characters)
Leonard, 1928: 314 (records)
Brimley, 1938: 149 (records)
Leschen, 1988a: 231 (fungal hosts)
Leschen et al., 1990: 283, Figs 7, 19, 20 (characters, records)
Campbell, 1991: 125 (records)
Downie & Arnett, 1996: 366 (records)

Majka et al., 2011: 89 (distribution)
Webster et al., 2012: 244 (records, habitat)
DISTRIBUTION. Canada: Manitoba, New Brunswick, Nova Scotia, Ontario, Quebec; United States: Arkansas, Connecticut, Florida, Georgia, Indiana, Louisiana, Maine, Massachussetts, Michigan, Missouri, New Hampshire, New York, North Carolina, Oklahoma, Pennsylvania, Rhode island, Texas, Vermont, Virginia.

***Scaphisoma corbetti* Löbl**

Scaphisoma corbetti Löbl, 1986c: 361, Figs 16–18. Holotype male, MHNG; type locality: India: Corbett National Park near Garjia, 450 m, Kumaon.
Löbl, 1992a: 531 (records)
DISTRIBUTION. India: Uttarakhand; Nepal.

***Scaphisoma corcyricum* Löbl**

Scaphosoma corcyricum Löbl, 1964b: 1, Figs 1–3. Holotype male, HNHM; type locality: Greece: Corfu.
Löbl, 1967d: 35, Figs 4a, b (characters)
Löbl, 1970c: 749, Figs 21, 22 (characters, records)
Tamanini, 1970: 18 (characters)
Löbl & Leschen, 2003b: 40, Figs 18, 73, Map 21 (characters, records)
DISTRIBUTION. Croatia; Cyprus; Greece: Corfu; Turkey; New Zealand [introduced].

***Scaphisoma cortesaguilari* Fierros-López**

Scaphisoma cortesaguilari Fierros-López, 2006a: 59, Figs 1b, 2b, 3b, 4d-f, 6b. Holotype male, CZUG; type locality: Mexico: Jalisco, Mascota, El Atajo 1440 m.
DISTRIBUTION. Mexico.

***Scaphisoma crassipes* Achard**

Scaphosoma crassipes Achard, 1923: 115. Lectotype female, NHML: 115; type locality: Japan: Miyanoshita.
Miwa & Mitono, 1943: 541 (characters)
Nakane, 1955c: 56 (characters)
Löbl, 1967c: 132 (characters, records)
Löbl, 1970c: 777, Figs 58, 59 (implicite lectotype fixation, characters, records)
DISTRIBUTION. Japan: Honshu.

***Scaphisoma cribripenne* (Pic)**

Scutoscaphosoma cribripenne Pic, 1923b: 17. Lectotype male, MNHN; type locality: Vietnam: Tonkin, Lac Tho.

Löbl, 1981d: 158, Fig. 5 (lectotype designation, transfer to *Scaphisoma*, characters, records)
Löbl, 1990b: 564 (records)
DISTRIBUTION. Thailand, Vietnam.

Scaphisoma cribrosum Pic
Scaphosoma cribrosum Pic 1928b: 42. Holotype, MRAC; type locality: Democratic Republic of the Congo: Haut-Uelé, Watsa.
DISTRIBUTION. Democratic Republic of the Congo.

Scaphisoma cruciatum Champion
Scaphosoma cruciatum Champion, 1927: 275. Holotype female, NHML; type locality: India: Haldwani District.
Löbl, 1992a: 540, Fig. 26 (characters)
DISTRIBUTION. India: Uttarakhand (Kumaon).

Scaphisoma cubense Reitter
Scaphisoma cubense Reitter, 1880a: 48. Syntypes, MNHN; type locality: Cuba.
DISTRIBUTION. Cuba.

Scaphisoma cursor Löbl
Scaphisoma cursor Löbl, 1975b: 279, Figs 12, 13. Holotype male, MHNG; type locality: East Malasia: Sarawak, Semengoh Forest Reserve, 11 mi SW Kuching.
DISTRIBUTION. East Malasia: Sarawak.

Scaphisoma curtipenne (Pic)
Baeocera curtipennis Pic, 1916c: 19. Lectotype male, MNHN; type locality: West Malaysia: Perak [Malacca].
Achard, 1920c: 58 (copy of original description)
Pic, 1920g: 189 (as *Baeocera, characters*)
Löbl, 1973a: 156, Figs 11, 12 (lectotype fixed by inference, transfer to *Scaphisoma*, characters)
DISTRIBUTION. West Malaysia.

Scaphisoma cuspidatum Löbl
Scaphisoma cuspidatum Löbl, 1990b: 585, Figs 123–125. Holotype male, MHNG; type locality: Thailand: Doi Inthanon, 1650 m, Chiang Mai.
DISTRIBUTION. Thailand.

Scaphisoma cuyunon Löbl & Ogawa

Scaphisoma cuyunon Löbl & Ogawa, 2016b: 1400, Figs 9, 146, 147. Holotype male, MHNG; type locality: Philippines: Palawan, Conception, sea level.

DISTRIBUTION. Philippines: Palawan.

Scaphisoma dakotanum Fall

Scaphisoma dakotana Fall, 1910: 116. Syntypes, MCZC; type locality: USA: Bismarck, Dakota.

DISTRIBUTION. United States: North Dakota.

Scaphisoma danielae Löbl

Scaphisoma danielae Löbl, 1982a: 332, Figs 5, 6. Holotype male, MHNG; type locality: Japan: Omogo, 700 m, Ehime Pref.

DISTRIBUTION. Japan: Shikoku.

Scaphisoma dayak Löbl

Scaphisoma dayak Löbl, 1982f: 8, Figs 5–7. Holotype male, MHNG; type locality: East Malaysia: Sarawak, Kuala Bok.

DISTRIBUTION. East Malaysia: Sarawak.

Scaphisoma debile Löbl

Scaphisoma debile Löbl, 1980c: 392, Figs 27, 28. Holotype male, NZAC; type locality: Fiji: Nandarivatu, 950 m, Viti Levu.

DISTRIBUTION. Fiji.

Scaphisoma decorsei Pic

Scaphosoma decorsei Pic, 1948b: 72. Holotype, MNHN; type locality: Central African Republic: Chari, Fort Archambolt.

DISTRIBUTION. Central African Republic.

Scaphisoma decorum Löbl

Scaphisoma decorum Löbl, 1977e: 45, Figs 21, 66, 67. Holotype male, ANIC; type locality: Australia: Queensland, 2 mi west Kuranda.

DISTRIBUTION. Australia: Queensland.

Scaphisoma deharvengi Löbl & Ogawa

Scaphisoma deharvengi Löbl & Ogawa, 2016b: 1368, Figs 80–82. Holotype male, MHNG; type locality: Philippines: Luzon, Mountain Prov., Mount Data Lodge, 2200–2300 m.

DISTRIBUTION. Philippines: Luzon.

Scaphisoma delamarei **Pic**

Scaphosoma delamarei Pic, 1948b: 72. Holotype, MNHN; type locality: Ivory Coast: Banco.

DISTRIBUTION. Ivory Coast.

Scaphisoma delictum **Löbl**

Scaphisoma delictum Löbl, 1981d: 158, Figs 6, 7. Holotype male, HNHM; type locality: Vietnam: Ninh Binh, Cuc Phuong.
Löbl, 1990b: 564 (records)

DISTRIBUTION. Thailand; Vietnam.

Scaphisoma densepunctatum **Löbl & Ogawa**

Scaphisoma densepunctatum Löbl & Ogawa, 2016b: 1369, Figs 83, 84. Holotype male, SMNS; type locality: Philippines: Mindanao, 30 km NW of Maramag, Bagongsilang, 1700 m.

DISTRIBUTION. Philippines: Mindanao.

Scaphisoma dentipenne **Löbl**

Scaphisoma dentipenne Löbl, 1971a: 250, Fig. 3. Holotype male, ZMUC; type locality: Philippines: Mantalingajan, Pinigisan, 600 m.
Löbl & Ogawa, 2016b: 1442 (characters)

DISTRIBUTION. Philippines: Palawan.

Scaphisoma desertorum **Casey**

Scaphisoma desertorum Casey, 1893: 530. Syntypes, USNM; type localities: USA: Williams, Arizona; El Paso, Texas.

DISTRIBUTION. United States: Arizona, Texas.

Scaphisoma diabolum **Löbl**

Scaphisoma diabolum Löbl, 1986a: 158, Fig. 22, 23. Holotype male, MHNG; type locality: India: Sukna, 200 m, Darjeeling District.
Löbl, 1992a: 534 (records)

DISTRIBUTION. India: West Bengal (Darjeeling District); Nepal.

Scaphisoma diaphanum **Löbl**

Scaphisoma diaphanum Löbl, 1973b: 313, Figs 7, 8. Holotype male, NHMW; type locality: New Caledonia: Mt. Koghis.
Löbl, 1977i: 818 (records)
Löbl, 1981a: 361 (characters, records)

DISTRIBUTION. New Caledonia.

***Scaphisoma digitale* Löbl**
Scaphisoma digitale Löbl, 2001: 181, Figs 1–5. Holotype male, NMEC; type locality: Vietnam: Tam Dao, NNW Hanoi, 800–900 m.
DISTRIBUTION. Vietnam.

***Scaphisoma dilatatum* Löbl**
Scaphisoma dilatatum Löbl, 2003a: 68. Holotype male, NHMW; type locality: China: Lugi Lake, Luo Shui, Yunnan.
DISTRIBUTION. China: Yunnan.

***Scaphisoma discolor* Löbl**
Scaphisoma discolor Löbl, 1977b: 42, Figs 62, 63. Holotype male, SAMA; type locality: Australia: Tambourine Mt., Queensland.
DISTRIBUTION. Australia: New South Wales, Queensland, Victoria.

***Scaphisoma discretum* Löbl**
Scaphisoma discretum Löbl, 1986a: 145, Fig.4. Holotype male, MHNG; type locality: India: Manas Lild Life Sanctuary, 200 m, Assam.
DISTRIBUTION. India: Assam, Meghalaya.

***Scaphisoma dispar* Löbl**
Scaphosoma dispar Löbl, 1970b: 125, Fig. 1. Holotype male, MNHN; type locality: Philippines: Luzon, Los Banos.
Löbl & Ogawa, 2016b: 1420, Figs 10, 194, 195 (characters, records)
DISTRIBUTION. Philippines: Luzon.

***Scaphisoma disparides* Löbl & Ogawa**
Scaphisoma disparides Löbl & Ogawa, 2016b: 1420, Figs 11, 196–198. Holotype male, MHNG; type locality: Philippines: Luzon, Ilocos Norte, Bangui.
DISTRIBUTION. Philippines: Palawan, Mindanao.

***Scaphisoma dissimile* Löbl**
Scaphisoma dissimile Löbl, 1975a: 391, Figs 30, 31. Holotype male, HNHM; type locality: Papua New Guinea: Stephansort, Astrolabe Bay.
DISTRIBUTION. Papua New Guinea.

***Scaphisoma dissymmetricum* Löbl & Ogawa**
Scaphisoma dissymmetricum Löbl & Ogawa, 2016b: 1405, Figs 155, 156. Holotype male, MHNG; type locality: Philippines: Palawan, Sabang, trail to Underground River, sea level.
DISTRIBUTION. Philippines: Palawan.

Scaphisoma distans Löbl

Scaphisoma distans Löbl, 1975a: 822, Figs 7, 8. Holotype male, SAMA; type locality: Fiji: Ovalau.
Löbl, 1980c: 390, Figs 20, 21 (characters, records)
Note. Erroneously Fig. no "22" in Löbl, 1980c: 390.
DISTRIBUTION. Fiji.

Scaphisoma distinctoides Löbl & Ogawa

Scaphisoma distinctum Löbl & Ogawa, 2016b: 1421, Figs 199, 200. Holotype male, SMNS; type locality: Philippines: Mindanao, 30 km NW of Maramag, Bagogsilang, 1700 m.
Scaphisoma distinctoides Löbl & Ogawa, 2017: 415, Pl. 82; replacement name for *Scaphisoma distinctum* Löbl & Ogawa, 2016b (nec *Scaphisoma distinctum* Blatchley, 1910; in synonymy with *Baeocera apicalis* LeConte, 1860).
DISTRIBUTION. Philippines: Mindanao.

Scaphisoma distinguendum Oberthür

Scaphisoma distinguendum Oberthür, 1883: 14. Syntypes, MNHN; type locality: Ethiopia ["Abyssinie"].
DISTRIBUTION. Ethiopia.

Scaphisoma diversicorne Löbl

Scaphisoma diversicorne Löbl, 2003b: 158, Figs 4–7. Holotype male, NHMW; type locality: India: Meghalaya, Nokrek NP, 1100 m, Garo Hills.
DISTRIBUTION. India: Meghalaya.

Scaphisoma diversum Löbl & Ogawa

Scaphisoma diversum Löbl & Ogawa, 2016b: 1356, Figs 53, 54. Holotype male, EUMJ; type locality: Philippines: Luzon, Mountain Prov., Mt. Data, 2300 m.
DISTRIBUTION. Philippines: Luzon.

Scaphisoma dives Löbl

Scaphisoma dives Löbl, 1990b: 575, Figs 100–193. Holotype male, MHNG; type locality: Thailand: Doi Suthep, north slope, 1550 m, Chiang Mai.
DISTRIBUTION. Thailand.

Scaphisoma dohertyi Pic

Scaphosoma dohertyi Pic, 1915a: 24. Lectotype male, MNHN; type locality: West Malaysia: Perak, Malacca.
Achard, 1920c: 57 (copy of original description)

Löbl, 1981f: 107, Fig. 5 (lectotype designation, characters, records)
Löbl, 1986a: 174 (records)
Löbl, 1990b: 556 (records)
Löbl, 2000: 607 (records)
Löbl, 2015a: 109 (records)

DISTRIBUTION. China: Yunnan; India: Meghalaya, West Bengal (Darjeeling District); Indonesia: Java, Bali, Sumbawa; East Malaysia, West Malaysia; Thailand, Vietnam.
Note. Replace Taiwan by Thailand in Löbl, 1981f: 107.

Scaphisoma dumosum **Löbl**

Scaphisoma dumosum Löbl, 2000: 652, Figs 85–89. Holotype male, SMNS; type locality: China: Yunnan, above Dali, 2000–2200 m.
Löbl, 2003a: 62 (records)

DISTRIBUTION. China: West Sichuan, Yunnan.

Scaphisoma duplex **Löbl & Ogawa**

Scaphisoma duplex Löbl & Ogawa, 2016b: 1430, Figs 219, 220. Holotype male, MHNG; type locality: Philippines: Luzon, Lagunas Prov., Mt. Banahaw near school, about 1 km from Kinabuhayan, 500 m.

DISTRIBUTION. Philippines: Luzon.

Scaphisoma duplicatum **Löbl**

Scaphisoma duplicatum Löbl, 1972b: 89, Figs 25, 26. Holotype male, ZMUB; type locality: Philippines: Masbate, Aroroy.

DISTRIBUTION. Philippines: Masbate.

Scaphisoma duryi **Leng & Mutchler**

Scaphisoma carolinae Dury, 1911: 275. Syntypes, CMNH; type locality: USA: Balsam, North Carolina.
Scaphisoma duryi Leng & Mutchler, 1927: 25, replacement name for *Scaphisoma carolinae* Dury, 1911 (nec *Scaphisoma carolinae* Casey, 1893).

DISTRIBUTION. United States: North Carolina.

Scaphisoma dusunum **Löbl**

Scaphisoma dusunum Löbl, 1987a: 104, Figs 31.33. Holotype male, MHNG; type locality: East Malaysia: Sabah, Kinabalu Nat. Park.

DISTRIBUTION. East Malaysia: Sabah.

Scaphisoma dybasi Löbl

Scaphisoma dybasi Löbl, 1981b: 75, Figs 4, 5. Holotype male, FMNH; type locality: Micronesia: Mt. Tagpochau, 380 m, Saipan.

DISTRIBUTION. Micronesia: S. Mariana Islands.

Scaphisoma echinatum Löbl

Scaphisoma echinatum Löbl, 1986a: 198, Figs 84–87. Holotype male, MHNG; type locality: India: West Bengal, Algarah, 1800 m.

DISTRIBUTION. India: West Bengal (Darjeeling District).

Scaphisoma edentatum Löbl

Scaphisoma edentatum Löbl, 2012b: 179, Figs 11–14. Holotype male, NMPC; type locality: West Malayasia: Tanah Rata village, near Gunung Jasat, 1470–1705 m, Cameron Highlands, Pahang.

DISTRIBUTION. West Malaysia.

Scaphisoma egenum Löbl

Scaphisoma egenum Löbl, 1990b: 569, Figs 90, 91. Holotype male, MHNG; type locality: Thailand: Khao Yai Nat. Park, 750–850 m.

DISTRIBUTION. Thailand.

Scaphisoma egregium Löbl

Scaphisoma egregium Löbl, 1971c: 967, Figs 30, 31. Holotype male, MHNG; type locality: Sri Lanka: above Wellawaya, 300 m.

DISTRIBUTION. Sri Lanka.

Scaphisoma elgonense Pic

Scaphosoma elgonense Pic, 1946: 84. Syntypes, MNHN; type locality: Kenya: Mt. Elgon [Koptaweli Valley, 2300 m].

DISTRIBUTION. Kenya.

Scaphisoma elongatum Waterhouse

Scaphisoma elongatum Waterhouse, 1879: 533. Syntypes, NHML; type locality: Brazil: Rio de Janeiro.

DISTRIBUTION. Brazil.

Scaphisoma elpis Löbl & Ogawa

Scaphisoma elpis Löbl & Ogawa, 2016b: 1384, Figs 114, 115. Holotype male, MHNG; type locality: Philippines: Palawan, Sabang, 50–100 m.

DISTRIBUTION. Philippines: Palawan.

Scaphisoma emeicum Löbl

Scaphisoma emeicum Löbl, 2000: 613, Figs 3, 4. Holotype male, MHNG; type locality: China: Mt. Emei, 1700 m, Sichuan.

DISTRIBUTION. China: Sichuan.

Scaphisoma endroedyi Löbl

Scaphisoma endroedyi Löbl, 1986d: 465, Figs 1, 2. Holotype male, ANIC; type locality: Australia: Queensland, Fraser Is.

DISTRIBUTION. Australia: Queensland.

Scaphisoma erichsoni (Matthews)

Baeocera erichsoni Matthews, 1888: 169. Lectotype male, NHML; type locality: Mexico: Pinos Altos, Chihuahua.

Löbl, 1992b: 383, Figs 7, 8 (lectotype designation, transfer to *Scaphisoma*, characters)

DISTRIBUTION. Mexico.

Scaphisoma erythraeum Pic

Scaphosoma erythraeum Pic, 1933b: 119. Holotype, MCSN; type locality: Eritrea: Adi Ugri.

DISTRIBUTION. Eritrea.

Scaphisoma excellens Löbl

Scaphisoma excellens Löbl, 1981a: 378, Figs 41 42. Holotype male, NZAC; type locality: New Caledonia: Pic d'Amoa near Poindimié, 500 m.

DISTRIBUTION. New Caledonia.

Scaphisoma exiguum (Casey)

Scaphiomicrus exiguus Casey, 1900: 60. Holotype, USNM; type locality: United States: Oregon.

Hatch, 1957: 282 (characters)

DISTRIBUTION. United States: Oregon.

Scaphisoma eximium Löbl

Scaphisoma eximium Löbl, 1969a: 3, Figs 4–7. Holotype male, MHNG; type locality: New Caledonia: St. Louis.

Löbl, 1973b: 329 (records)

Löbl, 1977i: 818 (records)

Löbl, 1981a: 372 (characters, records)

DISTRIBUTION. New Caledonia.

Scaphisoma expandum Löbl

Scaphisoma expandum Löbl, 2011d: 183, Figs 1–3. Holotype male, NMEC; type locality: West Bhutan: Chiley-La, 3000–3500 m, Paro Prov.

DISTRIBUTION. Bhutan.

Scaphisoma falciferum Löbl

Scaphisoma falciferum Löbl, 1986c: 359, Figs 14, 15. Holotype male, MHNG; type locality: Pakistan: Nathia Gali, 2500 m, Hazara.

Löbl, 1990a: 121 (records)

Löbl, 1992a: 534 (records)

Löbl, 2000: 608 (records)

DISTRIBUTION. China: Yunnan; India: Himachal Pradesh, Uttarakhand (Kumaon); Nepal; Pakistan.

Scaphisoma fastum Löbl

Scaphisoma fastum Löbl, 1990b: 552, Figs 62–64. Holotype male, BPBM; type locality: Thailand: Chiang Mai, Chiang Dao, 450 m.

DISTRIBUTION. Thailand.

Scaphisoma fatuum Löbl

Scaphisoma fatuum Löbl, 1992a: 545, Fig. 133. Holotype male, MHNG; type locality: Nepal: Kali Gandaki Khola, 1500–1700 m, Myandi District.

DISTRIBUTION. Nepal.

Scaphisoma favens Löbl

Scaphisoma favens Löbl, 1990b: 577, Figs 108–110. Holotype male, MHNG; type locality: Thailand: Doi Suthep, 1180 m, Chiang Mai.

DISTRIBUTION. Thailand.

Scaphisoma feai (Löbl)

Macroscaphosoma feai Löbl, 1970b: 130, Figs 7, 9. Holotype male, MNHN; type locality: Equatorial Guinea: Bioko [Fernando Po], Punta Fraites.

Leschen & Löbl, 2005 (transfer from *Macroscaphosoma*)

DISTRIBUTION. Equatorial Guinea: Bioko.

Scaphisoma fenestratum Löbl

Scaphisoma fenestratum Löbl, 2003c: 184, Figs 3–4. Holotype male, HNHM; type locality: Papua New Guinea: Mt. Wilhelm Field Station.

DISTRIBUTION. Papua New Guinea.

***Scaphisoma fernshawense* Blackburn**

Scaphisoma fernshawense Blackburn, 1903: 99. Lectotype female, NHML; type locality: Australia: Victoria, Fernshaw.

Löbl, 1977b: 35, Figs 18, 50, 51 (lectotype fixed by inference, characters, records)

DISTRIBUTION. Australia: Victoria.

***Scaphisoma festivum* Löbl**

Scaphisoma festivum Löbl, 1975a: 396, Figs 38, 39. Holotype male, MHNG; type locality: Papua New Guinea: near Friedrich-Wilhems-Hafen [Madang], Shering Peninsula.

DISTRIBUTION. Papua New Guinea.

***Scaphisoma fibrosum* Löbl**

Scaphisoma fibrosum Löbl, 2000: 640, Figs 56, 57. Holotype male, MHNG; type locality: China: Yunnan, Xishuangbanna.

DISTRIBUTION. China: Yunnan.

***Scaphisoma fijianum* Löbl**

Scaphisoma fijianum Löbl, 1977i: 824, Figs 11, 12. Holotype male, BPBM; type locality: Fiji: Nandarivatu, 3700 ft, Viti Levu.

Löbl, 1980c: 394, Figs 31, 32 (characters, records)

DISTRIBUTION. Fiji.

***Scaphisoma filium* Löbl**

Scaphisoma filium Löbl, 1973b: 321, Figs 17, 18. Holotype male, NHMW; type locality: New Caledonia: Table d'Union, 600–800 m.

Löbl, 1981a: 375 (characters, records)

DISTRIBUTION. New Caledonia.

***Scaphisoma flagellulum* Prokofiev**

Scaphisoma flagellulum Prokofiev, 2003: 1, Fig. 2. Holotype male, SIEE; type locality: Vietnam: Binh Phuoc Prov., Phuoc Long District, Bu Gia Map National Park.

DISTRIBUTION. Vietnam.

***Scaphisoma flavapex* Achard**

Scaphosoma flavapex Achard, 1921b: 87. Syntypes, NBCL/NMPC; type localities: Indonesia: Java, Smeroe, Ranoe Koembala, Groote Meer [Semeru, Ranoe Koembolo, Great Lake].

DISTRIBUTION. Indonesia: Java.

***Scaphisoma flavescens* (Casey)**

Scaphiomicrus flavescens Casey, 1900: 59. Syntypes, USNM; type locality: USA: Michigan.

DISTRIBUTION. United States: Michigan.

***Scaphisoma flavofasciatum* Löbl**

Scaphisoma flavofasciatum Löbl, 1986a: 162, Fig. 28. Holotype female, HNHM; type locality: India: West Bengal, Ghoom, 2200 m.

DISTRIBUTION. India: West Bengal (Darjeeling District).

***Scaphisoma flavonotatum* Pic**

Scaphosoma flavonotatum Pic, 1905a: 128. Syntypes, not traced, ?MNHN; type locality: Algeria: Djidjelli.

Scaphosoma flavonotatum var. *nigricolor* Pic, 1920a: 5. Lectotype female, MNHN; type locality: Algeria: Forêt d'Aschrit.

Scaphosoma reitteri var. *vaulogeri* Pic, 1920d: 13. Lectotype male, MNHN; type locality: Algeria, Forêt d'Aschrit.

Normand, 1934: 45 (records)
Löbl, 1965d: 337, Fig. 6 (characters)
Löbl, 1969a: 486 (records)
Tamanini, 1969a: 484 (characters)
Löbl, 1970c: 788, Figs 72, 73 (lectotype designations for *S. flavonotatum* var. *nigricolor* and *S. reitteri* var. *vaulogeri*, synonymy with *S. flavonotatum*, characters, records)
Tamanini, 1970: 19, Figs 4F, 10E (characters, records)
Löbl, 1974a: 61 (records)
Löbl, 1989a: 11 (records)
Dauphin, 2004: 259 (records)

DISTRIBUTION. Algeria; France: Corse; Gibraltar; Italy: Sardinia; Tunisia.

***Scaphisoma flexuosum* Löbl**

Scaphisoma flexuosum Löbl, 1986a: 167, Fig. 34. Holotype male, MHNG; type locality: India: Manas Wild Life Sanctuary, 200 m, Assam.

Löbl, 1986c: 348 (records)

DISTRIBUTION. India: Assam, Uttarakhand (Kumaon).

***Scaphisoma forcipatum* Champion**

Scaphosoma forcipatum Champion, 1927: 276, Fig. 1. Holotype male, NHML; type locality: India: Uttarakhand, Nainital, 7000–8600 ft.

Löbl, 1986a: 198, Figs 88–91 (characters, records)

Löbl, 1986c: 348 (records)
Löbl, 1992a: 537 (records)
Löbl, 2000: 608 (records)

DISTRIBUTION. China: Yunnan; India: Himachal Pradesh, Uttarakhand, West Bengal (Darjeeling District); Nepal; Pakistan.

***Scaphisoma foveatum* Löbl**

Scaphisoma foveatum Löbl, 1987a: 91, Figs 8, 9. Holotype male, NHML; type locality: East Malaysia: Sarawak, Gunung Mulu Nat. Park, lower montane forest.

DISTRIBUTION. East Malaysia: Sarawak.

***Scaphisoma franzi* Löbl**

Scaphisoma franzi Löbl, 1973b: 323, Figs 21, 22. Holotype male, NHMW; type locality: New Caledonia: Tindou near Hienghène.
Löbl, 1981a: 372 (characters, records)

DISTRIBUTION. New Caledonia.

***Scaphisoma fratellum* Löbl**

Scaphisoma fratellum Löbl, 1992a: 548, Fig. 136. Holotype male, SMNS; type locality: Nepal: above Yamputhin, left bank of Kabeli khola, 1800–2000 m, Taplejung District.

DISTRIBUTION. Nepal.

***Scaphisoma frater* Löbl**

Scaphisoma frater Löbl, 1977f: 48, Figs 15, 16. Holotype male, MHNG; type locality: La Réunion: N.D. de la Paix, Plaine des Cafres.

DISTRIBUTION. La Réunion.

***Scaphisoma fraterculum* Löbl**

Scaphosoma fraterculum Löbl, 1986a: 147, Figs 6, 7. Holotype male, MHNG; type locality: India: Meghalaya, Mawphlang, 1800 m, Khasi Hills.
Löbl, 1992a: 532 (records)
Löbl, 2005: 180 (records)

DISTRIBUTION. India: Meghalaya, West Bengal (Darjeeling District); Nepal.

***Scaphisoma frontale* Löbl**

Scaphisoma frontale Löbl, 2003c: 186, Figs 7, 8. Holotype male, HNHM; type locality: Papua New Guinea: Mt. Wilhelm, Imbuka Ridge, 3600 m.

DISTRIBUTION. Papua New Guinea.

Scaphisoma fulcratum Löbl

Scaphisoma fulcratum Löbl, 1992a: 543, Fig. 129. Holotype male, MHNG; type locality: Nepal: 2 km south Godawari, 1700 m, Patan District.

DISTRIBUTION. India: Himachal Pradesh; Nepal.

Scaphisoma funebre Löbl

Scaphisoma funebre Löbl, 1981a: 368, Figs 28, 29. Holotype male, NZAC; type locality: New Caledonia: Mt. Koghis, 550 m.

DISTRIBUTION. New Caledonia.

Scaphisoma funereum Löbl

Scaphisoma funereum Löbl, 1977e: 31, Figs 44, 45. Holotype male, SAMA; type locality: Australia: National Park, New South Wales.

Löbl & Leschen, 2003b: 40, Figs 19, 19, 71, 72, Map 22 (characters, records, fungal hosts)

DISTRIBUTION. Australia: New South Wales, Queensland; New Zealand.

Scaphisoma funiculatum Löbl

Scaphisoma funiculatum Löbl, 1980a: 101, Figs 13, 14. Holotype male, MHNG; type locality: Taiwan: Fenchihu, 1400 m.

DISTRIBUTION. Taiwan.

Scaphisoma furcatum Löbl & Ogawa

Scaphisoma furcatum Löbl & Ogawa, 2016b: 1422, Figs 201–203. Holotype male, EUMJ; type locality: Philippines: Mindanao, Eagle Centre, 1100 m, Baracatan, north slope of Mt. Apo.

DISTRIBUTION. Philippines: Mindanao.

Scaphisoma furcigerum Löbl & Ogawa

Scaphisoma furcigerum Löbl & Ogawa, 2016b: 1423, Figs 13, 204–206. Holotype male, FMNH; type locality: Philippines: Mindanao, Cotabato Prov., Burungkot, Upi, 1500 ft.

DISTRIBUTION. Philippines: Mindanao.

Scaphisoma furcillatum Löbl & Ogawa

Scaphisoma furcillatum Löbl & Ogawa, 2016b: 1424, Figs 14, 207, 208. Holotype male, SMNS; type locality: Philippines: Mindanao, Misamis occ., Don Victoriano, 1700 m.

DISTRIBUTION. Philippines: Mindanao.

Scaphisoma fuscum Löbl

Scaphisoma fuscum Löbl, 1975b: 283, Figs 18, 19. Holotype male, MHNG; type locality: Indonesia: Sumatra, Fort de Kock [Bukittinggi], 920 m.

DISTRIBUTION. Indonesia: Sumatra.

Scaphisoma gallienii Pic

Scaphosoma gallienii Pic, 1920b: 3. Syntypes, MNHN; type locality: Madagascar.

DISTRIBUTION. Madagascar.

Scaphisoma galloisi Achard

Scaphosoma galloisi Achard, 1923: 114. Lectotype male, MNHN; type locality: Japan: Chuzenji.

Scaphosoma harmandi Achard, 1923: 114. Lectotype male, MNHN; type locality: Japan: Tokyo.

Miwa & Mitono, 1943: 543 (characters)

Nakane, 1955c: 56 (characters)

Chûjô, 1961: 5 (records)

Nakane, 1963b: 79 (characters)

Löbl, 1970c: 766, Figs 44, 45 (lectotype fixed by inference for *S. galloisi* and *S. harmandi*, synonymy of *S. harmandi* with *S. galloisi*, characters, records)

Hisamatsu, 1977: 194 (records)

DISTRIBUTION. Japan: Honshu, Tsushima, Kyushu.

Scaphisoma garomontium Löbl

Scaphisoma garomontium Löbl, 1986a: 168, Figs 35, 36. Holotype male, MHNG; type locality: India: Meghalaya, Rongrengiri, 400 m, Garo Hills.

DISTRIBUTION. India: Meghalaya.

Scaphisoma geminatum Löbl

Scaphisoma geminatum Löbl, 1986a: 212, Figs 112–114. Holotype male, MHNG; type locality: India: Meghalaya, below Cherrapunjee, 1200 m, Khasi Hills.

Löbl, 2000: 608 (records)

Löbl, 2003a: 62 (records)

DISTRIBUTION. China: Fujian, Guangdong, Jiangxi; India: Meghalaya.

Scaphisoma gentile Löbl

Scaphisoma gentile Löbl, 1982a: 329, Figs 3, 4. Holotype male, MHNG; type locality: Japan: Omogo, 700 m, Ehime Pref.

DISTRIBUTION. Japan: Honshu, Shikoku.

Scaphisoma germanni Löbl

Scaphisoma germanni Löbl, 2012c: 87, Figs 1–5. Holotype male, MHNG; type locality: India: South Andaman Is., Chiriyatapu.

DISTRIBUTION. India: Andaman Islands.

Scaphisoma gestroi Reitter

Scaphisoma gestroi Reitter, 1881b: 140. Lectotype male, MNHN; type locality: Australia: Queensland, Somerset.

Löbl, 1977e: 49, Figs 72, 73 (lectotype designation, characters)

DISTRIBUTION. Australia: Queensland.

Scaphisoma glabrellum Löbl & Ogawa

Scaphisoma glabrellum Löbl & Ogawa, 2016b: 1425, Figs 209–211. Holotype male, SMNS; type locality: Philippines: Mindanao, 30 km NW of Maramag, Bagongsilang, 1700 m.

DISTRIBUTION. Philippines: Mindanao.

Scaphisoma glabripenne Löbl

Scaphisoma glabripenne Löbl, 1977e: 34, Figs 48, 49. Holotype male, SAMA; type locality: Australia: Lord Howe Island.

DISTRIBUTION. Australia: Lord Howe Island.

Scaphisoma gomyi Löbl

Scaphisoma gomyi Löbl, 1977f: 46, Figs 13, 14. Holotype male, MHNG; type locality: La Réunion: Plaine des Chicots, below the Gîte, 1800 m.

Lecoq, 2015: 209 (records)

DISTRIBUTION. La Réunion.

Scaphisoma goudoti Achard

Scaphosoma goudoti Achard, 1920i: 240. Syntypes, NMPC; type locality: Madagascar.

DISTRIBUTION. Madagascar.

Scaphisoma gracilendum Löbl

Scaphisoma gracilendum Löbl, 1990b: 545, Figs 51, 52. Holotype male, MHNG; type locality: Thailand: Khao Yai Nat. Park, near Headquarters, 750–850 m.

Löbl, 1997: 104 (misspelled *gracilentum*)

DISTRIBUTION. Thailand.

Scaphisoma gracilicorne Achard

Scaphosoma gracilicorne Achard, 1920d: 129. Lectotype male, MNHN; type locality: Indonesia: Sumatra.
Löbl, 2015a: 109, Figs 99–102 (lectotype designation, characters)
DISTRIBUTION. Indonesia: Sumatra.

Scaphisoma grande Pic

Scaphosoma grande Pic, 1920f: 24. Lectotype female, MCSN; type locality: Myanmar: Tenasserim [Mte fra Meekalam e Kyeat].
Pic, 1921a: 165 (characters)
Löbl, 1973a: 153 (lectotype fixed by inference, characters)
DISTRIBUTION. Myanmar.

Scaphisoma grouvellei Achard

Scaphosoma grouvellei Achard, 1920d: 130. Syntypes, MNHN; type locality: Indonesia: Java, Mts Tengger.
Note. Achard (l.c.) gave as altitude "4000 m", instead of 4000 feets.
DISTRIBUTION. Indonesia: Java.

Scaphisoma guatemalense Matthews

Scaphisoma guatemalense Matthews, 1888: 175. Syntypes, NHML; type locality: Guatemala: Zapote, Capetillo.
DISTRIBUTION. Guatemala.

Scaphisoma hadrops Löbl

Scaphosoma hadrops Löbl, 1965a: 57, Fig.10. Holotype male, MNHN; type locality: Japan:
Nemova.
Löbl, 1970c: 774, Figs 54, 55 (characters)
DISTRIBUTION. Japan: Hokkaido.

Scaphisoma haemorrhoidale Reitter

Scaphisoma haemorrhoidale Reitter, 1877: 369. Lectotype male, HNHM; type locality: Japan.
Scaphosoma sinense Pic, 1920b: 5 [credited to Portevin]. Lectotype male, MNHN; type locality: China: Yunnan, Nankin.
Scaphosoma lewisi Achard, 1923: 112. Lectotype male, NMPC; type locality: Japan: Kobe.
Scaphosoma haemorrhoidale var. *plagipenne* Achard, 1923: 112. Lectotype female, NMPC; type locality: Japan: Kyoto.

Scaphosoma lautum Löbl, 1965e: 30, Figs 3, 3a, b. Holotype male, MNHN; type locality: China: Fujian, Kuatun.

Miwa & Mitono, 1943: 540 (characters)
Nakane, 1955c: 55 (characters)
Chûjô, 1961: 5 (records)
Nakane, 1963b: 79 (characters)
Löbl, 1965h: 3 (*S. lewisi* as aberration of *S. haemorrhoidale*)
Löbl, 1965e: 25, Figs 1, 1a, b (characters and lectotype designation for *S. sinense*)
Löbl, 1966b: 134 (characters, records)
Löbl, 1967b: 110 (records)
Löbl, 1970c: 791, Figs 76, 77 (lectotype fixed by inference for *S. haemorrhoidale* var. *plagipenne* and *S. lewisi*, characters, records)
Löbl, 1977g: 165 (records)
Löbl, 1982b: 105 (records)
Iablokoff-Khnzorian, 1985: 141 (records)
Li, 1992: 62 (records)
Löbl, 2000: 608 (synonymy of *S. sinense* and *S. lautum* with *S. haemorrhoidale*, characters, records)

DISTRIBUTION. China: Beijing, Fujian, Hubei, Jiangsu, Liaoning, Yunnan; Japan: Honshu, Kyushu, Ryukyus, Shikoku, Tsushima; North Korea; Russia: Far East.

Scaphisoma hajeki Löbl

Scaphisoma hajeki Löbl, 2012b: 176, Figs 5–7. Holotype male, NMPC; type locality: West Malaysia: Batu, 590 m, Cameron Highlands, Perak.

DISTRIBUTION. West Malaysia.

Scaphisoma hamatum Löbl & Ogawa

Scaphisoma hamatum Löbl & Ogawa, 2016b: 1406, Figs 157–159. Holotype male, FMNH; type locality: Philippines: Mindanao, Cotabato Prov., 50 km. N. of Parang, 500 ft.

DISTRIBUTION. Philippines: Mindanao.

Scaphisoma hanseni Löbl & Leschen

Scaphisoma hanseni Löbl & Leschen, 2003b: 41, Figs 17, 69, 70, Map 23. Holotype male, NZAC; type locality: New Zealand: WD, Okuku Reserve.

DISTRIBUTION. New Zealand.

Scaphisoma hapiroense **Löbl**

Scaphisoma hapiroense Löbl, 1968b: 420, Fig. 1. Holotype female, ZMPA; type locality: North Korea, Mjohjang-san Mts, Hapiro Valley, Hjangsan.
Löbl, 1970c: 793 (characters)

DISTRIBUTION. North Korea.

Scaphisoma hastatum **Löbl**

Scaphisoma hastatum Löbl, 1977e: 49, Figs 74, 75. Holotype male, QMBA; type locality: Australia: New South Wales, Barrington House, via Salisbury.

DISTRIBUTION. Australia: New South Wales.

Scaphisoma hawkeswoodi Prokofiev

Scaphisoma hawkeswoodi Prokofiev, 2003: 2, Figs 4–9. Holotype male, SIEE; type locality: Vietnam: Binh Phuoc Prov., Phuoc Long District, Bu Gia Map National Park.

DISTRIBUTION. Vietnam.

Scaphisoma heishuiense **Löbl**

Scaphisoma heishuiense Löbl, 2000: 638, Figs 47–50. Holotype male, MHNG; type locality: China: Yunnan, Heishui, 35 km north Lijiang.

DISTRIBUTION. China: Yunnan.

Scaphisoma heissi **Löbl**

Scaphisoma heissi Löbl, 1982f: 12, Figs 10, 11. Holotype male, MHNG; type locality: Indonesia: Sumatra, Brastagi.

DISTRIBUTION. Indonesia: Sumatra.

Scaphisoma helferi **Pic**

Scaphosoma helferi Pic, 1930b: 58. Syntypes, NMPC; type locality: Myanmar: Tenasserim.

DISTRIBUTION. Myanmar.

Scaphisoma hexameroides **Löbl & Ogawa**

Scaphisoma hexameroides Löbl & Ogawa, 2016b: 1402, Figs 148, 149. Holotype male, EUMJ; type locality: Philippines: Palawan, Olanguan, 0–50 m.

DISTRIBUTION. Philippines: Palawan.

Scaphisoma hexamerum **Löbl & Ogawa**

Scaphisoma hexamerum Löbl & Ogawa, 2016b: 1369, Figs 85, 86. Holotype male, MHNG; type locality: Philippines: Luzon, Lagunas Prov., Mt. Banahaw, above Kinabuhayan, 600–700 m.

DISTRIBUTION. Philippines: Luzon.

Scaphisoma hiekei Löbl

Scaphisoma hiekei Löbl, 1972b: 93, Figs 32, 33. Holotype male, ZMUB; type locality: Philippines: Palawan, Binaluan.
Löbl & Ogawa, 2016b: 1408 (characters, records)
DISTRIBUTION. Philippines: Palawan.

Scaphisoma hisamatsui Löbl

Scaphisoma hisamatsui Löbl, 1982a: 328, Figs 1, 2. Holotype male, MHNG; type locality: Japan: Shiroyama, 200 m, Matsuyama, Ehime Pref.
DISTRIBUTION. Japan: Shikoku.

Scaphisoma hospitator Löbl

Scaphisoma hospitator Löbl, 2004a: 347, Figs 1–3. Holotype male, MHNG; type locality: India: Mahadeo Hills, 5.5 km SW Pachmarhi, Tridhara, 900 m, Mahdya Pradesh.
DISTRIBUTION. India: Mahdya Pradesh.

Scaphisoma humerosum Reitter

Scaphisoma humerosum Reitter, 1880a: 48. Syntypes, MNHN; type locality: Venezuela, Caracas.
DISTRIBUTION. Venezuela.

Scaphisoma hybridum Boheman

Scaphisoma hybridum Boheman, 1851: 558. Syntypes, NHRS; type locality: South Africa: "in tractibus fluvii Gariepis".
DISTRIBUTION. South Africa.

Scaphisoma idaanum Löbl

Scaphisoma idaanum Löbl, 1987a: 102, Figs 28–30. Holotype male, NHML; type locality: East Malaysia: Sarawak, Gunung Mulu Nat. Park, near Base Camp, 50–100 m.
DISTRIBUTION. East Malaysia: Sarawak.

Scaphisoma ilonggo Löbl & Ogawa

Scaphisoma ilonggo Löbl & Ogawa, 2016b: 1358, Figs 59–62 Holotype male, FMNH; type locality: Philippines: Mindanao, Davao Prov., Meran, east slope of Mt. Apo, 6000 ft.
DISTRIBUTION. Philippines: Mindanao.

Scaphisoma imitator Löbl

Scaphisoma imitator Löbl, 1986a: 149, Fig. 10. Holotype male, MHNG; type locality: India: Sevoke, 200 m, Darjeeling District.

DISTRIBUTION. India: Meghalaya, West Bengal (Darjeeling District).

Scaphisoma immodicum Löbl

Scaphisoma immodicum Löbl, 1986c: 363, Figs 19–21. Holotype male, NHMW; type locality: India: Baragaran near Katrain, Himachal Pradesh.

Löbl, 1992a: 538 (records)

DISTRIBUTION. India: Himachal Pradesh; Nepal.

Scaphisoma immundum Reitter

Scaphisoma immundum Reitter, 1880a: 47. Syntypes, MNHN; type locality: Colombia: Carthagena.

DISTRIBUTION. Colombia.

Scaphisoma impar Löbl

Scaphisoma impar Löbl, 1971c: 969, Figs 34, 35. Holotype male, MHNG; type locality: Sri Lanka: Inginiyagala.

DISTRIBUTION. Sri Lanka.

Scaphisoma impolitum Löbl

Scaphisoma impolitum Löbl, 1986a: 183, Figs 60–62. Holotype male, MHNG; type locality: India: Meghalaya, Rogrengiri, 400 m, Garo Hills.

DISTRIBUTION. India: Meghalaya.

Scaphisoma impressipenne Pic

Scaphosoma impressipenne Pic, 1947a: 10. Syntypes, MNHN; type locality: Democratic Republic of the Congo: Ouellé.

DISTRIBUTION. Democratic Republic of the Congo.

Scaphisoma impunctatum Reitter

Scaphisoma impunctatum Reitter, 1880a: 46. Lectotype female, MNHN; type locality: USA: Missouri.

Scaphisoma obesula Casey, 1893: 531. Lectotype male, USNM; type locality: USA: Enterprise Lav, Florida.

Leschen, 1988a: 227, Figs 1–6, 7, -9 (records, immature stages, natural history, fungal hosts)

Leschen et al., 1990: 285, Figs 1, 8, 23–25 (lectotype designations for *S. impunctatum* and *S. obesula*, synonymy, characters, records, fungal hosts)

Downie & Arnett, 1996: 366 (records)

Sikes, 2004: 105 (distribution)

DISTRIBUTION. United States: Arkansas, Florida, Iowa, Massachussetts, Mississippi, Missouri, Oklahoma, Rhode Island.

Scaphisoma imuganense Löbl

Scaphisoma imuganense Löbl, 1972b: 91, Figs 27, 28. Holotype male, ZMUB; type locality: Philippines: Luzon, Imugan.
Löbl & Ogawa, 2016b: 1391 (characters, records)

DISTRIBUTION. Philippines: Luzon.

Scaphisoma inaequale Löbl

Scaphisoma inaequale Löbl, 1977e: 38, Figs 54, 55. Holotype male, QMBA; type locality: Australia: Queensland, Carrai Plateau, via Kemsey.

DISTRIBUTION. Australia: Queensland.

Scaphisoma incertum (Pic)

Toxidium incertum Pic, 1954b: 39. Syntypes, MRAC type locality: between Stanleyville [Kisangani] and Kilo.
Löbl & Leschen, 2010: 90, Figs 21, 22 (transfer to *Scaphisoma*, characters)

DISTRIBUTION. Democratic Republic of the Congo.

Scaphisoma incisum Löbl

Scaphisoma incisum Löbl, 2000: 642, Figs 58–61. Holotype male, MHNG; type locality: China: Yunnan, Lijiang, 1800 m.

DISTRIBUTION. China: Yunnan.

Scaphisoma incomptum Löbl

Scaphisoma incomptum Löbl, 1976a: 8, Figs 1, 2. Holotype male, NBCL; type locality: Indonesia: Oosthaven [Pandjang, South Sumatra].

DISTRIBUTION. Indonesia: Sumatra.

Scaphisoma inconspicuum Casey

Scaphisoma inconspicua Casey, 1893: 530. Holotype, USNM; type locality: USA: Florida.

DISTRIBUTION. United States: Florida.

Scaphisoma inconventum Löbl & Ogawa

Scaphisoma inconventum Löbl & Ogawa, 2016b: 1391, Figs 125, 126. Holotype male, MHNG; type locality: Philippines: Luzon, Mount Data, near Lodge.

DISTRIBUTION. Philippines: Luzon.

Scaphisoma incurvum **Löbl**

Scaphisoma incurvum Löbl, 1990b: 576, Figs 104–107. Holotype male, MHNG; type locality: Thailand: Kaeng Krachan Nat. Park, 300–400 m, 25–30 km from Headquarters.

DISTRIBUTION. Thailand.

Scaphisoma indistinctum **Pic**

Scaphisoma indistinctum Pic, 1928b: 43. Holotype, MRAC; type locality: Democratic Republic of the Congo, Tengo Katanta, Manyema.

DISTRIBUTION. Democratic Republic of the Congo.

Scaphisoma indra **Löbl**

Scaphisoma indra Löbl, 1986a: 185, Figs 63–67. Holotype male, MHNG; type locality: India: between Teesta and Rangpo, 11 km from Teesta, 350 m, Darjeeling District.

Löbl, 1992a: 535 (records)

DISTRIBUTION. India: West Bengal (Darjeeling District); Nepal.

Scaphisoma indubium **Löbl**

Scaphosoma indubium Löbl, 1965a: 50, Figs 5, 5a. Holotype male, MNHN; type locality: Japan: Kumanotaira near Karuizawa.

Löbl, 1967c: 132 (records)

Löbl, 1970c: 778, Figs 60, 61 (characters, records)

Hisamatsu, 1977: 194 (records)

Iablokoff-Khnzorian, 1985: 141 (records)

Li, 2015: 38, Fig. 3 (records)

DISTRIBUTION. China: Heilongjiang; Japan: Honshu; Russia: Far East.

Scaphisoma indutum **Löbl**

Scaphisoma indutum Löbl, 1977e: 43, Figs 20, 64, 65. Holotype male, SAMA; type locality: Australia: Cradle Mts, Tasmania.

DISTRIBUTION. Australia: Victoria, Tasmania.

Scaphisoma ineptum **Löbl**

Scaphisoma ineptum Löbl, 1987a: 89, Figs 6, 7. Holotype male, EUMJ; type locality: East Malaysia: Kinabalu Nat. Park, Headquarters.

DISTRIBUTION. East Malaysia: Sabah.

Scaphisoma inexspectatum Löbl & Ogawa

Scaphisoma inexspectatum Löbl & Ogawa, 2016b: 1408, Figs 15, 162, 163. Holotype male, FMNH; type locality: Philippines: Mindanao, Cotabato Prov., Burungkot, Upi, 1500 ft.

DISTRIBUTION. Philippines: Mindanao.

Scaphisoma infirmum Löbl

Scaphisoma infirmum Löbl, 2003c: 185, Figs 5, 6. Holotype male, HNHM; type locality: Papua New Guinea: Mt. Wilhelm Field Station.

DISTRIBUTION. Papua New Guinea.

Scaphisoma inflexum Löbl & Ogawa

Scaphisoma inflexum Löbl & Ogawa, 2016b: 1409, Figs 146–149. Holotype male, SMNS; type locality: Philippines: Leyte, Visca, north Baybay, 200–500 m.

DISTRIBUTION. Philippines: Leyte.

Scaphisoma inhospitale Löbl

Scaphisoma inhospitale Löbl, 1990a: 118, Figs 1–3. Holotype male, MHNG; type locality: India: Himachal Pradesh, Khadjiar east of Dalhousie, 1950 m.

DISTRIBUTION. India: Himachal Pradesh.

Scaphisoma innotatum Pic

Scaphosoma innotatum Pic, 1926d: 46. Lectotype male, MNHN; type locality: Vietnam: Tonkin, Hoa Binh.
Löbl, 1981f: 110, Fig. 8 (lectotype designation, characters, records)
Löbl, 1986a: 201, Figs 92–94 (characters, records)
Löbl, 1986c: 348 (records)
Löbl, 1990b: 582 (records)
Löbl, 1992a: 537 (records)
Löbl, 2000: 609 (records)

DISTRIBUTION. China: Yunnan; India: Meghalaya, Uttarakhand (Garhwal); Nepal; Thailand; Vietnam.

Scaphisoma inopinatum Löbl

Scaphosoma inopinatum Löbl, 1967b: 105, Figs 1, 5–7. Holotype male, SNMC; type locality: Russia: Chita, Transbaikal.
Löbl, 1967d: 34, Figs 3a-c (characters, records)
Kofler, 1968: 42, Figs 1C, 2C (characters, records)
Tamanini, 1969a: 488 (records)

Tamanini, 1969b: 376, Figs 8F, G (characters, records, fungal hosts)
Kofler, 1970: 57 (records)
Löbl, 1970a: 13, Figs 9, 10 (characters, records)
Löbl, 1970c: 740, Figs 5, 6 (characters, records)
Tamanini, 1970: 22, Figs 9F, G, 10L (characters, records, fungal hosts)
Freude, 1971: 347, Fig. 3: 2 (characters)
Palm, 1971: 66 (records)
Israelson, 1971: 66 (records)
Strand, 1975: 10 (records)
Holzschuh, 1977: 30 (records)
Burakowski et al., 1978: 235 (records)
Iablokoff-Khnzorian, 1985: 141 (records)
Kahlen, 1987: 120 (distribution)
Löbl, 1994: 50 (records)
Krasutskij, 1996a: 97 (records, fungal hosts)
Nikitsky et al., 1996b: 24 (fungal hosts)
Krasutskij, 1997a: 307 (fungal hosts)
Krasutskij, 1997b: 774 (fungal hosts, habitat)
Kofler, 1998: 648 (fungal hosts)
Krasutskij, 2000: 80 (fungal hosts)
Löbl & Růžička, 2000: 289 (records)
Silfverberg, 2004: 38 (records)
Nikitsky & Schigel, 2004: 9 (fungal hosts)
Krasutskij, 2005: 37–42 (fungal hosts)
Borowski et al., 2005: 44 (records)
Borowski, 2006: 54, 58 (records, fungal hosts)
Byk et al., 2006: 337 (habitat, records)
Štourač & Rébl, 2009: 121 (records, fungal hosts)
Krasutskij, 2010: 374 (fungal hosts)
Schigel, 2011: 332, 341 (fungal hosts)
Wojas, 2016: 141 (records, fungal hosts)
Ostrovsky, 2016: 381 (records)
Süda, 2016: 66 (records)
Hamed & Vancl, 2016: 79 (records)
Nikitsky et al., 2016: 138 (records)
Vlasov & Nikitsky, 2017: 6 (records, fungal hosts)

Distribution. Austria; Belarus; Bosnia and Herzegovina; Czech Republic; Estonia; Finland; Germany; Hungary; Italy; Latvia; Lithuania; Mongolia; Netherlands; Norway; Poland; Russia: European including Caucasus, Siberia, Transbaikal, Far East; Slovenia; Sweden; Switzerland; Ukraine; Uzbekistan.

Scaphisoma inopportunum Löbl & Ogawa

Scaphisoma inopportunum Löbl & Ogawa, 2016b: 1392, Figs 16, 127, 128. Holotype male, MHNG; type locality: Philippines: Luzon, Lagunas Prov., Mt. Banahaw, ca 1 km from Kinabuhayan, 500 m.

DISTRIBUTION. Philippines: Luzon.

Scaphisoma inornatum Löbl

Scaphisoma inornatum Löbl, 1975a: 411, Figs 67, 68. Holotype male, MCSN; type locality: Indonesia: Western New Guinea, Hatam.

DISTRIBUTION. Indonesia: Western New Guinea.

Scaphisoma inquietum Löbl

Scaphisoma inquietum Löbl, 1992a: 547, Fig. 135. Holotype male, MHNG; type locality: Nepal: Ghoropani Pass, north slope, 2700 m, Parbat District.

DISTRIBUTION. Nepal.

Scaphisoma instabile Lea

Scaphisoma instabile Lea, 1926: 280. Holotype male, NHML; type locality: Australia: Huon R., Tasmania.

Löbl, 1977e: 21, Figs 26, 27 (characters, records)

DISTRIBUTION. Australia: New South Wales, Tasmania, Victoria, Western Australia.

Scaphisoma insulanum Löbl

Scaphisoma insulanum Löbl, 1982b: 102, Figs 2, 3. Holotype male, PCMS; type locality: Japan: Hatsuno, Amami-ôshima.

DISTRIBUTION. Japan: Ryukyus, Amami-ôshima.

Scaphisoma insulare Vinson

Scaphosoma insulare Vinson, 1943: 191, Figs 2, 12. Syntypes, NHML/MIMM; type locality: Mauritius: Macabé.

DISTRIBUTION. Mauritius.

Scaphisoma interjectum Löbl

Scaphisoma interjectum Löbl, 1992a: 544, Fig. 130. Holotype male, MHNG; type locality: Nepal: Chichile, above Ahale, 2300 m, Sankhuwasabha District.

DISTRIBUTION. Nepal.

Scaphisoma invalidum Löbl

Scaphisoma invalidum Löbl, 1992a: 552, Figs 142, 143. Holotype male, MHNG; type locality: Nepal: Chichila, 2200, Sankhuwasabha District.

DISTRIBUTION. Nepal.

Scaphisoma invertum Löbl

Scaphisoma invertum Löbl, 2000: 612, Figs 1, 2. Holotype male, MHNG; type locality: China: Guangxi, 10 km south Longsheng, ca 1000 m.

DISTRIBUTION. China: Guangxi, Yunnan.

Scaphisoma invisum Löbl

Scaphisoma invisum Löbl, 1990b: 545, Figs 49–50. Holotype male, MHNG; type locality: Thailand: Khao Yai Nat. Park, near Phliu Waterfalls, 150–300 m.

DISTRIBUTION. Thailand.

Scaphisoma iridescens Löbl & Ogawa

Scaphisoma iridescens Löbl & Ogawa, 2016b: 1402, Figs 150, 152. Holotype male, FMNH; type locality: Philippines: Mindanao, Mainit Riv., Mt Apo, 7000 ft.

DISTRIBUTION. Philippines: Mindanao.

Scaphisoma irideum Löbl

Scaphisoma irideum Löbl, 2012f: 312, Figs 11–13. Holotype male, NMEC; type locality: Indonesia: Maluku Islands, Halmahera Is., Central Weda Selatan, District Loleo, S env. Tilope vill.
Löbl, 2015b: 186 (records)

DISTRIBUTION. Indonesia: Maluku Islands: Halmahera Is.

Scaphisoma iriomotense Löbl

Scaphisoma iriomotense Löbl, 1977g: 163, Figs 1, 2. Holotype male, NSMT; type locality: Japan: foot of Mt. Goza, Iriomote-jima.

DISTRIBUTION. Japan: Ryukyus: Iriomote-jima.

Scaphisoma irregulare Löbl

Scaphisoma irregulare Löbl, 1975b: 276, Figs 7–9. Holotype male, MCST; type locality: Indonesia: Sumatra, Ajer Mantcior [near Padangpanjang].

DISTRIBUTION. Indonesia: Sumatra.

Scaphisoma irruptum Löbl

Scaphisoma irruptum Löbl, 2000: 648, Figs 74–76, 81. Holotype male, MHNG; type locality: China: Yunnan, Xishuangbanna.

DISTRIBUTION. China: Yunnan.

Scaphisoma italicum **Tamanini**

Scaphosoma italicum Tamanini, 1955: 15, Figs 27–34. Holotype male, MZUN; type locality: Italy: Val Santicelli, Massiccio del Pollino.

Löbl, 1965g: 734 (records)
Löbl, 1967d: 35, Figs 5a, b (characters)
Tamanini, 1969a: 487, Figs 7, 11 (characters, records)
Tamanini, 1969b: 370, Figs 5H, 6C, D, (characters, fungal hosts)
Tamanini, 1970: 24, Figs 5H, 6C, D, 8A, D, 10N (characters, records, fungal hosts)
Löbl, 1970c: 749, Figs 19, 20 (characters, records)
Angelini, 1986: 55 (records)
Angelini, 1987: 22 (records)

DISTRIBUTION. Albania; Italy; Montenegro.

Scaphisoma jaccoudi **Löbl**

Scaphisoma jaccoudi Löbl, 1975b: 285, Figs 23, 24. Holotype male, MHNG; type locality: West Malaysia: Fraser's Hill, 4200 ft., Selangor.

DISTRIBUTION. West Malaysia.

Scaphisoma jacobsoni **Löbl**

Scaphisoma jacobsoni Löbl, 1975b: 287, Figs 25, 26. Holotype male, MNHN; type locality: Indonesia: Sumatra, Fort de Kock [Bukittinggi], 920 m.

Löbl, 1979b: 328 (records)
Löbl, 1982f: 5 (records)
Löbl, 1990b: 556 (records)
Löbl, 2015a: 111 (records)

DISTRIBUTION. Indonesia: Bali, Java, Sumatra; East Malaysia: Sarawak; Thailand.

Scaphisoma jado **Löbl**

Scaphisoma jado Löbl, 1992a: 551, Figs 142, 143. Holotype male, MNHN; type locality: Nepal: Malemchi, 2800 m, Sindhupalcok District.

DISTRIBUTION. Nepal.

Scaphisoma jaliscanum **Fierros-López**

Scaphisoma jaliscanum Fierros-López, 2006a: 55, Figs 1a, 2a, 3a, 4a-c, 6a. Holotype male, CZUG; type locality: Mexico: Jalisco, Casimiro Castillo, Arroyo Tacubaya, 600 m.

DISTRIBUTION. Mexico.

***Scaphisoma janczyki* Löbl**

Scaphosoma janczyki Löbl, 1965a: 52, Figs 6, 6a, b. Holotype male, MNHN; type locality: Japan: near Tokyo and Alps of Nikko.

Scaphosoma ignotum Löbl, 1965a: 53, Figs 7, 7a, b. Holotype male, MNHN; type locality: Japan: Tokyo and Alps of Nikko.

Löbl, 1970c: 761, Figs 36, 37 (synonymy of *S. ignotum* with *S. janczyku*, characters)

DISTRIBUTION. Japan: Honshu.

***Scaphisoma jankodadai* Löbl & Ogawa**

Scaphisoma jankodadai Löbl & Ogawa, 2016b: 1426, Figs 212, 213. Holotype male, MHNG; type locality: Philippines: Palawan, above San Rafael, ca 300 m.

DISTRIBUTION. Philippines: Palawan.

***Scaphisoma japonicum* Löbl**

Scaphosoma japonicum Löbl, 1965a: 47, Fig. 3. Holotype female, MNHN; type locality: Japan: Kumonotaira near Karuizawa.

Löbl, 1967b: 110, Figs 11, 12 (characters, records)

Löbl, 1967c: 131 (characters, records)

Löbl, 1970c: 779, Figs 62, 63 (characters, records)

Iablokoff-Khnzorian, 1985: 141 (records)

Löbl, 2000: 609 (records)

Note. The holotype is in the original description erroneously stated to be a male.

DISTRIBUTION. China: Liaoning; Japan: Honshu, Shikoku, Hachijô-jima; Russia: Far East.

***Scaphisoma javanum* Löbl**

Scaphisoma javanum Löbl, 1979b: 326, Figs 13–15. Holotype male, MZBI; type locality: Indonesia: Java, Bogor, Kebun Raya.

Löbl, 1982f: 5, Fig. 1 (characters, records)

Löbl, 1990b: 566 (records)

Löbl & Ogawa, 2016b: 1427, Fig. 12 (characters, records)

DISTRIBUTION. Indonesia: Java; East Malasia: Sarawak; Philippines: Leyte, Luzon, Mindanao, Palawan; Thailand.

***Scaphisoma jeanneli jeanneli* Pic**

Scaphosoma jeanneli Pic, 1946: 84. Syntypes, MNHN; type locality: Kenya: Mt. Elgon [1880 m].

DISTRIBUTION. Kenya.

Scaphisoma jeanneli chappuisi **Pic**

Scaphosoma jeanneli var. *chappuisi* Pic, 1946: 84. Syntypes, MNHN; type locality: Kenya: Mt. Elgon [Suam Fishing Hut, east Mt. Elgon, 2400 m].

DISTRIBUTION. Kenya.

Scaphisoma jelineki **Löbl**

Scaphosoma jelineki Löbl, 1965a: 56, Figs 9, 9a, b. Holotype male, MNHN; type locality: Japan: near Tokyo and Alps of Nikko.

Löbl, 1970c: 770, Figs 48, 49 (characters)

Löbl, 1993: 39, Fig. 5 (characters, records)

DISTRIBUTION. Japan: Honshu; Russia: Far East.

Scaphisoma joachimschmidti **Löbl**

Scaphisoma joachimschmidti Löbl, 2005: 177, Figs 1–3. Holotype male, NMEC; type locality: Nepal: Bara Pokhari Lekh, Chhandi Khola valley, 2000–2300 m, Manaslu Mts.

DISTRIBUTION. Nepal.

Scaphisoma jocosum **Oberthür**

Scaphisoma jocosum Oberthür, 1883: 15. Holotype female, MNHN; type locality: Australia: King Georg's Sound, West Australia.

Löbl, 1977b: 40, Figs 19, 58, 59 (characters, records)

DISTRIBUTION. Australia: New South Wales, South Australia, Victoria, West Australia.

Scaphisoma kalabitum **Löbl**

Scaphisoma kalabitum Löbl, 1987a: 95, Figs 14–17. Holotype male, MHNG; type locality: East Malaysia: Sabah, Mount Kinabalu Nat. Park.

DISTRIBUTION. East Malaysia: Sabah.

Scaphisoma kali **Löbl**

Scaphisoma kali Löbl, 1979a: 108, Figs 29, 30. Holotype male, MHNG; type locality: India: above Aliyar Dam, 1150 m, Anaimalai Hills, Tamil Nadu.

DISTRIBUTION. India: Kerala, Tamil Nadu.

Scaphisoma kanchi **Löbl**

Scaphisoma kanchi Löbl, 1992a: 558, Figs 156, 157. Holotype male, MHNG; type locality: Nepal: Arun Valley below Num, 1050 m, Sankhuwasabha District.

DISTRIBUTION. Nepal.

Scaphisoma karen Löbl

Scaphisoma karen Löbl, 1990b: 589, Figs 132–134. Holotype male, MHNG; type locality: Thailand: Doi Inthanon, 1650 m, Chiang Mai.

DISTRIBUTION. Thailand.

Scaphisoma kashmirense Achard

Scaphosoma kashmirense Achard, 1920i: 240. Lectotype male, MNHN; type locality: Kashmir.

Scaphosoma ebeninum Champion, 1927: 278. Holotype female, NHML; type locality: India: Pindar Valley, Almora, 8,000–11,000 ft.

Löbl, 1970c: 764, Figs 40, 41 (lectotype fixed by inference for *S. kashmirense*, characters, records)

Löbl, 1986c: 345 (records)

Löbl, 1992a: 539 (synonymy of *S. ebeninum* with *S. kashmirense*, characters, records)

DISTRIBUTION. India: Himachal Pradesh, Kashmir; Nepal; Pakistan.

Scaphisoma kaszabianum Löbl

Scaphisoma kaszabianum Löbl, 1986a: 161, Figs 26, 27. Lectotype male, HMNH; type locality: India: Ghoom, 2200 m, Darjeeling District.

Löbl, 1992a: 533 (records)

Löbl, 2005: 180 (records)

DISTRIBUTION. India: West Bengal (Darjeeling District); Nepal.

Scaphisoma katantanum Pic

Scaphosoma katantanum Pic, 1928b: 43. Syntypes, MRAC; type locality: Democratic Republic of the Congo, Tengo Katanta [Manyema].

DISTRIBUTION. Democratic Republic of the Congo.

Scaphisoma katinganum Löbl

Scaphisoma katinganum Löbl, 1987a: 101, Figs 25–27. Holotype male, NHML; type locality: East Malaysia: Sarawak, Kerangas, Gunung Mulu Nat. Park, near Camp 5.

DISTRIBUTION. East Malaysia: Sarawak.

Scaphisoma kenyanum Pic

Scaphosoma kenyanum Pic, 1946: 84. Syntypes, MNHN; type locality: Kenya: Kikuyu.

DISTRIBUTION. Kenya.

***Scaphisoma khao* Löbl**

Scaphisoma khao Löbl, 1990b: 587, Figs 126–128. Holotype male, MHNG; type locality: Thailand: Khao Yai Nat. Park, near Headquarters, 750–850 m.

DISTRIBUTION. Thailand.

***Scaphisoma khasianum* Löbl**

Scaphisoma khasianum Löbl, 1986a: 170, Figs 37–39. Holotype male, NHML; type locality: India: Meghalaya, Mawphlang, 1800 m, Khasi Hills.

DISTRIBUTION. India: Meghalaya.

***Scaphisoma khmer* Löbl**

Scaphisoma khmer Löbl, 2001: 182, Figs 6–8. Holotype male, NMEC; type locality: Cambodia, Ban Lok.

DISTRIBUTION. Cambodia.

***Scaphisoma kibuyense* Pic**

Scaphosoma kibuyense Pic, 1955: 53. Holotype, MRAC; type locality: Rwanda, Rubengera, 1800 m, terr. Kibuye.

DISTRIBUTION. Rwanda.

***Scaphisoma kinabaluum* Löbl**

Scaphisoma kinabaluum Löbl, 1987a: 97, Figs 18–21. Holotype male, EUMJ; type locality: East Malaysia: Sabah, Mount Kinabalu Nat. Park, Headquarters, 1700 m.

DISTRIBUTION. East Malaysia: Sabah.

***Scaphisoma klapperichi* Löbl**

Scaphisoma klapperichi Löbl, 1980a: 114, Figs 34, 35. Holotype male, MHNG; type locality: Taiwan: Fenchihu.

Note. In text under Figs 34 and 35 read: *S. klapperichi* (not *S. asper*)

DISTRIBUTION. Taiwan.

***Scaphisoma kodadai* Löbl & Ogawa**

Scaphisoma kodadai Löbl & Ogawa, 2016b: 1370, Figs 87, 88. Holotype male, MHNG; type locality: Philippines: Luzon, Lagunas Prov., Mt. Banahaw above Kinabuhayan, 600–700 m.

DISTRIBUTION. Philippines: Luzon.

Scaphisoma kuscheli **Löbl**

Scaphisoma kuscheli Löbl, 1980c: 389, Figs 17, 18. Holotype male, NZAC; type locality: Fiji: Nandarivatu, 900 m, Viti Levu.

DISTRIBUTION. Fiji.

Scaphisoma kuschelianum **Löbl**

Scaphisoma kuschelianum Löbl, 1981a: 376, Figs 39, 40. Holotype male, NZAC; type locality: New Caledonia: Mt. Panié, 250 m.

DISTRIBUTION. New Caledonia.

Scaphisoma lacustre **(Casey)**

Scaphiomicrus lacustris Casey, 1900: 59. Syntypes, USNM; type locality: USA: Lake Superior.
Campbell, 1991: 125 (records)

DISTRIBUTION. Canada: Ontario; United States: Lake Superior.

Scaphisoma laetum **Matthews**

Scaphisoma laetum Matthews, 1888: 172. Syntypes, NHML; type locality: Guatemala: Zapote.

DISTRIBUTION. Guatemala.

Scaphisoma laevigatum **Löbl**

Scaphisoma laevigatum Löbl, 1970c: 755, Figs 29, 30. Holotype male, MHNG; type locality: Japan: Honshu, Hyogo.
Löbl, 1980a: 103, Fig. 15 (characters, records)
Löbl, 1993: 38 (characters, records)
Löbl, 2000: 609 (records)

DISTRIBUTION. China: Yunnan; Japan: Honshu; Russia: Far East; Taiwan.

Scaphisoma laeviusculum **Reitter**

Scaphosoma laeviusculum Reitter, 1898: 314. Lectotype male, HNHM; type locality: Azerbaijan: Lenkoran.
Jakobson, 1910: 637 (records)
Löbl, 1964b: 4 (characters)
Löbl, 1964c: Fig. 2 (characters)
Kofler, 1970: 57, Figs 1–3 (characters)
Löbl, 1970c: 751, Figs 23, 24 (lectotype fixed by inference, characters, records)
Iablokoff-Khnzorian, 1985: 141 (records)
Löbl & Ogawa, 2016c: 38 (records)

DISTRIBUTION. Armenia; Azerbaijan; Iran; "Russia mer."; Turkmenistan.

Scaphisoma laminatum Löbl

Scaphisoma laminatum Löbl, 1972b: 97, Figs 44, 45. Holotype male, ZMUB; type locality: Philippines: Panay, Port Banga near Capiz.
Löbl & Ogawa, 2016b: 1407 (characters, records)

DISTRIBUTION. Philippines: Leyte, Luzon, Mindanao, Panay.

Scaphisoma lannaense Löbl

Scaphisoma lannaenese Löbl, 1990b: 558, Figs 72, 73. Holotype male, MHNG; type locality: Thailand: Khao Yai Nat. Park, near Headquarters, 750–850 m.

DISTRIBUTION. Thailand.

Scaphisoma lateapicale Pic

Scaphosoma lateapicale Pic, 1926d: 45. Syntypes, MNHN; type locality: Vietnam: Tonkin, Lac Tho.

DISTRIBUTION. Vietnam.

Scaphisoma latenigrum Pic

Scaphosoma latenigrum Pic, 1946: 83. Syntypes, MNHN; type locality: Kenya: Kikuyu.

DISTRIBUTION. Kenya.

Scaphisoma laterufum Pic

Scaphosoma laterufum Pic, 1954b: 36. Holotype, MRAC; type locality: Democratic Republic of the Congo, Kundelungu Massif.

DISTRIBUTION. Democratic Republic of the Congo.

Scaphisoma latipenne Pic

Pseudoscaphosoma latipenne Pic, 1921a: 166. Syntypes, MCSN; type locality: Myanmar: Bhamò.

DISTRIBUTION. Myanmar.

Scaphisoma latitarse Löbl

Scaphisoma latitarse Löbl, 2012f: 313, Figs 14–18. Holotype male, NHMW; type locality: Indonesia: southeast Sulawesi, Kendari Airport, 30 km west of Kendari.
Ogawa, 2015: 83, Figs 4-14k. l (characters, records)

DISTRIBUTION. Indonesia: Sulawesi.

***Scaphisoma latro* Löbl**

Scaphisoma latro Löbl, 2000: 618, Figs 12, 13. Holotype male, MHNG; type locality: China: Hubei, Shennonglia Nat. Res., 2000–2200 m

DISTRIBUTION. China: Hubei.

***Scaphisoma leai* Löbl**

Scaphisoma leai Löbl, 1977e: 32, Figs 46, 47. Holotype male, SAMA; type locality: Australia: Summit of Mt. Gower, Lord Howe Island.

DISTRIBUTION. Australia: Lord Howe Island.

***Scaphisoma lepesmei* Pic**

Scaphosoma lepesmei Pic, 1942: 1. Syntypes, MNHN; type locality: Cameroon [SE Mt. Cameroon, 1800–2000 m].

DISTRIBUTION. Cameroon.

***Scaphisoma lepidum* Löbl**

Scaphisoma lepidum Löbl, 1990b: 548, Figs 53, 54. Holotype male, MHNG; type locality: Thailand: Khao Yai Nat. Park, near Headquarters, 750–850 m.

DISTRIBUTION. Thailand.

***Scaphisoma leucopyga* Champion**

Scaphosoma leucopyga Champion, 1927: 276. Syntypes, NHML; type locality: India: R. Sarda Gorge, Kumaon.

Scaphosoma kaszabi Löbl, 1965f: 267, Figs 1, 2. Holotype male, MHNG; type locality: Afghanistan: Kamu, Bashgul Valley, 1500 m.

Löbl, 1970c: 784, Figs 66, 67 (as *S. kaszabi*, characters, records)

Löbl, 1979c: 784 (characters)

Löbl, 1986a: 166, Fig. 33 (synonymy of *S. kaszabi* with *S. leucopyga*, characters, records)

Löbl, 1986c: 347 (records)

Löbl, 1992a: 534, 587, Fig. 35 (characters, records)

DISTRIBUTION. Afghanistan; India: Assam, Himachal Pradesh, Uttarakhand, West Bengal (Darjeeling District); Nepal; Pakistan.

***Scaphisoma liberum* Pic**

Scaphosoma liberum Pic, 1918: 1. Syntypes, MNHN; type locality: Cameroon [Musake].

DISTRIBUTION. Cameroon.

Scaphisoma lienhardi Löbl & Ogawa

Scaphisoma lienhardi Löbl & Ogawa, 2016b: 1445, Fig. 258. Holotype female, MHNG; type locality: Philippines: Palawan, Roxas region of Matalangao, 50 m.

DISTRIBUTION. Philippines: Palawan.

Scaphisoma liliput Löbl & Ogawa

Scaphisoma liliput Löbl & Ogawa, 2016b: 1359. Holotype male, MHNG; type locality: Philippines: Luzon, Lagunas Prov., Mt. Makiling, 600 m.

DISTRIBUTION. Philippines: Luzon.

Scaphisoma liliputanum Löbl

Scaphisoma liliputanum Löbl, 1977i: 823, Figs 9, 10. Holotype male, SAMA; type locality: Fiji: Viti Levu.

Löbl, 1980c: 390, Figs 19, 22 [erroneously Fig. no "20"] (characters, records)

Löbl, 1981b: 73 (characters, records)

DISTRIBUTION. Fiji; Micronesia: Caroline Islands.

Scaphisoma limbatum Erichson

Scaphisoma limbatum Erichson, 1845: 11. Lectotype male, ZMUB; type locality: Hungary [Banat, Romania].

Gutfleisch, 1859: 222 (characters)

Redtenbacher, 1872: 335 (characters, records)

Reitter, 1880d: 45 (characters)

Kuthy, 1897: 86 (records)

Ganglbauer, 1899: 344 (characters)

Jakobson, 1910: 637 (records)

Roubal, 1930: 298 (records)

Horion, 1949: 255 (as *Caryoscapha*, records)

Tamanini, 1969b: 361, Figs 4A-E (as *Caryoscapha*, characters)

Tamanini, 1969c: 131, Figs 23, 41, 44 (as *Caryoscapha*, characters)

Tamanini, 1970: 10, Figs 3A-E (as *Caryoscapha*, characters, records)

Löbl, 1970a: 15, Figs 4, 17 (as *Caryoscapha*, characters, records)

Löbl, 1970c: 794, Figs 2, 78, 79 (as *Caryoscapha*, lectotype designation, characters, records)

Franz, 1970: 267 (as *Caryoscapha*, records)

Freude, 1971: 347, Figs 4: 1 (as *Caryoscapha*, characters)

Burakowski et al., 1978: 237 (as *Caryoscapha*, records, habitat)

Iablokoff-Khnzorian, 1985: 42, Fig. 2d (as *Caryoscapha*, characters)

Löbl, 1987c: 389 (as *Caryoscapha*, characters)
Kompantsev & Potoskaya, 1987: 97, 99, Fig. 5 (as *Caryoscapha*, immature stages, fungal hosts, natural history)
Nikitsky et al., 1996: 24 (as *Caryoscapha*, fungal hosts)
Krasutskij, 1996a: 96 (as *Caryoscapha*, records, fungal hosts)
Krasutskij, 1997a: 307 (as *Caryoscapha*, records, fungal hosts)
Nikitsky et al., 1998: 7 (as *Caryoscapha*, fungal hosts)
Tsinkevich, 2004: 19 (records, fungal hosts)
Leschen & Löbl, 2005: 25 (transfer to *Scaphisoma*)
Mateleshko, 2005: 128 (fungal hosts)
Borowski, 2006: 54, 58 (records, fungal hosts)
Löbl, 2011c: 111 (records)
Hebda & Zając, 2013: 605 (records)
Li, 2015: 40, Fig. 6 (as *Caryoscapha*, characters, records)
Semenov, 2017: 203 (records)
Vlasov & Nikitsky, 2017: 4 (records, fungal hosts)

DISTRIBUTION. Austria; Belarus; Bosnia and Herzegovina; China: Heilongiang, Jilin; Croatia; Czech Republic; Finland, France; Germany; Hungary, Italy; Poland; Romania; Russia: European to Far East; Serbia; Slovakia; Slovenia; Sweden; Ukraine.

Note. Records from Finland and Sweden in Thomson, 1862: 128 and in Sahlberg, 1889: 81 may refer to *S. subalpinum* Reitter (e.g., West, 1942: 120), four specimens in coll. Thomson identified as *limbatum* are *S. assimile assimile* Erichson and *S. balcanicum* Tamanini (tested pers.).

Scaphisoma lineare Löbl & Ogawa

Scaphisoma lineare Löbl & Ogawa, 2016b: 1361, Figs 65–68. Holotype male, MHNG; type locality: Philippines: Luzon, Lagunas Prov., Mt. Makiling, summit road, 600 m.

DISTRIBUTION. Philippines: Luzon.

Scaphisoma lineatopunctatum (Pic)

Pseudoscaphosoma lineatopunctatum Pic, 1916d: 7. Lectotype female, MNHN; type locality: Indonesia: Kalimantan [Martapura].
Löbl, 2015c: 132 (lectotype designation, characters)

DISTRIBUTION. Indonesia: Kalimantan; East Malaysia: Sabah.

Scaphisoma linum Löbl

Scaphisoma linum Löbl, 2000: 628, Figs 31, 32. Holotype male, MHNG; type locality: China: Sichuan, Lipizing env., near Shimien.

DISTRIBUTION. China: Sichuan.

Scaphisoma loebli **Tamanini**

Scaphisoma loebli Tamanini, 1969b: 372, Figs 7B, E, 8A-E. Holotype male, MSNM; type locality: Italy: Milano.
Löbl, 1970c: 747, Figs 15, 16 (characters, records)
Tamanini, 1970: 24 Figs 8B, E, 9A-E (characters, records, fungal hosts)
Löbl, 1974a: 61 (records)

DISTRIBUTION. Croatia; France; Italy; Switzerland: Ticino; Turkey.

Scaphisoma lombokense **Löbl**

Scaphisoma lombokense Löbl, 1986b: 90, Fig. 6. Holotype male, MHNG; type locality: Indonesia: Lombok, Sesaot.

DISTRIBUTION. Indonesia: Lombok.

Scaphisoma longicolle **Matthews**

Scaphisoma longicolle Matthews, 1888: 177. Syntypes, NHML; type localities: British Honduras: R. Hondo; Mexico: Tabasco, Teapa and Frontera.

DISTRIBUTION. Belize; Mexico.

Scaphisoma longicorne **Löbl**

Scaphisoma longicorne Löbl, 1977e: 55, Figs 84, 85. Holotype male, NHML; type locality: Australia: New South Wales, Wingham.

DISTRIBUTION. Australia: New South Wales.

Scaphisoma longiusculum **Löbl**

Scaphisoma longiusculum Löbl, 2012b: 178, Figs 8–10. Holotype male, NMPC; type locality: West Malaysia: Tanah Rata village, near Gunung Jasat, 1470–1705 m, Cameron Highlands, Pahang.

DISTRIBUTION. West Malaysia.

Scaphisoma loriai **Löbl**

Scaphisoma loriai Löbl, 1975a: 397, Figs 42, 43. Holotype male, MCST; type locality: Papua New Guinea, Bujako [Bujakori, 9°40'S, 147°45'E].

DISTRIBUTION. Papua New Guinea.

Scaphisoma lucens **Löbl**

Scaphisoma lucens Löbl, 1977e: 36, Figs 52, 63. Holotype male, QMBA; type locality: Australia: Queensland, Crystal Cascades via Cairns.

DISTRIBUTION. Australia: Queensland.

***Scaphisoma luctans* Löbl**

Scaphisoma luctans Löbl, 1986a: 147, Figs 8, 9. Holotype male, MHNG; type locality: India: Manas Wild Life Sanctuary, 200 m, Assam.

DISTRIBUTION. India: Assam, West Bengal (Darjeeling District).

***Scaphisoma luctuosum* Löbl**

Scaphisoma luctuosum Löbl, 1986a: 175, Figs 43–45. Holotype male, MHNG; type locality: India: Meghalaya, Sonsag, 400 m, Garo Hills.

DISTRIBUTION. India: Meghalaya.

***Scaphisoma lunabianum* Löbl & Ogawa**

Scaphisoma lunabianum Löbl & Ogawa, 2016b: 1371, Figs 17, 89, 90. Holotype male, FMNH; type locality: Philippines: Mindanao, Davao Prov., Lake Linau, north slope of Mt. Apo, 6000 ft.

DISTRIBUTION. Philippines: Mindanao.

***Scaphisoma lunatum* Matthews**

Scaphisoma lunatum Matthews, 1888: 173, Pl. 4, Fig. 13. Syntypes, NHML; type locality: Nicaragua: Chontales.

Fierros-López, 2006b: 41, Figs 2d, 3d (characters, records; note: illustations suggest two distinct species)

DISTRIBUTION. Costa Rica; Nicaragua.

***Scaphisoma luteipes* Oberthür**

Scaphisoma luteipes Oberthür, 1883: 15. Syntypes, MNHN; type locality: Panama: Matachin.

DISTRIBUTION. Panama.

***Scaphisoma luteomaculatum* Pic**

Scaphosoma luteomaculatum Pic, 1915e: 5. Lectotype male, MNHN; type locality: Indonesia: Java, Palabuan.

Scaphisoma dansalanense Löbl, 1972b: 95, Figs 36–38. Holotype male, ZMUB; type locality: Philippines: Mindanao, Dansalan near Lanao.

Scaphosoma sapitense var. *infasciatum* Achard, 1920d: 131. Lectotype female, MNHN; type locality: Indonesia: Lombok, Sapit.

Pic, 1921a: 165 (records)

Achard, 1920d: 130 (records)

Löbl, 2014a: 55 (records)

Löbl, 2015a: 111, Figs 103–106 (lectotype designations for *S. luteomaculatum* and var. *infasciatum*, synonymy of *S. dansalanense* and *infasciatum* with *S. luteomaculatum*, characters, records)
Löbl & Ogawa, 2016b: 1427, Fig. 18 (characters, records)

DISTRIBUTION. Indonesia: Bali, Buru, Java, Lombok, Sumatra, Sumbawa. Myanmar; Philippines: Leyte, Luzon, Mindanao. Palawan.

Scaphisoma luteopygidiale (Pic)

Scutoscaphosoma luteopygidiale Pic, 1947c: 3. Lectotype male, MNHN; type locality: Philippines: Palawan, Binaluan.
Löbl, 1970b: 128, Fig. 5 lectotype fixed by inference, transfer to *Scaphisoma*, characters)
Löbl, 1981d: 163, Fig. 15 (characters)
Löbl & Ogawa, 2016b: 1410, Figs 19, 167 (characters, records)

DISTRIBUTION. Philippines: Palawan.

Scaphisoma luzonicum Pic

Scaphosoma luzonicum Pic, 1926a: 1. Lectotype male, MNHN; type locality: Philippines: Luzon, Los Banos.
Löbl, 1972b: 107, Figs 41–43 (misspelled *luconicum*, lectotype designation, characters)
Löbl & Ogawa, 2016b: 1428, Figs 214, 215 (characters, records)

DISTRIBUTION. Philippines: Luzon.

Scaphisoma maculatum (Pic)

Pseudoscaphosoma maculatum Pic, 1920e: 96. Syntypes, ?MCSN, not traced; type locality: East Malaysia: Sarawak.

DISTRIBUTION. East Malaysia: Sarawak.

Scaphisoma maculiger Löbl

Scaphisoma maculiger Löbl, 1975b: 289, Figs 27–29. Holotype male, ZMUC; type locality: India: Kar, Nicobar.
Löbl, 1986a: 175 (records)
Löbl, 1992a: 588, Fig. 37 (characters)

DISTRIBUTION. India: Assam, Nicobar.

Scaphisoma maculosum Löbl & Ogawa

Scaphisoma maculosum Löbl & Ogawa, 2016b: 1372, Figs 20, 91, 92. Holotype male, MHNG; type locality: Philippines: Palawan, Cabayugan near Lion's Cave, sea level.

DISTRIBUTION. Philippines: Palawan.

***Scaphisoma madagascariense* Achard**

Scaphosoma madagascariense Achard, 1920i: 241. Syntypes, NMPC; type locality: Madagascar: Suberbieville [Maevatanana District].

DISTRIBUTION. Madagascar.

***Scaphisoma mahense* Scott**

Scaphosoma mahense Scott, 1922: 224, Pl. 20, Fig. 17, Pl. 21, Fig. 21. Holotype, NHML; type locality: Seychelles: Cascade Estate, Mahé.

Vinson, 1943: 197, Figs 8, 18 (characters, records)

DISTRIBUTION. Mauritius; Seychelles.

***Scaphisoma maindroni* Achard**

Scaphosoma maindroni Achard, 1920i: 240. Lectotype male, NMPC; type locality: India: Pondichéry [Coromantel].

Scaphosoma mutatum Champion, 1927: 276. Lectotype male, NHML; type locality: India: W. Almora, Kumaon.

Löbl, 1979a: 102, Fig. 24 (lectotype designation for *S. maindroni*, characters, records)

Löbl, 1981f: 105 (records)

Löbl, 1984c: 996 (records)

Löbl, 1986a: 151 (records)

Löbl, 1986c: 346 (synonymy of *S. mutatum* with *S. maindroni*, records)

Löbl, 1990b: 549 (records)

Löbl, 1992a: 534 (lectotype designation for *S. mutatum*, records)

Löbl, 2000: 609 (records)

DISTRIBUTION. China: Guizhou, Hong Kong, Yunnan; India: Assam, Himachal Pradesh, Kerala, Pondichery, Uttarakhand, Tamil Nadu; Nepal; Pakistan; Thailand; Vietnam.

***Scaphisoma malaccanum* (Pic)**

Baeocera malaccana Pic, 1915b: 32. Lectotype male, MNHN; type locality: West Malaysia: Perak, Malacca.

Achard, 1920c: 58 (characters, copy of Pic's description)

Pic, 1920g: 189 (in *Baeocera*, characters)

Löbl, 1973a: 155, Figs 7–10 (lectotype fixed by inference, transfer to *Scaphisoma*, characters)

DISTRIBUTION. West Malaysia.

Scaphisoma malayanum Löbl

Scaphisoma malayanum Löbl, 1986b: 99, Figs 17, 18. Holotype male, MHNG; type locality: West Malaysia: Sungei Buloh, Kuala Lumpur.

DISTRIBUTION. West Malaysia.

Scaphisoma malignum Löbl

Scaphisoma malignum Löbl, 1986a: 182, Figs 56, 57. Holotype male, MHNG; type locality: India: Sevoke, 200 m, Darjeeling District.

Löbl, 1986c: 347 (records)

Löbl, 1992a: 535 (records)

DISTRIBUTION. India: Assam, Meghalaya, Uttarakhand (Garhwal), West Bengal (Darjeeling District); Nepal.

Scaphisoma maramag Löbl & Ogawa

Scaphisoma maramag Löbl & Ogawa, 2016b: 1373, Figs 93, 94. Holotype male, SMNS; type locality: Philippines: Mindanao, 30 km NW Maramag, Bagongsilang, 1700 m.

DISTRIBUTION. Philippines: Mindanao.

Scaphisoma marshallae Löbl

Scaphisoma marshallae Löbl, 1987a: 94, Figs 12, 13. Holotype male, NHML; type locality: East Malaysia: Sarawak, Gunung Mulu Nat. Park, near Camp 5.

DISTRIBUTION. East Malaysia: Sarawak.

Scaphisoma mascareniense Vinson

Scaphosoma mascareniense Vinson, 1943: 190, Figs 1, 11. Holotype, NHML; type locality: Mauritius: Les Mares.

DISTRIBUTION. Mauritius.

Scaphisoma mauritiense Vinson

Scaphosoma mauritiense Vinson, 1943: 196, Figs 7, 17. Syntypes, NHML; type localities: Mauritius: Macabé forest; Les Mares, The Pouce, Moka, Mt. Cocotte.

DISTRIBUTION. Mauritius.

Scaphisoma mediofasciatum Reitter

Scaphosoma mediofasciatum Reitter, 1908: 32. Syntypes, NHMW; type locality: Tanzania: Mt. Bomole and Amani.

DISTRIBUTION. Tanzania.

***Scaphisoma medium* Löbl**

Scaphisoma medium Löbl, 2003c: 188, Figs 9, 10. Holotype male, HNHM; type locality: Papua New Guinea: Mt. Wilhelm, Imbuka Ridge, 3600 m.

DISTRIBUTION. Papua New Guinea.

***Scaphisoma mendax* Löbl**

Scaphisoma mendax Löbl, 1980a: 112, Figs 31, 32. Holotype male, MHNG; type locality: Taiwan: Kosempo.

DISTRIBUTION. Taiwan.

***Scaphisoma meracum* Löbl**

Scaphisoma meracum Löbl, 1990b: 579, Figs 114–116. Holotype male, MHNG; type locality: Thailand: Doi Suthep, 1400 m, Chiang Mai.

DISTRIBUTION. Thailand.

***Scaphisoma michaeli* Löbl**

Scaphisoma michaeli Löbl, 2003a: 70, Figs 14–17. Holotype male, ZMUB; type locality: China: Shaanxi, Qinling Shan, pass on road Zhouzhi-Foping, north slope, 1990 m, 105 km SW Xi'an.

DISTRIBUTION. China: Shaanxi.

***Scaphisoma migrator* Löbl**

Scaphisoma migrator Löbl, 2000: 643, Figs 62–65. Holotype male, MHNG; type locality: China: Shaanxi, east Xian, Mt. Huashan, 500 m.
Löbl, 2003a: 62 (records)

DISTRIBUTION. China: Hubei, Shaanxi, Sichuan.

***Scaphisoma mimicum* Löbl**

Scaphisoma mimicum Löbl, 1986a: 151, Fig. 11. Holotype male, MHNG; type locality: India: Sevoke, 200 m, Darjeeling District.

DISTRIBUTION. India: Meghalaya, West Bengal (Darjeeling District).

***Scaphisoma minax* Löbl**

Scaphisoma minax Löbl, 1986a: 209, Figs 109–111. Holotype male, MHNG; type locality: India: Meghalaya, below Cherrapunjee, 1200 m, Khasi Hills.
Löbl, 1992a: 538 (records)
Löbl, 2005: 180 (records)

DISTRIBUTION. India: Meghalaya, West Bengal (Darjeeling District); Nepal.

Scaphisoma mindanaosum **Pic**

Scaphosoma mindanaosum Pic, 1926b: 2. Lectotype female, MNHN; type locality: Philippines: Mindanao, P. Bango.

Löbl, 1972b: 107, Figs 39, 40 (lectotype designation, characters)
Löbl, 1981f: 108 (characters, records)
Löbl, 2015c: 134, Figs 4, 5 (characters, records)
Löbl & Ogawa, 2016b: 1429, Fig. 21 (characters, records)

DISTRIBUTION. Indonesia: Kalimantan, Sumatra; Laos; Philippines: Luzon, Mindanao, Palawan; Vietnam.

Scaphisoma minus **Löbl**

Scaphosoma minutissimum Pic, 1951c: 1100. Holotype, MNHN; type locality: Senegal: Casamance, Bignona.

Scaphisoma minus Löbl, 1997: xii; replacement name for *Scaphosoma minutissimum* Pic, 1951 (nec *Scaphisoma minutissimum* Champion, 1927).

DISTRIBUTION. Senegal.

Scaphisoma minutipenis **Löbl & Ogawa**

Scaphisoma minutipenis Löbl & Ogawa, 2016b: 1374, Figs 95, 96. Holotype male, FMNH; type locality: Philippines: Mindanao, Cotabato Prov., Burungkot, Upi, 1500 ft.

DISTRIBUTION. Philippines: Mindanao.

Scaphisoma minutissimum **Champion**

Scaphosoma minutissimum Champion, 1927: 278. Lectotype male, NHML; type locality: India: Uttarakhand, R. Sarda Gorge, Kumaon.

Löbl, 1990b: 546 (as cf. *S. minutissimum*, characters, records)
Löbl, 1992a: 538 (lectotype designation, characters)
Löbl, 2000: 609 (records)

DISTRIBUTION. China: Yunnan; India: Uttarakhand (Kumaon); Thailand.

Scaphisoma minutulum **Löbl**

Scaphosoma minutissimum Pic, 1937: 207. Holotype, ZMUH; type locality: Costa Rica: Finca La Caja near San José.

Scaphisoma minutulum Löbl, 1997: xii; replacement name for *Scaphosoma minutissimum* Pic, 1937 (nec *Scaphisoma minutissimum* Champion, 1927).

DISTRIBUTION. Costa Rica.

Scaphisoma minutum Achard

Scaphosoma minutum Achard, 1920f: 363. Lectotype male, NMPC; type locality: India: West Bengal, Kurseong, Darjeeling District.
Löbl, 1986a: 146, Fig. 5 (lectotype designation, characters, records)

DISTRIBUTION. India: West Bengal (Darjeeling District).

Scaphisoma mirandum Löbl

Scaphisoma mirandum Löbl, 1990b: 548, Fig. 55. Holotype male, MHNG; type locality: Thailand: Khao Yai Nat. Park, near Headquarters.

DISTRIBUTION. Thailand.

Scaphisoma mirum Löbl & Ogawa

Scaphisoma mirum Löbl & Ogawa, 2016b: 1375, Figs 22, 97, 98. Holotype male, FMNH; type locality: Philippines: Mindanao, Davao Prov., east slope of Mt. McKinley, 6000 ft.

DISTRIBUTION. Philippines: Mindanao.

Scaphisoma mocquerysi Achard

Scaphosoma mocquerysi Achard, 1920i: 241. Syntypes, NMPC/MNHN; type locality: Madagascar: Antongil Bay.

DISTRIBUTION. Madagascar.

Scaphisoma modestum Löbl

Scaphisoma modestum Löbl, 1981a: 365, Figs 24, 25. Holotype male, NZAC; type locality: New Caledonia: Mt. Rembai, 800 m.

DISTRIBUTION. New Caledonia.

Scaphisoma modicum Löbl

Scaphisoma modicum Löbl, 1984c: 1003, Fig. 11. Holotype male, MHNG; type locality: Myanmar: Kalaw, 1300 m, Shan State.

DISTRIBUTION. Myanmar.

Scaphisoma modiglianii Pic

Scaphosoma modiglianii Pic, 1920e: 95. Syntypes, MCSN; type locality: Indonesia: Sumatra, Si Rambé [near Balige].

DISTRIBUTION. Indonesia: Sumatra.

Scaphisoma montanellum Vinson

Scaphosoma montanellum Vinson, 1943: 192, Figs 3, 13. Holotype, NHML; type locality: Mauritius: The Pouce.

Scaphosoma montanellum var. *fasciatum* Vinson, 1943: 194. Syntypes, NHML; type locality: Mauritius: Mt. Cocotte.
Löbl, 1977f: 43, Figs 11, 12 (implicit synonymy of var. *fasciatum* with *S. montanellum*, characters, records)
Lecoq, 2015: 210 (records)
DISTRIBUTION. Mauritius; La Réunion.

Scaphisoma monticola **(Löbl)**
Caryoscapha monticola Löbl, 1987c: 389, Figs 8, 9. Holotype male, NHMB; type locality: Nepal: Modi Kola, Banthandi-Landrung, 1600–2500 m.
Löbl, 1992a: 562 (records)
Leschen & Löbl, 2005: 25 (transfer to *Scaphisoma*)
DISTRIBUTION. India: West Bengal (Darjeeling District); Western Nepal.

Scaphisoma montivagum **Löbl & Ogawa**
Scaphisoma montivagum Löbl & Ogawa, 2016b: 1376, Figs 99, 100. Holotype male, MHNG; type locality: Philippines: Luzon, Mount Data Lodge, 2200–2300 m.
DISTRIBUTION. Philippines: Luzon.

Scaphisoma morosum **Löbl**
Scaphisoma morosum Löbl, 1990b: 589, Figs 129–131. Holotype male, MHNG; type locality: Thailand: road to Wab Pang An, 50 km NE Chiang Mai, 900 m.
Löbl, 2000: 609 (records)
DISTRIBUTION. China: Yunnan; Myanmar; Thailand.

Scaphisoma mucronatum **Löbl**
Scaphisoma mucronatum Löbl, 1980c: 392, Figs 29, 30. Holotype male, NZAC; type locality: Fiji: Nandrau, 750 m, Viti Levu.
DISTRIBUTION. Fiji.

Scaphisoma murphyi **Löbl**
Scaphisoma murphyi Löbl, 1981d: 162, Fig. 9. Holotype male, MHNG; type locality: West Malaysia: Fraser's Hill, Pahang.
DISTRIBUTION. West Malaysia.

Scaphisoma murutum **Löbl**
Scaphisoma murutum Löbl, 1987a: 92, Figs 10, 11. Holotype male, MHNG; type locality: East Malaysia: Pangi, Sabah.
DISTRIBUTION. East Malaysia: Sabah, Sarawak.

Scaphisoma mussardi Löbl

Scaphisoma mussardi Löbl, 1971c: 979, Figs 48, 49. Holotype male, MHNG; type locality: Sri Lanka: Kandy, ca 600 m.
Löbl, 1979a: 107 (records)
DISTRIBUTION. India: Tamil Nadu; Sri Lanka.

Scaphisoma mutator Löbl

Scaphisoma mutator Löbl, 2000: 644, Figs 66–69. Holotype male, MHNG; type locality: China: Sichuan, Wolong Nat. Res., 1500 m.
DISTRIBUTION. China: Shaanxi, Sichuan.

Scaphisoma nabiluanum Löbl & Ogawa

Scaphisoma nabiluanum Löbl & Ogawa, 2016b: 1377, Figs 23, 101, 102. Holotype male, NMNS; type locality: Philippines: Luzon, Ifugao Prov., Mt. Polis, 1900 m.
DISTRIBUTION. Philippines: Luzon.

Scaphisoma nakanei Löbl

Scaphisoma nakanei Löbl, 1980a: 107, Figs 22, 23. Holotype male, NSMT; type locality: Taiwan: Nanshanshi, 800 m, Nantou Hsien.
Löbl, 1981f: 105 (records)
Rougemont, 1996: 12 (records)
Löbl, 2000: 610 (records)
DISTRIBUTION. China: Hong Kong, Jiangsu; Taiwan; Vietnam.

Scaphisoma nanellum Löbl & Ogawa

Scaphisoma nanellum Löbl & Ogawa, 2016b: 1360, Figs 63, 64. Holotype male, FMNH; type locality: Philippines: Leyte, Tarragona.
DISTRIBUTION. Philippines: Leyte.

Scaphisoma nanulum Löbl

Scaphisoma nanulum Löbl, 1973b: 314, Figs 9, 10. Holotype male, NHMW; type locality: New Caledonia: Tiouandé near Hienghène.
Löbl, 1981a: 369 (characters, records)
DISTRIBUTION. New Caledonia.

Scaphisoma napu Löbl

Scaphisoma napu Löbl, 1983a: 289, Figs 7, 8. Holotype male, MHNG; type locality: Indonesia: Sulawesi, Makale.
Ogawa, 2015: 83, Figs 4-14i, j (characters, records)
DISTRIBUTION. Indonesia: Sulawesi.

Scaphisoma natalense Pic

Scaphosoma natalense Pic, 1937: 207. Holotype, ZMUH; type locality: South Africa: KwaZulu-Natal, Pietermaritzburg, Fort Napier.

DISTRIBUTION. South Africa: KwaZulu-Natal.

Scaphisoma neboissi Löbl

Scaphisoma neboissi Löbl, 1977e: 21, Figs 28, 29. Holotype male, SAMA; type locality: Australia: South Australia, Pelican Lagoon, Muston, Kangaroo Island.

DISTRIBUTION. Australia: South Australia.

Scaphisoma nebulosoides Löbl

Scaphisoma nebulosum Löbl, 1986a: 195, Figs 78–80. Holotype male, MHNG; type locality: India: West Bengal, between Ghoom and Lopchu, 13 km from Ghoom, 2000 m, Darjeeling District.

Scaphisoma nebulosoides Löbl, 1997: xii; replacement name for *Scaphisoma nebulosum* Löbl, 1986 (nec *Scaphisoma nebulosum* Matthews, 1888).

DISTRIBUTION. India: West Bengal (Darjeeling District).

Scaphisoma nebulosum Matthews

Scaphisoma nebulosum Matthews, 1888: 173. Syntypes, NHML; type locality: Guatemala: Zapote, Capetillo.

DISTRIBUTION. Guatemala.

Scaphisoma necopinum Löbl

Scaphisoma necopinum Löbl, 1986a: 206, Figs 102–104. Holotype male, MHNG; type locality: India: Mahanadi near Kurseong, 1200 m, Darjeeling District.
Löbl, 1992a: 536 (records)

DISTRIBUTION. India: Sikkim, West Bengal (Darjeeling District); Nepal.

Scaphisoma nefastum Löbl

Scaphisoma nefastum Löbl, 1986a: 196, Figs 81–83. Holotype male, MHNG; type locality: India: West Bengal, between Ghoom and Lopchu, 13 km from Ghoom, 2000 m, Darjeeling District.
Löbl, 1986c: 348 (records)
Löbl, 1992a: 537 (record)

DISTRIBUTION. India: West Bengal (Darjeeling District), Nepal.

Scaphisoma neglectum Löbl

Scaphisoma neglectum Löbl, 2003a: 69, Figs 11–13. Holotype male, ZIBC; type locality: China: Beijing, Dongling Mts, Xiaolongmen, Da Nam Gou, 1500 m.

DISTRIBUTION. China: Beijing.

Scaphisoma negligens **Löbl**

Scaphisoma negligens Löbl, 1986a: 193, Figs 75.77. Holotype male, MHNG; type locality: India: Meghalaya, Mawphlang, 1800 m, Khasi Hills.

DISTRIBUTION. India: Meghalaya.

Scaphisoma neotropicale **Matthews**

Scaphisoma neotropicale Matthews, 1888: 176. Syntypes, NHML; type localities: Mexico: Teapa; Guatemala: Calderas; Panama: Volcan de Chiriqui.

DISTRIBUTION. Guatemala; Mexico; Panama.

Scaphisoma nepalense **Löbl**

Scaphisoma nepalense Löbl, 1992a: 557, Figs 154, 155. Holotype male, MHNG; type locality: Nepal: Pokhare NE Barahbise, 2700 m, Sindhupalcok District.

Löbl, 2005: 180 (records)

DISTRIBUTION. Nepal.

Scaphisoma nevermanni **Pic**

Scaphosoma nevermanni Pic, 1937: 206. Holotype, ZMUH; type locality: Costa Rica: Farm Hamburg at Reventazon, near Las Mercedes.

DISTRIBUTION. Costa Rica.

Scaphisoma niasense **(Pic)**

Pseudoscaphosoam niasense Pic, 1915b: 31. Syntypes, MNHN; type locality: Indonesia: Nias.

Achard, 1920d: 131 (records)

Löbl, 1975b: 270 (transfer to *Scaphisoma*)

DISTRIBUTION. Indonesia: Nias, Sumatra.

Scaphisoma nietneri **Löbl**

Scaphisoma nietneri Löbl, 1971c: 969, Figs 36, 37. Holotype male, MHNG; type locality: Sri Lanka: 3 km northeast Puliyan Kulam.

Löbl, 1990b: 552, Figs 60, 61 (characters, records)

DISTRIBUTION. Sri Lanka; Thailand.

Scaphisoma nigroapicale **Pic**

Scaphosoma nigroapicale Pic, 1955: 53. Syntypes, MRAC/MNHN; type locality: Rwanda: Rubengera, 1900 m, terr. Kibuye.

DISTRIBUTION. Rwanda.

***Scaphisoma nigrofasciatum* Pic**

Scaphosoma nigrofasciatum Pic, 1915b: 31. Lectotype male, MNHN; type locality: India: Tamil Nadu, Chambaganor.

Achard, 1915c: 292 (as synonym of *S. binotatum* Achard, 1915)

Pic, 1916e: 49 (as a form of *S. binotatum* Achard, 1915, or as valid species)

Scott, 1922: 222, Pl. 20, Fig. 18, Pl. 21, Fig. 19 (as forme of *S. pictum*, characters)

Vinson, 1943: 189, Figs 9, 10a, b (as *S. pictum*, characters)

Löbl, 1971c: 974, Figs 42, 43 (lectotype fixed by inference, characters, records)

Löbl, 1977f: 42 (records)

Löbl, 1979a: 107 (records)

Löbl, 1986c: 347 (records)

Löbl, 1990a: 121 (records)

Löbl, 1992a: 534 (records)

Deepthi et al., 2004 (fungal hosts)

Lecoq, 2015: 210 (records)

Singh & Sharma, 2016: 213 (fungal hosts)

DISTRIBUTION. India: "Bengalien", Goa, Himachal Pradesh, Kerala, Tamil Nadu, Uttarakhand; Mauritius; Nepal; La Réunion; Seychelles; Sri Lanka.

***Scaphisoma nigroplagatum* Achard**

Scaphosoma nigroplagatum Achard, 1920c: 56. Syntypes, NMPC/MNHN; type localities: West Malaysia: Perak, Myanmar: Rangoon.

Achard, 1920d: 129 (records)

Pic, 1921a: 165 (misspelled *nigroplagiatum*, records)

DISTRIBUTION. East Malayasia; West Malaysia; Myanmar.

***Scaphisoma nigrum* Löbl**

Scaphisoma nigrum Löbl, 1986b: 91, Figs 7–9. Holotype male, MHNG; type locality: West Malaysia: Cameron Highlands.

DISTRIBUTION. West Malaysia.

***Scaphisoma nilgiriense* Löbl**

Scaphisoma nilgiriense Löbl, 2003d: 94, Figs 1, 2. Holotype male, MHNG; type locality: India: Nilgiri Hills, 10 km SW Thiashola Reserve forest near Carrington Estate, ca 2100 m, Tamil Nadu.

DISTRIBUTION. India: Tamil Nadu.

Scaphisoma nima **Löbl**

Scaphisoma nima Löbl, 1992a: 550, Fig. 139. Holotype male, MHNG; type locality: Nepal: Kalopani, 2550 m, Mustang District.
Löbl, 2005: 181 (records)
DISTRIBUTION. Nepal.

Scaphisoma nishikawai **Löbl & Ogawa**

Scaphisoma nishikawai Löbl & Ogawa, 2016b: 1431, Figs 221–223. Holotype male, EUMJ; type locality: Philippines: Mindanao, Sultan Kadarat Prov., Kraan.
DISTRIBUTION. Philippines: Mindanao.

Scaphisoma nitidulum **Zayas**

Scaphosoma nitidula Zayas, 1988: 26, Fig. 17. Syntypes, FZCH; type locality: Cuba.
DISTRIBUTION. Cuba.

Scaphisoma notatum **Löbl**

Scaphisoma notatum Löbl, 1986a: 154, Figs 16, 17. Holotype male, MHNG; type locality: India: Mawphlang, 1800 m, Khasi Hills.
Löbl, 1986c: 346 (records)
Löbl, 1990a: 121 (records)
Löbl, 1992a: 532, 588, Fig. 38 (characters, records)
Löbl, 2000: 610 (records)
Löbl, 2003a: 62 (records)
DISTRIBUTION. China: Hubei, Shaanxi, Sichuan, Yunnan; India: Himachal Pradesh, Meghalaya, Uttarakhand (Kumaon), West Bengal (Darjeeling District); Nepal; Pakistan.

Scaphisoma notulum **Fauvel**

Scaphosoma notula Fauvel, 1903: 292. Lectotype female, ISNB; type locality: New Caledonia: Nouméa.
Löbl, 1969a: 2, Fig. 3 (lectotype fixed by inference, characters)
Löbl, 1981a: 367, Figs 26, 27 (characters, records)
DISTRIBUTION. New Caledonia.

Scaphisoma novaecaledonicum **Löbl**

Scaphisoma novaecaledonicum Löbl, 1977i: 818, Figs 1, 2. Holotype male, BPBM; type locality: New Caledonia: Mt. Koghi.
Löbl, 1981a: 369 (characters)
DISTRIBUTION. New Caledonia.

***Scaphisoma novicum* Blackburn**

Scaphisoma novicum Blackburn, 1891: 91. Lectotype male, NHML; type locality: Australia: Alpine District, Victoria.

Blackburn, 1903: 99 (characters)

Löbl, 1977e: 29, Figs 38, 39 (lectotype fixed by inference, characters, records)

DISTRIBUTION. Australia: South Australia, Tasmania, Victoria.

***Scaphisoma nugator* (Casey)**

Scaphiomicrus nugator Casey, 1900: 59. Syntypes, USNM; type locality: USA: Keokuk, Iowa.

DISTRIBUTION. United States: Iowa.

***Scaphisoma obenbergeri* Löbl**

Scaphosoma obenbergeri Löbl, 1963b: 273, Figs 1, 2. Holotype male, NMPC; type locality: "Cp. or." [East Carpathian, Ukraine].

Löbl, 1964d: 50 (records)

Löbl, 1966b: 131 (records)

Palm, 1966: 44, Fig. C (characters)

Löbl, 1967b: 110 (records)

Kofler, 1968: 41, Figs 1F, 2F (characters, records)

Tamanini, 1969b: 364, Figs 4G, H (characters)

Tamanini, 1970: 16, Figs 3G, H, 10C (characters)

Kofler, 1970: 58 (records)

Löbl, 1970a: 13, Figs 12, 24 (characters, records)

Löbl, 1970c: 769, Figs 46, 47 (characters, records)

Franz, 1970: 267 (records)

Freude, 1971: 346, Fig. 3: 8 (characters)

Löbl, 1974a: 61 (records)

Holzschuh, 1977: 31 (records)

Burakowski et al., 1978: 237 (records)

Iablokoff-Khnzorian, 1985: 141 (records)

Schillhammer, 1996: 230 (records)

Kofler, 1998: 651 (fungal hosts)

Geiser, 1999: 380 (records)

Byk, 2001: 354 (records)

Matějíček & Boháč, 2003: 131 (records)

Borowski, 2006: 54–58 (records, fungal hosts)

Byk et al., 2006: 337 (habitat, records)

Konvička et al., 2009: 8 (records)

Monsevičius, 2013: 27 (records)

Solodovnikov, 2016: 127 (records)
Hamed & Vancl, 2016: 79 (records)
Frisch, 2017: 245, Fig. 1 (characters, distribution, records)

DISTRIBUTION. Austria; Belarus; Bosnia and Herzegovina; Bulgaria; Czech Republic; Georgia; Germany; Hungary; Italy; Lithuania; Poland; Romania; Slovakia; Switzerland; Ukraine.

***Scaphisoma obliquemaculatum* Motschulsky**

Scaphisoma obliquemaculatum Motschulsky, 1863: 435. Lectotype female, ZMUM; type locality: Sri Lanka: Nura-Ellia [Nuwara Eliya].

Pseudoscaphosoma atrithorax Pic, 1916d: 7. Lectotype female, MNHN; type locality: Indonesia: Kalimantan [Martapura].

Scaphosoma rufomaculatum Pic, 1921c: 5. Lectotype male, MNHN; type locality: Indonesia: Sumatra [Padar Mardang].

Scaphosoma luteoapicale Pic, 1923b: 17. Lectotype male, MNHN; type locality: Vietnam: Tonkin [Lac Tho].

Löbl, 1971c: 984, Figs 54, 55 (lectotype designation for *S. obliquemaculatum*, characters, records)
Löbl, 1975b: 273 (lectotype fixations by inference for *S. rufomaculatum* and *S. luteoapicale*, synonymy with *S. obliquemaculatum*, records)
Löbl, 1977c: 221 (records)
Löbl, 1977f: 43 (records)
Löbl, 1990b: 573 (records)
Löbl, 2015a: 111 (records)
Löbl, 2015c: 135 (lectotype designation for *P. atrithorax*, synonymy with *S. obliquemaculatum*, records)
Lecoq, 2015: 211 (records)

DISTRIBUTION. Indonesia: Java, Kalimantan, Sulawesi, Sumatra, Sumbawa; East Malaysia: Sarawak; Mascarene Archipelago; Sri Lanka; Thailand; Vietnam.

***Scaphisoma oblongum* Pic**

Scaphosoma oblongum Pic, 1916b: 4. Lectotype female, MNHN; type locality: Indonesia: Nias [Lahago].

Löbl, 2015c: 136 (lectotype designation)

DISTRIBUTION. Indonesia: Nias.

***Scaphisoma obscurum* Löbl & Ogawa**

Scaphisoma obscurum Löbl & Ogawa, 2016b: 1378, Figs 103, 104. Holotype male, MHNG; type locality: Philippines: Palawan, Sabang, sea level.

DISTRIBUTION. Philippines: Palawan.

Scaphisoma occidentale Champion

Scaphisoma occidentale Champion, 1913: 69. Holotype, NHML; type locality: Mexico: Omiltene, Guerero, 8000 ft.

DISTRIBUTION. Mexico.

Scaphisoma ochropenne Löbl & Ogawa

Scaphisoma ochropenne Löbl & Ogawa, 2916b: 1410, Figs 24, 168–170. Holotype male, FMNH; type locality: Philippines: Mindanao, Davao Prov., Meran, east slope of Mt. Apo, 6000 ft.

DISTRIBUTION. Philippines: Mindanao.

Scaphisoma onychionum Löbl

Scaphisoma onychiunum Löbl, 1986a: 165, Figs 29–32. Holotype male, MHNG; type locality: India: Mawphlang, 1800 m, Khasi Hills.

Löbl, 1997: 122 (misspelled *onychiorum*)

DISTRIBUTION. India: Meghalaya.

Scaphisoma opacum Löbl & Ogawa

Scaphisoma opacum Löbl & Ogawa, 2016b: 1383, Figs 25, 112, 113. Holotype male, FMNH; type locality: Philippines: Mindanao, Davao Prov., east slope of Mt. McKinley, 3200 ft.

DISTRIBUTION. Philippines: Mindanao.

Scaphisoma operosum Löbl

Scaphisoma operosum Löbl, 1990b: 556, Figs 69–71. Holotype male, MHNG; type locality: West Malaysia: Cameron Highlands.

DISTRIBUTION. West Malaysia.

Scaphisoma opochti Fierros-López

Scaphisoma opochti Fierros-López, 2006a: 61, Figs 1c, 2c, 3c, 4g-i, 6c. Holotype male, CZUG; type locality: Mexico: Morelos, Tlayacapan, San José de los Laureles, Camino a Amatlán, 1721 m.

DISTRIBUTION. Mexico.

Scaphisoma oppositum Löbl

Scaphisoma oppositum Löbl, 2000: 626, Figs 27, 28. Holotype male, MHNG; type locality: China: Yunnan, Gaoligong Shan, 90 km west Baoshan.

DISTRIBUTION. Yunnan.

Scaphisoma ornatipenne **Löbl**

Scaphisoma ornatipenne Löbl, 1975a: 400, Figs 46, 48, 49. Holotype male, HNHM; type locality: Papua New Guinea: Mc. Adams Park, Wau.

DISTRIBUTION. Papua New Guinea.

Scaphisoma ornatum **Fall**

Scaphisoma ornata Fall, 1910: 117. Lectotype female, MCZC; type locality: USA: Mobile, Alabama.

Leschen et al., 1990: 281, Figs 4, 14–16 (lectotype designation, characters, records)

DISTRIBUTION. United States: Alabama, Oklahoma.

Scaphisoma oviforme **Löbl & Ogawa**

Scaphisoma oviforme Löbl & Ogawa, 2016b: 1393, Figs 129, 130. Holotype male, FMNH; type locality: Philippines: Mindanao, Cotabato Prov., Burungkot, Upi, 1500 ft.

DISTRIBUTION. Philippines: Mindanao.

Scaphisoma palawanum **Pic**

Scaphosoma palawanum Pic, 1926b: 1. Lectotype male, MNHN; type locality: Philippines: Palawan, Binaluan.

Löbl, 1972b: 103, Figs 23, 24 (lectotype designation, characters)

Löbl & Ogawa, 2016b: 1394, Figs 26, 131, 132 (characters, records)

DISTRIBUTION. Palawan.

Scaphisoma paliferum **Löbl**

Scaphisoma paliferum Löbl, 1984c: 999, Figs 6–8. Holotype male, MHNG; type locality: Myanmar: Kalawa, 1300 m, Shan State.

Löbl, 1990b: 566 (records)

DISTRIBUTION. Myanmar; Thailand.

Scaphisoma pallens **Löbl**

Scaphisoma pallens Löbl, 1981a: 370, Figs 30, 31. Holotype male, NZAC; type locality: New Caledonia: Mt. Koghi, 500 m.

DISTRIBUTION. New Caledonia.

Scaphisoma palposum **Löbl**

Scaphisoma palposum Löbl, 2011c: 110. Holotype female, MHNG; type locality: Taiwan: Hsinchu Co., near Hsinkuang vill, Henhan Township, km 48 road no. 60, ca 1800 m.

DISTRIBUTION. Taiwan.

***Scaphisoma palu* Löbl**

Scaphisoma palu Löbl, 1983a: 290, Figs 9, 10. Holotype male, MHNG; type locality: Indonesia: Sulawesi, Rontepao.

Ogawa, 2015: 84, Figs 4-14g, h (characters, records)

DISTRIBUTION. Indonesia: Sulawesi.

***Scaphisoma palumboi palumboi* (Ragusa)**

Baeocera palumboi Ragusa, 1892: 255. Lectotype female, MZUC; type locality: Italy: Castelvetrano.

Tamanini, 1969b: 370, Figs 6E-H (lectotype designation, transfer from *Baeocera*, characters)

Löbl, 1970c: 744, Figs 9, 10 (characters, records)

Tamanini, 1970: 20, Figs 6E-H, 7A, 10H (characters)

Poggi, 1983: 156 (records)

DISTRIBUTION. Italy: Sicily.

***Scaphisoma palumboi erratum* Löbl**

Scaphosoma erratum Löbl, 1965d: 334, Figs 1, 3, 4. Holotype male, NMPC; type locality: Algeria: Dj. Edough Bugeaud.

Normand, 1934: 45 (as *S. agaricinum*, records)

Löbl, 1966b: 132 (records)

Löbl, 1967d: 34, Figs 2a, b (characters)

Tamanini, 1969a: 487, Figs 8, 12 (characters)

Tamanini, 1969b: Fig. 6I

Löbl, 1970c: 745, Figs 13, 14 (downgraded to subspecies, characters, records)

Löbl, 1974a: 61 (records)

Löbl, 1989a: 11 (records)

DISTRIBUTION. Algeria; Gibraltar; Morokko; Tunisia.

***Scaphisoma palumboi ruffoi* Tamanini**

Scaphisoma ruffoi Tamanini, 1969a: 483, Figs 1–4. Holotype male, MSNV; type locality: Italy: Sorgano-Sardo.

Löbl, 1970c: 745, Figs 11, 12 (downgraded to subspecies, characters, records)

Tamanini, 1970: 22, Figs 7B-D, 10I (as species, characters)

DISTRIBUTION. Italy: Sardinia.

***Scaphisoma pandanum* Löbl & Ogawa**

Scaphisoma pandanum Löbl & Ogawa, 2016b: 1394, Figs 27, 133, 134. Holotype male, MHNG; type locality: Philippines: Mindoro, North Pandan Island, sea level.

DISTRIBUTION. Philippines: Mindoro.

***Scaphisoma papuanum* Löbl**

Scaphisoma papuanum Löbl, 1975a: 392, Figs 32, 33. Holotype male, MHNG; type locality: Papua New Guinea: Sattelberg [Sattelburg].

DISTRIBUTION. Papua New Guinea.

***Scaphisoma papuum* (Löbl)**

Metalloscapha papua Löbl, 1975a: 384, Figs 19–25. Holotype male, NHML; type locality: Papua New Guinea: Mt. Abilala, ca 9000 ft., Finistere Mts., Morobe District.

Löbl, 2003c: 182 (transfer to *Scaphisoma*)

DISTRIBUTION. Papua New Guinea.

***Scaphisoma paraboleti* Löbl**

Scaphisoma paraboleti Löbl, 1993: 38, Figs 3, 4. Holotype male, MHNG; type locality: Russian Far East: Kamenushka, east Ussuriysk.

DISTRIBUTION. Russia: Far East.

***Scaphisoma parasolutum* Löbl**

Scaphisoma parasolutum Löbl, 2000: 631, Figs 35, 36. Holotype male, MHNG; type locality: China: Yunnan, Mengyang Nat. Res., ca 500 m.

DISTRIBUTION. China: Yunnan.

***Scaphisoma paravarium* Löbl**

Scaphisoma paravarium Löbl, 2000: 638, Figs 51–55. Holotype male, MHNG; type locality: China: Yunnan, Gaoligong Mts., 2200–2500.

DISTRIBUTION. China: Yunnan.

***Scaphisoma pauliani* Pic**

Scaphisoma subconvexum var. *pauliani* Pic, 1942: 1 (nec *Scaphisoma subconvexum* Pic, 1926). Syntypes, MNHN; type locality: Cameroun, SE Mt. Cameroon, 1800–2000 m.

Löbl, 1997: xii (*S. pauliani* as available, to replace the homonymous *S. subconvexum* Pic, 1930; see under *S. semialutaceum* Pic, 1930).

DISTRIBUTION. Cameroon.

***Scaphisoma pauloatrum* Pic**

Scaphosoma pauloatrum Pic, 1954b: 36. Holotype, MRAC; type locality: Democratic Republic of the Congo, Kundelungi Massif.

DISTRIBUTION. Democratic Republic of the Congo.

Scaphisoma pecki **Löbl**

Scaphisoma pecki Löbl, 1982h: 21. Holotype female, MHNG; type locality: Japan: Ishizuchi Nat. Park, Omogo Valley, 700 m, Ehime Pref.

DISTRIBUTION. Japan: Shikoku.

Scaphisoma penangense **Löbl**

Scaphisoma penangense Löbl, 1986b: 95, Fig. 13. Holotype male, MHNG; type locality: West Malaysia: Penang.

DISTRIBUTION. West Malaysia.

Scaphisoma peninsulare **Horn**

Scaphisoma peninsulare Horn, 1894: 363. Syntypes, MCZC; type locality: Mexico: Sierra dw la Laguna, Baja California.

DISTRIBUTION. Mexico: Baja California.

Scaphisoma peraffine **Löbl**

Scaphisoma peraffine Löbl, 1986a: 187, Figs 68, 69. Holotype male, MHNG; type locality: India: Manas Wild Life Sactuary, 200 m, Assam.
Löbl, 1986c: 347 (records)

DISTRIBUTION. India: Assam, Uttarakhand (Kumaon).

Scaphisoma peraffirmatum **Löbl**

Scaphisoma peraffirmatum Löbl, 2015b: 177, Figs 35–38. Holotype male, SMNS; type locality: Indonesia: Maluku Islands, Halmahera Is., 28 km south Tobelo, Toguliua, 200 m.

DISTRIBUTION. Indonesia: Maluku Islands: Halmahera Is.

Scaphisoma perbrincki **Löbl**

Scaphisoma perbrincki Löbl, 1971c: 965, Figs 28, 29. Holotype male, MHNG; type locality: Sri Lanka: Diyaluma Falls, ca 400 m.

DISTRIBUTION. India: Goa; Sri Lanka.

Scaphisoma perdecorum **Löbl**

Scaphisoma perdecorum Löbl, 2015b: 179, Figs 39–41. Holotype male, SMNS; type locality: Indonesia: Maluku Islands, Halmahera Is., Tobelo (SW).

DISTRIBUTION. Indonesia: Maluku Islands: Halmehera Is.

Scaphisoma peregrinum **Löbl**

Scaphisoma peregrinum Löbl, 2015b: 184, Figs 48–51. Holotype male, SMNS; type locality: Indonesia: Maluku Islands, Ternate Island, Marikurubu, Gn. Gamalama, 700–1500 m.

DISTRIBUTION. Indonesia: Maluku Islands: Ternate Is.

Scaphisoma perelegans **Blackburn**

Scaphisoma perelegans Blackburn, 1903: 98. Holotype female, NHML; type locality: Australia: Victoria.
Löbl, 1977e: 27, Figs 16, 34, 35 (characters, records)
DISTRIBUTION. Australia: Victoria.

Scaphisoma perfectum **Löbl**

Scaphisoma perfectum Löbl, 2015b: 174, Figs 28, 29. Holotype male, SMNS; type locality: Indonesia: Maluku Islands, Morotai Is., W. Daruba, Raja, 100 m.
DISTRIBUTION. Indonesia: Maluku Islands: Morotai Is.

Scaphisoma perforatum **Pic**

Scaphosoma perforatum Pic, 1926a: 322. Syntypes, MNHN; type locality: Vietnam: Tonkin, Hoa Binh.
DISTRIBUTION. Vietnam.

Scaphisoma perkinsi **Scott**

Scaphisoma perkinsi Scott, 1908: 534. Syntypes, NHML; type localities: USA: Hawaii, Oahu, Mokuleiia, Waianae and Kaala Mts.
Löbl, 1981b: 74, Fig. 3 (characters, records)
Morimoto, 1985: 256 (records)
DISTRIBUTION. Hawaii; Japan: Ogasawara.

Scaphisoma perleve **Löbl**

Scaphisoma perleve Löbl, 2015b: 173, Figs 24, 25. Holotype male, SMNS; type locality: Indonesia: Maluku Islands, Halmahera Is., Ibu, Kamp. Baru, Gn. Alon, 800 m.
DISTRIBUTION. Indonesia: Maluku Islands: Halmahera Is.

Scaphisoma perminutum **Löbl**

Scaphisoma perminutum Löbl, 2015b: 171, Figs 20–23. Holotype male, SMNS; type locality: Indonesia: Maluku Islands, Ternate Island, Marikurubu, Gn. Gamalama, 700–1500 m.
DISTRIBUTION. Indonesia: Maluku Islands: Ternate Is.

Scaphisoma permixtum **Löbl**

Scaphisoma permixtum Löbl, 2015b: 182, Figs 46, 47. Holotype male, SMNS; type locality: Indonesia: Maluku Islands, Halmahera Is., 28 km south Tobelo, Togoliua, ca 200 m.
DISTRIBUTION. Indonesia: Maluku Islands: Halmahera Is.

***Scaphisoma permutatum* Löbl**

Scaphisoma permutatum Löbl, 2015b: 177, Figs 32–34. Holotype male, SMNS; type locality: Indonesia: Maluku Islands, Halmahera Is., 28 km south Tobelo, 100 m.

DISTRIBUTION. Indonesia: Maluku Islands: Halmahera Is.

***Scaphisoma perpusillum* Löbl**

Scaphisoma perpusillum Löbl, 1973b: 316, Figs 11, 12. Holotype male, NHMW; type locality: New Caledonia: Tiouandé near Hienghène.
Löbl, 1974b: 407 (records)
Löbl, 1977i: 818 (records)
Löbl, 1981a: 366 (characters, records)

DISTRIBUTION. New Caledonia.

***Scaphisoma perrieri* Achard**

Scaphosoma perrieri Achard, 1920i: 240. Syntypes, NMPC; type locality: Madagascar: Suberbieville [Maevatanana District].

DISTRIBUTION. Madagascar.

***Scaphisoma persimilans* Löbl**

Scaphisoma persimilans Löbl, 2015b: 174, Figs 26, 27. Holotype male, SMNS; type locality: Indonesia: Maluku Islands, Halmahera Is., Sidangoli, Batu putih, 100 m.

DISTRIBUTION. Indonesia: Maluku Islands: Halmahera Is.

***Scaphisoma persimile* Löbl**

Scaphisoma persimile Löbl, 1977f: 49, Figs 17, 18. Holotype male, MHNG; type locality: La Réunion: Plaine des Chicots, below the Gîte, 1800 m.
Lecoq, 2015: 211 (records)

DISTRIBUTION. La Réunion.

***Scaphisoma pertubator* Löbl**

Scaphisoma pertubator Löbl, 2015b: 176, Figs 30, 31. Holotype male, SMNS; type locality: Indonesia: Maluku Islands, Ternate Is., Marikurubu, Gn. Gamalama, 300–600 m.

DISTRIBUTION. Indonesia: Maluku Islands: Ternate Isl.

***Scaphisoma peterseni* Löbl**

Scaphisoma peterseni Löbl, 1971a: 248, Figs 1, 2. Holotype male, ZMUC; type locality: Papua New Guinea: New Ireland, Lemkamin, 900 m.

DISTRIBUTION. Papua New Guinea: New Ireland.

Scaphisoma phalacroide **Pic**

Scaphosoma phalacroide Pic, 1920e: 96. Holotype male, MCSN; type locality: Brazil: Santos.

DISTRIBUTION. Brazil.

Scaphisoma philippinense **Oberthür**

Scaphisoma philippinense Oberthür, 1883: 14. Lectotype female, MNHN; type locality: Philippines: Kingua [?Kinga].

Heller, 1917: 45 (characters)

Löbl, 1971a: 249 (characters)

Löbl & Ogawa, 2016b: 1442 (lectotype designation, characters)

DISTRIBUTION. Philippines: ?Luzon.

Scaphisoma phungi **Pic**

Scaphosoma phungi Pic, 1922a: 1. Syntypes, MNHN, lost; type locality: Vietnam: Tonkin.

Pic, 1925a: 195 (transfer to *Macrobaeocera*, characters)

Löbl, 1975b: 270 (transfer to *Scaphisoma*, characters)

DISTRIBUTION. Vietnam.

Scaphisoma piceicolle piceicolle **Pic**

Scaphosoma piceicolle Pic, 1930c: 176. Syntypes, MACN; type locality: Argentina: Tucuman.

DISTRIBUTION. Argentina.

Scaphisoma piceicolle boxi **Pic**

Scaphosoma piceicolle var. *boxi* Pic, 1930c: 176. Syntypes, MBRA/MNHN; type locality: Argentina: Tucuman.

DISTRIBUTION. Argentina.

Scaphisoma piceonotatum **Pic**

Scaphosoma piceonotatum Pic, 1930c: 89. Syntypes, MRAC/MNHN; type locality: Democratic Republic of the Congo: Matenda.

DISTRIBUTION. Democratic Republic of the Congo.

Scaphisoma pici **Löbl**

Baeocera discoidalis Pic, 1916: 7. Lectotype female, MNHN; type locality: Indonesia: Kalimantan, Martapura.

Scaphosoma pici Löbl, 1979a: 102; replacement name for *Baeocera discoidalis* Pic, 1916 (nec *Scaphisoma binotatum* var. *discoidale* Achard, 1915).

Löbl, 1973a: 158 (lectotype fixed by inference for *B. discoidalis*, transfer to *Scaphisoma*, characters)
Löbl, 2015c: 135 (note)
DISTRIBUTION. Indonesia: Kalimantan.

Scaphisoma pictum Motschulsky
Scaphisoma pictum Motschulsky, 1863: 435. Lectotype male, ZMUM; type locality: Sri Lanka: Nura-Ellia [Nuwara Eliya].
Scott, 1922: 222 (missidentified *S. nigrofasciatum* Pic)
Vinson, 1943: 189 (missidentified *S. nigrofasciatum* Pic)
Löbl, 1971c: 976, Figs 44, 45 (lectotype designation, characters, records)
Löbl, 1979a: 107 (records)
DISTRIBUTION. India: Kerala, Tamil Nadu; Sri Lanka.

Scaphisoma pinnigerum Löbl
Scaphisoma pinnigerum Löbl, 1992a: 555, Fig. 152. Holotype male, MHNG; type locality: Nepal: Phulcoki near Godawari, 1700 m, Patan District.
DISTRIBUTION. Nepal.

Scaphisoma planum Löbl
Scaphisoma planum Löbl, 1977e: 24, Figs 30, 31. Holotype male, SAMA; type locality: Australia: Meirose, South Australia.
DISTRIBUTION. Australia: South Australia, Victoria.

Scaphisoma pocsi Löbl
Scaphisoma pocsi Löbl, 1981f: 110, Figs 9, 10. Holotype male, HNHM; type locality: Vietnam: Cuc Phuong Bong.
DISTRIBUTION. Vietnam.

Scaphisoma politum Macleay
Scaphisoma politum Macleay, 1871: 156. Lectotype female, ANIC; type locality: Australia: Queensland, Gayndah.
Löbl, 1977e: 27, Figs 36, 37 (lectotype designation, characters, records)
DISTRIBUTION. Australia: New South Wales, Queensland.

Scaphisoma portevini Pic
Scaphosoma portevini Pic, 1920b: 5. Lectotype male, MNHN; type locality: China [Yunnan].
Scaphosoma taliense Achard, 1920a: 328. Syntypes, MNHN/NMPC; type locality: China: Yunnan, Tali.

Achard, 1920d: 129 (synonymy of *S. taliense* with *S. portevini*)
Löbl, 1965e: 29, Figs 2, 2a, b (lectotype designation for *S. portevini*, characters)
Löbl, 2000: 610 (records)
Löbl, 2003a: 63 (records)
DISTRIBUTION. China: Anhui, Guangxi, Sichuan, Yunnan; India: Mahdya Pradesh; Japan; Korea.

Scaphisoma poussereaui **Löbl**
Scaphisoma poussereaui Löbl, 2015d: 368, Figs 1–3. Holotype male, MNHN; type locality: La Réunion: chemin de Centure, Maison Boyer.
DISTRIBUTION. La Réunion.

Scaphisoma praesigne **Löbl**
Scaphisoma praesigne Löbl, 1992a: 542, Figs 127, 128. Holotype male, MHNG; type locality: Nepal: Gokarna, 1400 m, Kathmandu District.
DISTRIBUTION. Nepal.

Scaphisoma prehensor **Champion**
Scaphosoma prehensor Champion, 1927: 277. Holotype male, NHML; type locality: India: Haldwani, Kumaon.
Löbl, 1986a: 203, Figs 95–97 (characters, records)
Löbl, 1986c: 348
Löbl, 1990a: 121 (records)
Löbl, 1992a: 536 (records)
DISTRIBUTION. India: Himachal Pradesh, Meghalaya, Uttarakhand (Kumaon), West Bengal (Darjeeling District); Nepal.

Scaphisoma pressum **Löbl**
Scaphisoma pressum Löbl, 1990b: 571, Figs 92–96. Holotype male, MHNG; type locality: Thailand: Khao Yai Nat. Park, near Headquarters, 750–850 m.
Löbl, 2000: 610 (records)
DISTRIBUTION. China: Yunnan; Thailand.

Scaphisoma promtum **Löbl**
Scaphisoma promptum Löbl, 1977e: 54, Figs 24, 82, 83. Holotype male, QMBA; type locality: Australia: Bunya Mts, Queensland.
DISTRIBUTION. Australia: New South Wales, Queensland.

Scaphisoma propinquum Löbl

Scaphisoma propinquum Löbl, 1977b: 50. Holotype male, NHML; type locality: Australia: Port Darwin.

DISTRIBUTION. Australia: Northern Territory.

Scaphisoma prostratum Löbl

Scaphisoma prostratum Löbl, 2003a: 72, Figs 18–20. Holotype male, ZIBC; type locality: China: Jiangxi, Wuyi Shan, Huanggashan, 2100 m.

DISTRIBUTION. China: Jiangxi.

Scaphisoma pseudamabile Löbl

Scaphisoma pseudamabile Löbl, 1990b: 567, Figs 84–86. Holotype male, MHNG; type locality: Thailand: Khao Yai Nat. Park, near Headquarters, 750–850 m.

DISTRIBUTION. Thailand.

Scaphisoma pseudantennatum Löbl

Scaphisoma pseudantennatum Löbl, 2000: 625, Figs 24–26. Holotype male, MHNG type locality: China: Sichuan, Wolong Nat. Res., 1500 m.

DISTRIBUTION. China: Sichuan.

Scaphisoma pseudatrox Löbl

Scaphisoma pseudatrox Löbl, 1990b: 564, Figs 82, 83. Holotype male, MHNG; type locality: Thailand: Khao Sabap Nat. Park, near Phliu Waterfalls, 150–300 m.

DISTRIBUTION. Thailand.

Scaphisoma pseudodelictum Löbl

Scaphisoma pseudodelictum Löbl, 1986a: 180, Figs 54, 55. Holotype male, MHNG; type locality: India: Sevoke, 200 m, Darjeeling District.
Löbl, 1990b: 565 (records)
Löbl, 2000: 610 (records)
Löbl, 2003a: 63 (records)

DISTRIBUTION. China: Jiangxi, Yunnan; India: Assam, Meghalaya, West Bengal (Darjeeling District); Thailand.

Scaphisoma pseudofasciatum Löbl

Scaphisoma pseudofasciatum Löbl, 1975a: 390, Figs 28, 29. Holotype male, HNHM; type locality: Papua New Guinea: Sattelberg [Sattelburg], Huon Golf.

DISTRIBUTION. Papua New Guinea.

Scaphisoma pseudokalabitum **Löbl & Ogawa**

Scaphisoma pseudokalabitum Löbl & Ogawa, 2016b: 1411, Figs 28, 171–173. Holotype male, MHNG; type locality: Philippines: Palawan, Central, Cabayugan near Lion's Cave, sea level.

DISTRIBUTION. Philippines: Palawan.

Scaphisoma pseudorubellum **Löbl**

Scaphisoma pseudorubellum Löbl, 1977f: 45; Figs 7, 8. Holotype male, MHNG; type locality: La Réunion: Plaine des Marsouins.

Lecoq, 2015: 211 (records)

DISTRIBUTION. La Réunion.

Scaphisoma pseudorufum **Löbl**

Scaphisoma pseudorufum Löbl, 1986a: 143, Fig. 3. Holotype male, MHNG; type locality: India: Sukna, 200 m, Darjeeling District.

Löbl, 1992a: 531 (records)

Löbl, 2000: 611 (records)

Löbl, 2012b: 183 (records)

DISTRIBUTION. China: Guangxi, Yunnan; India: West Bengal (Darjeeling District); Nepal; West Malaysia.

Scaphisoma pseudosolutum **Löbl**

Scaphisoma pseudosolutum Löbl, 2000: 632, Figs 37–39, 41. Holotype male, MHNG; type locality: China: Hubei, Shennongjia Nat. Res., 2000–2200 m.

DISTRIBUTION. China: Hubei.

Scaphisoma pseudovarium **Löbl**

Scaphisoma pseudovarium Löbl, 2000: 636, Figs 43–46. Holotype male, MHNG; type locality: Yunnan, Gaoligong Mts., 2200–2500 m.

DISTRIBUTION. China: Yunnan.

Scaphisoma pulchellum **Löbl**

Scaphosoma ornatum Champion, 1927: 275. Holotype female, NHML; type locality: India: R. Sarda Gorge, Kumaon.

Scaphisoma pulchellum Löbl, 1986a: 165; replacement name for *Scaphosoma ornatum* Champion, 1927 (nec *Scaphisoma ornatum* Fall, 1910).

Löbl, 1986a: 165 (characters, records)

Löbl, 1986b: 89, Figs 4, 5 (characters, records)

Löbl, 1992a: 586, Fig. 32 (characters)

DISTRIBUTION. India: Uttarakhand (Kumaon), West Bengal (Darjeeling District); Singapore; Thailand.

***Scaphisoma pulchrum* Löbl & Ogawa**

Scaphisoma pulchrum Löbl & Ogawa, 2016b: 1432, Figs 228–230. Holotype male, FMNH; type locality: Philippines: Mindanao, Davao Province, Meran, east slope of Mt. Apo, 6000 ft.

DISTRIBUTION. Philippines: Mindanao.

***Scaphisoma punctaticolle* Löbl**

Scaphisoma punctaticolle Löbl, 1980a: 103, Figs 16, 17. Holotype male, NSMT; type locality: Taiwan: Rokkiri.

DISTRIBUTION. Taiwan.

***Scaphisoma punctatipenne* Pic**

Scaphosoma oblongum var. *punctatipenne* Pic, 1916b: 4. Lectotype female, MNHN; type locality: Indonesia: Kalimantan, Martapura.
Löbl, 2015c: 136 (lectotype designation, status, characters)

DISTRIBUTION. Indonesia: Kalimantan.

***Scaphisoma punctatissimum* Matthews**

Scaphisoma punctatissimum Matthews, 1888: 177. Syntypes, NHML; type locality: Guatemala: San Isidro.

DISTRIBUTION. Guatemala.

***Scaphisoma punctatum* (Pic)**

Pseudoscaphosoma punctatum Pic, 1915b: 31. Lectotype female, MNHN; type locality: Indonesia: Kalimantan [Martapura].
Löbl, 1975b: 270 (transfer to *Scaphisoma*)
Löbl, 2015c: 137 (lectotype designation, characters)

DISTRIBUTION. Indonesia: Kalimantan.

***Scaphisoma puncticolle* Matthews**

Scaphisoma puncticolle Matthews, 1888: 176. Syntypes, NHML; type locality: Panama: Volcan de Chiriqui 3000 ft.

DISTRIBUTION. Panama.

***Scaphisoma punctulatipenne* Löbl**

Scaphosoma punctatum Pic, 1946: 84. Syntypes, MNHN; type locality: Kenya: Mt. Elgon.

Scaphisoma punctulatipenne Löbl, 1997: xii; replacement name for *Scaphosoma punctatum* Pic, 1946 (secondary junior homonym with *Pseudoscaphosoma punctatum* Pic, 1915).

DISTRIBUTION. Kenya.

***Scaphisoma punctulatum* LeConte**

Scaphisoma punctulatum LeConte, 1860: 323. Holotype female, MCZC; type locality: USA: Georgia.

Casey, 1893: 526 (characters, records)

Blatchley, 1910: 496 (characters, records)

Leschen et al., 1990: 291, Figs 11, 37, 38 (characters, records)

Peck & Thomas, 1998: 46 (distribution)

DISTRIBUTION. United States: Florida, Georgia, Indiana, Missouri, Oklahoma.

***Scaphisoma pusillum* LeConte**

Scaphisoma pusillum LeConte, 1860: 323. Lectotype male, MCZC; type locality: USA: South Carolina.

Scaphiomicrus dimidiatus Casey, 1900: 59. Syntypes, USNM; type locality: USA: Boston Neck, Rhode Island.

Casey, 1893: 532 (characters, records)

Casey, 1900: 59 (records)

Fall, 1910: 119 (synonymy of *S. dimidiatus* with *S. pusillum*, characters)

Brimley, 1938: 149 (records)

Leschen et al., 1990: 281, Figs 5, 17, 18 (lectotype designation for *S. pusillum*, characters, records, fungal hosts)

Downie & Arnett, 1996: 366 (records)

DISTRIBUTION. United States: Arizona, Florida, Georgia, Iowa, Kansas, Massachusetts, Michigan, Missouri, New Hampshire, North Carolina, Oklahoma, Rhode Island, South Carolina, Virginia.

***Scaphisoma puthzi* Löbl**

Scaphisoma puthzi Löbl, 1986a: 153, Figs 14, 15. Holotype male, MHNG; type locality: India: Tura Peak, 700–900 m, Garo Hills.

DISTRIBUTION. India: Meghalaya.

***Scaphisoma quadratum* Oberthür**

Scaphisoma quadratum Oberthür, 1883: 13. Syntypes, MNHN; type locality: Republic of South Africa: Transvaal.

Reitter, 1889: 6 (probably member of *Baeoceridium*)

DISTRIBUTION. Republic of South Africa.

***Scaphisoma quadrifasciatum* Löbl**

Scaphisoma quadrifasciatum Löbl, 1986a: 160, Figs 24, 24a, 25. Holotype male, MHNG; type locality: India: between Algarah and Labha, 1900 m, Darjeeling District.

Löbl, 1986c: 345 (records)
Löbl, 1992a: 532, 587, Fig. 33 (characters, records)

DISTRIBUTION. India: Himachal Pradesh, Uttarakhand, West Bengal (Darjeeling District); Nepal; Pakistan.

***Scaphisoma quadrimaculatum* Pic**

Scaphosoma quadrimaculatum Pic, 1922a: 1. Lectotype female, MNHN; type locality: Vietnam: Tonkin, Lac Tho.
Löbl, 1981f: 105, Figs 1, 2 (lectotype fixed by inference, characters, records)
Löbl, 1986a: 160 (records)
Löbl & Ogawa, 2016b: 1379 (records)

DISTRIBUTION. India: Assam; Indonesia: Sumatra; Laos; Philippines. Palawan; Vietnam.

***Scaphisoma quadripunctatum* (Pic)**

Amalocera quadripunctata Pic, 1956b: 72. Lectotype male, HNHM; type locality: Papua New Guinea: Erima, Astrolabe Bay.
Löbl, 1975a: 405, Figs 58, 59 (lectotype fixed by inference, transfer to *Scaphisoma*, characters, records)

DISTRIBUTION. Papua New Guinea.

***Scaphisoma queenslandicum* Blackburn**

Scaphisoma queenslandicum Blackburn, 1903: 98. Lectotype female, NHML; type locality: Australia: Cairns, Queensland.
Löbl, 1977e: 40, Figs 60, 61 (lectotype fixed by inference, characters, records)

DISTRIBUTION. Australia: New South Wales, Queensland, Victoria.

***Scaphisoma raffrayi* Pic**

Scaphosoma raffrayi Pic, 1916c: 18. Syntypes, MNHN; type locality: Ethiopia.

DISTRIBUTION. Ethiopia.

***Scaphisoma ramosum* Löbl**

Scaphisoma ramosum Löbl, 1972b: 93, Fig. 31. Holotype male, ZMUB; type locality: Philippines: Masbate, Aroroy.
Löbl & Ogawa, 2016b: 1395 (note)

DISTRIBUTION. Philippines: Masbate.

***Scaphisoma rarum* Löbl**

Scaphisoma rarum Löbl, 1971c: 968, Figs 32, 33. Holotype male, MHNG; type locality: Sri Lanka: Nedunleni.

DISTRIBUTION. Sri Lanka.

***Scaphisoma rasum* Löbl**

Scaphisoma rasum Löbl, 1977e: 30, Figs 42, 43. Holotype male, SAMA; type locality: Australia: New South Wales.

DISTRIBUTION. Australia: New South Wales, Queensland.

***Scaphisoma remingtoni* Löbl**

Scaphisoma remingtoni Löbl, 1974b: 405, Figs 1, 2. Holotype male, FMNH; type locality: New Caledonia: 7 mi SE La Foa.
Löbl, 1977i: 818 (records)
Löbl, 1981a: 368 (characters, records)

DISTRIBUTION. New Caledonia.

***Scaphisoma renominatum* Löbl**

Amalocera suturalis Achard, 1920d: 128. Lectotype male, MNHN; type locality: East Malaysia: Banguey [Banggi Is.].
Scaphisoma renominatum Löbl, 1975b: 272; replacement name for *Amalocera suturalis* Achard, 1920 (nec *Scaphisoma suturale* LeConte, 1860).
Löbl, 1975b: 272 (lectotype designation, transfer to *Scaphisoma*, characters)

DISTRIBUTION. East Malaysia: Banggi Is.

***Scaphisoma repandum* Casey**

Scaphisoma repanda Casey, 1893: 525. Syntypes, USNM; type localities: USA: Iowa, Missouri, Massachusetts.
Campbell, 1991: 125 (records)
Webster et al., 2012: 246 (records, habitat)

DISTRIBUTION. Canada: New Brunswick, Ontario; United States: Iowa, Massachusetts, Missouri, New Hampshire.

***Scaphisoma reticulatum* Löbl**

Scaphisoma reticulatum Löbl, 1973b: 327, Figs 29, 30. Holotype male, NHMW; type locality: New Caledonia: Pic du Pin.
Löbl, 1981a: 374 (characters, records)

DISTRIBUTION. New Caledonia.

***Scaphisoma riedeli* Löbl**

Scaphisoma riedeli Löbl, 2014b: 62, Figs 1–5. Holotype male, SMNS; type locality: Indonesia: Western New Guinea, Baliem valley, Jiwika-Wandaku, 1700–2300 m.

DISTRIBUTION. Indonesia: Western New Guinea.

Scaphisoma robustum **Pic**

Scaphosoma robustum Pic, 1955: 52. Holotype, MRAC; type locality: Rwanda: Rubengera, 1900 m, terr. Kibuye.

DISTRIBUTION. Rwanda.

Scaphisoma rodolphei **Pic**

Scaphosoma rodolphei Pic, 1946: 84. Syntypes, MNHN; type locality: Ethiopia: Nanoropus.

DISTRIBUTION. Ethiopia.

Scaphisoma rougemonti **Löbl**

Scaphisoma rougemonti Löbl, 1984c: 996, Fig. 3. Holotype male, MHNG; type locality: Myanmar: Kalawa, 1300 m, Shan State.

Löbl, 1990b: 564 (records)

DISTRIBUTION. Myanmar; Thailand.

Scaphisoma rouyeri **Pic**

Scaphosoma (*Scutoscaphosoma*) *rouyeri* Pic, 1916b: 3. Lectotype male, MNHN; type locality: Indonesia: Java, Mt. Smerou [Mt. Semeru].

Scutoscaphosoma subovatum Pic, 1920f: 24. Lectotype male, MNHN; type locality: East Malaysia: [Sarawak].

Löbl, 1981d: 156, Figs 1–3, 17–19 (lectotype designations for *S. rouyeri* and *S. subovatum*, synonymy, transfer to *Scaphisoma*, characters, records)

Löbl, 1990b: 563 (records)

Löbl, 2015c: 138 (records)

DISTRIBUTION. Indonesia: Java, Kalimantan; East Malaysia: Sarawak; Thailand.

Scaphisoma ruandanum **Pic**

Scaphosoma ruandanum Pic, 1955: 52. Holotype, MRAC; type locality: Rwanda: Rubengera, 1900 m, terr. Kibuye.

DISTRIBUTION. Rwanda.

Scaphisoma rubellum **Vinson**

Scaphosoma rubellum Vinson, 1943: 195, Figs 6, 16. Holotype, NHML; type locality: Mauritius: Grand Basin.

DISTRIBUTION. Mauritius.

Scaphisoma rubens **Casey**

Scaphisoma rubens Casey, 1893: 529. Syntypes, USNM; type localities: USA: Massachusetts; Catskill Mts; Long Island, New York.

Bowditch, 1896: 5 (records)
Leonard, 1928: 314 (records)
Campbell, 1991: 125 (records)
Sikes, 2004: 105 (distribution)
Majka et al., 2011: 89 (distribution)
Webster et al., 2012: 246 (records, habitat)
Evans, 2014: 132 (characters)

DISTRIBUTION. Canada: New Brunswick, Nova Scotia, Ontario, Quebec; United States: Maine, Massachusetts, New Hampshire, New York, Rhode Island, Vermont.

Scaphisoma rubripenne **Löbl**

Scaphisoma rubripenne Löbl, 2003b: 155, Figs 1–3. Holotype male, NHMW; type locality: Nepal: Godawari.

DISTRIBUTION. Nepal.

Scaphisoma rubripes **Pic**

Scaphosoma rubripes Pic, 1920b: 5. Syntypes, MNHN; type locality: Brazil.
Note. Pic (l.c.) credited the species epithet to Portevin.

DISTRIBUTION. Brazil.

Scaphisoma rubrum **Reitter**

Scaphisoma rubrum Reitter, 1877: 370. Lectotype male, MNHN; type locality: Japan.

Scaphosoma ustulatum Achard, 1923: 112. Syntypes, NMPC; type localities: Japan: Kyoto, Nichigo.
Achard, 1923: 113 (characters)
Miwa & Mitono, 1943: 540 (characters)
Nakane, 1955c: 55 (characters)
Chûjô, 1961: 5 (records)
Nakane, 1963b: 80 (characters)
Löbl, 1965a: 46, Figs 2, 2a, b (lectotype designation for *S. rubrum*, synonymy of *S. ustulatum* with *S. rubrum*, characters)
Löbl, 1970c: 785, Figs 68, 69 (characters, records)
Hisamatsu, 1977: 195 (records)
Löbl, 1977g: 165 (records)
Löbl, 1981d: 163, Fig. 13 (characters)
Löbl, 1982b: 105 (records)
Morimoto, 1985: Pl. 45, Fig. 33 (characters)

DISTRIBUTION. Japan: Honshu, Kyushu, Shikoku, Ryukyu's, Tsushima.

Scaphisoma rufescens **(Pic)**

Pseudoscaphosoma punctatum var. *rufescens* Pic, 1920f: 24. Syntypes, MNHN; type locality: East Malaysia: Banguey [Banggi Is.].

Scutoscaphosoma distinctipenne Pic, 1923d: 195. Lectotype male, MNHN; type locality: Vietnam: Tonkin, Lac Tho [secondary homonym of *Pseudoscaphosoma niasense* var. *distinctipenne* Pic, 1920f: 24].

Löbl, 1981d: 157, Fig. 4 (lectotype designation for S. *distinctipenne* Pic, 1923, synonymy of *S. distinctipenne* with *S. rufescens*, transfer to *Scaphisoma*, status, characters, records)

Löbl, 1990b: 563 (records)

Löbl, 2000: 611 (records)

Löbl, 2012b: 183 (records)

Löbl, 2015a: 111 (records)

Löbl, 2015c: 138 (records)

Löbl & Ogawa, 2016b: 1412, Fig. 29 (characters, records)

DISTRIBUTION. China: Yunnan; East Malaysia: Sabah, Sarawak, Banggi Island; Singapore; Indonesia: Bali, Kalimantan; Philippines: Palawan; Thailand; Vietnam; West Malaysia.

Scaphisoma ruficeps **Pic**

Scaphosoma ruficeps Pic, 1916c: 18. Syntypes, MNHN; type locality: Madagascar.

DISTRIBUTION. Madagascar.

Scaphisoma ruficolle **(Pic)**

Pseudoscaphosoma ruficolle Pic, 1915e: 5. Lectotype, MNHN; type locality: East Malaysia: Banguey [Banggi Is.].

Pseudoscaphosoma ruficolle var. *maculipenne* Achard, 1920d: 131. Lectotype male, MNHN; type locality: East Malaysia: Banguey [Banggi Is.].

Pseudoscaphosoma niasense var. *distinctipenne* Pic, 1920f: 24. Lectotype male, MNHN; type locality: East Malaysia: Banguey [Banggi Is.].

Löbl, 1975b: 272 (lectotype fixations by inference for *P. ruficolle*, var. *maculipenne* and var. *distinctipenne*, their synonymies, transfer of *P. ruficolle* to *Scaphisoma*, characters)

DISTRIBUTION. East Malaysia: Sabah, Banggi Is.

Scaphisoma ruficolor **(Pic)**

Baeocera ruficolor Pic, 1916c: 19. Lectotype male, MNHN; type locality: Indonesia: Kalimantan, Martapura.

Pic, 1920g: 189 (as *Baeocera*, characters)

Löbl, 1973a: 156, Figs 13–15 (lectotype fixed by inference, transfer to *Scaphisoma*, characters)
Löbl, 2015c: 139 (records)
Distribution. Indonesia: Kalimantan.

Scaphisoma rufifrons Fairmaire
Scaphisoma rufifrons Fairmaire, 1898b: 394. Syntypes, MNHN; type locality: Madagascar: Suberbieville [in Maevatanana District].
Distribution. Madagascar.

Scaphisoma rufipenne (Pic)
Pseudoscaphosoma rufipenne Pic, 1916d: 7. Syntypes, MNHN; type locality: Borneo.
Distribution. "Borneo".

Scaphisoma rufithorax Pic
Scaphosoma rufithorax Pic, 1930a: 87. Syntypes, MRAC/MNHN; type locality: Democratic Republic of the Congo: Uluku, Buhunde.
Distribution. Democratic Republic of the Congo.

Scaphisoma rufoides Löbl
Pseudoscaphosoma rufum Pic, 1925a: 196. Lectotype female, MNHN; type locality: Vietnam: Tonkin, Lac Tho.
Scaphisoma rufoides Löbl, 1975b: 272; replacement name for *Pseudoscaphosoma rufum* Pic, 1925 (nec *Scaphisoma rufum* Achard, 1923).
Löbl, 1975b: 272 (lectotype fixed by inference, characters)
Distribution. Vietnam.

Scaphisoma rufonotatum Pic
Scaphisoma rufonotatum Pic, 1926e: 143. Syntypes, MNHN; type locality: Vietnam: Tonkin, Hoa Binh.
Löbl, 1990b: 543 (records)
Distribution. Thailand; Vietnam.

Scaphisoma rufulum LeConte
Scaphisoma rufulum LeConte, 1860: 323. Holotype, MCZC; type locality: USA: Junction of Colorado and Gila rivers, California.
Casey, 1893: 529 (characters, records)
Brimley, 1938: 149 (records)
Distribution. United States: Arizona, California, North Carolina.

Scaphisoma rufum Achard

Scaphosoma rufum Achard, 1923: 115. Lectotype female, BMHN; type locality: Japan: Nagasaki.
Miwa & Mitono, 1943: 542 (characters)
Nakane, 1955c: 56 (characters)
Chûjô, 1961: 5 (records)
Nakane, 1963b: 79 (characters; credited to Reitter)
Löbl, 1966b: 132, Figs 4, 5 (characters, records)
Löbl, 1970c: 756, Figs 31, 32 (lectotype fixed by inference, characters, records)
Löbl, 1972a: 117, Figs 1, 2 (characters, records)
Löbl, 1977g: 165 (records)
Hisamatsu, 1977: 194 (records)
Löbl, 1982b: 105 (records)
Morimoto, 1985: Pl. 45, Fig. 31 (characters)
Löbl, 1986a: 142, Fig. 1 (characters, records)
Löbl, 1986c: 345 (records)
Löbl, 1990b: 542 (records)
Löbl, 1992a: 531 (records)

DISTRIBUTION. China: Jiangsu; India: Himachal Pradesh, Meghalaya, West Bengal (Darjeeling District), Sikkim; Japan: Honshu, Kyushu, Ryukyus, Shikoku, Tsushima; North and South Korea; Nepal; Singapore; Thailand.

Scaphisoma rugosum Löbl

Scaphisoma rugosum Löbl, 1973b: 321, Figs 19, 20. Holotype male, NHMW; type locality: New Caledonia: Tindou near Hienghène.
Löbl, 1981a: 378 (characters)

DISTRIBUTION. New Caledonia.

Scaphisoma sadang Löbl

Scaphisoma sadang Löbl, 1983a: 286, Figs 3, 4. Holotype male, MHNG; type locality: Indonesia: Sulawesi, Makale.
Ogawa, 2015: 85, Figs 4-14c, d (characters, records)

DISTRIBUTION. Indonesia: Sulawesi.

Scaphisoma sagax Löbl & Ogawa

Scaphisoma sagax Löbl & Ogawa, 2016b: 1380, Figs 105–107. Holotype male, MHNG; type locality: Philippines: Palawan, Cabayugan near Lyons Cave, sea level.

DISTRIBUTION. Philippines: Palawan.

Scaphisoma sakaii **Löbl**

Scaphisoma sakaii Löbl, 1988b: 1133, Figs 1, 2. Holotype male, NSMT; type locality: Japan: Okinawa, Kunigani-son, Hiji-Hiji Falls.

DISTRIBUTION. Japan: Ryukyus.

Scaphisoma sakaiorum **Löbl**

Scaphisoma sakaiorum Löbl, 1987a: 99, Figs 22–24. Holotype male, MHNG; type locality: East Malaysia: Sabah, Kinabalu Nat. Park, Headquarters.

DISTRIBUTION. East Malaysia: Sabah.

Scaphisoma sapitense **Pic**

Scaphosoma sapitense Pic, 1915b: 31. Lectotype female, MNHN; type locality: Indonesia: Lombok, Sapit.
Achard, 1920d: 130 (records)
Löbl, 2015a: 113, Figs 197-111 (lectotype designation, characters)

DISTRIBUTION. Indonesia: Lombok.

Scaphisoma sasagoense **Löbl**

Scaphosoma sasagoensis Löbl, 1965a: 49, Figs 4, 4a. Holotype male, MNHN; type locality: Japan: Mountains of Sasago, near Kôfu.
Löbl, 1970c: 782, Figs 64, 64a, 65 (characters)

DISTRIBUTION. Japan: Honshu.

Scaphisoma satoi **Löbl**

Scaphisoma satoi Löbl, 1982f: 14, Fig. 12. Holotype male, PCMS; type locality: East Malaysia: Sarawak, Kuala Bok.

DISTRIBUTION. East Malaysia: Sarawak.

Scaphisoma scabiosum **Löbl**

Scaphisoma scabiosum Löbl, 1986a: 191, Figs 72–73. Holotype male, MHNG; type locality: India: Meghalaya, Nongpoh, 700 m, Khasi Hills.
Löbl, 1990b: 571 (records)

DISTRIBUTION. India: Meghalaya; Thailand.

Scaphisoma scapulare **Löbl & Ogawa**

Scaphisoma scapulare Löbl & Ogawa, 2016b: 1395, Figs 30, 135, 136. Holotype male, EUMJ; type locality: Philippines: Luzon, Mountain Prov., Mt. Puguis, 1900 m, near Bontoc.

DISTRIBUTION. Philippines: Luzon.

Scaphisoma schmidti Pic

Scaphosoma schmidti Pic, 1937: 206. Holotype, ZMUH; type locality: Costa Rica: Finca La Caja near San José.

DISTRIBUTION. Costa Rica.

Scaphisoma schoutedeni schoutedeni Pic

Scaphosoma schoutedeni Pic, 1928b: 42. Syntypes, MRAC/MNHN; type localities: Democratic Republic of the Congo: Thsisika, Kasai; Bulobo; Moto; Abimva, Haut-Uélé; Masinga.

DISTRIBUTION. Democratic Republic of the Congo.

Scaphisoma schoutedeni atropygum Pic

Scaphosoma schoutedeni var. *atropygum* Pic, 1930a: 88. Syntypes, MRAC/MNHN; type locality: Democratic Republic of the Congo: Masua, Lubutu.

DISTRIBUTION. Democratic Republic of the Congo.

Scaphisoma scurrile Löbl & Ogawa

Scaphisoma scurrile Löbl & Ogawa, 2016b: 1388, Figs 120, 122. Holotype male, MHNG; type locality: Philippines: Palawan, Olangoan, 18 km NE San Rafael, sea level.

DISTRIBUTION. Philippines: Palawan.

Scaphisoma segne Löbl

Scaphisoma segne Löbl, 1990b: 568, Figs 87–89. Holotype male, MHNG; type locality: Thailand: Chiang Mai, Doi Suthep, south slope, 1450 m.

Löbl, 2000: 611 (records)

DISTRIBUTION. China: Sichuan, Yunnan, Zhejiang; Thailand.

Scaphisoma semialutaceum Pic

Scaphosoma subconvexum race *semialutaceum* Pic, 1930a: 88. Syntypes, MRAC/MNHN; type locality: Democratic Republic of the Congo: Biruwe, Buhunde.

Note: *Scaphosoma subconvexum* Pic, 1930a: 88 is a primary junior homonym with *Scaphisoma subconvexum* Pic, 1926. *Scaphisoma subconvexum* Pic, 1930 is a permanently invalid. Its "race" *semialutaceum* was used as an available substitute name in Löbl (1997: xii, 133).

DISTRIBUTION. Democratic Republic of the Congo.

Scaphisoma semibreve Löbl

Scaphisoma semibreve Löbl, 2014a: 57, Figs 14, 15. Holotype male, MCZC; type locality: Indonesia: Maluku Islands, Tanimbar, Yamdena.

DISTRIBUTION. Indonesia: Maluku Islands: Yamdena.

Scaphisoma semiopacum Fall

Scaphisoma semiopaca Fall, 1910: 117. Holotype, MCZC; type locality: USA: Luling.

DISTRIBUTION. United States: Texas.

Scaphisoma seorsum (Löbl)

Caryoscapha seorsum Löbl, 1965b: 1, Figs 1, 2. Holotype male, MNHN; type locality: Japan: Chuzenji.

Löbl, 1967c: 132 (records)

Löbl, 1970c: 795, Figs 80, 81 (characters, records)

Leschen & Löbl, 2005: 25 (transfer to *Scaphisoma*)

Löbl, 2011c: 111 (records)

DISTRIBUTION. Japan: Hokkaido, Honshu, ?Taiwan.

Scaphisoma serosum Löbl

Scaphisoma serosum Löbl, 2000: 614, Figs 5, 6. Holotype male, MHNG; type locality: China: Sichuan, Wolong Nat. Res., 1000 m.

DISTRIBUTION. China: Sichuan.

Scaphisoma serpens Löbl

Scaphisoma serpens Löbl, 2000: 630, Figs 33, 34. Holotype male, ZMUB; type locality: China: Shaanxi, Qin Lin Shan, 93 km southwest Xian.

DISTRIBUTION. China: Shaanxi.

Scaphisoma sesaotense Löbl

Scaphisoma sesaotense Löbl, 1986b: 93, Figs 10–12. Holotype male, MHNG; type locality: Indonesia: Lombok, Sesaot.

Löbl, 2015a: 113 (records)

DISTRIBUTION. Indonesia: Lombok, Sumbawa.

Scaphisoma sexuale Löbl

Scaphisoma sexuale Löbl, 1972b: 86, Figs 17, 18. Holotype male, ZMUB; type locality: Philippines: Palawan, Binaluan.

Löbl & Ogawa, 2016b: 1386 (characters, records)

DISTRIBUTION. Philippines: Palawan.

Scaphisoma siamense Löbl

Scaphisoma siamense Löbl, 1990b: 583, Figs 117–119. Holotype male, MHNG; type locality: Thailand: Doi Inthanon, near Forestry Department, 1250 m, Chiang Mai.

DISTRIBUTION. Thailand.

Scaphisoma signaticolle Löbl & Ogawa

Scaphisoma signaticolle Löbl & Ogawa, 2016b: 1433, Figs 231–233. Holotype male, MHNG; type locality: Philippines: Luzon, Mountain Prov., Mount Data Lodge, 2200–2300 m.

DISTRIBUTION. Philippines: Luzon.

Scaphisoma signum Löbl

Scaphisoma signum Löbl, 2000: 616, Figs 7, 8. Holotype male, MHNG; type locality: China: Sichuan, Wolong Nat. Res., 1500 m.

DISTRIBUTION. China: Sichuan.

Scaphisoma sikkimense Löbl

Scaphisoma sikkimense Löbl, 1992a: 559, Figs 161, 162. Holotype male, NSMT; type locality: India: Sikkim, Choka, 3050 m.

DISTRIBUTION. India: Sikkim.

Scaphisoma silhouettae Scott

Scaphosoma silhouettae Scott, 1922: 223, Pl. 21, Fig. 20. Holotype male, NHML; type locality: Seychelles: Silhouette.

DISTRIBUTION. Seychelles.

Scaphisoma simillimum Löbl

Scaphisoma simillimum Löbl, 1970c: 748, Figs 17, 18. Holotype male, MHNG; type locality: Turley: Artvin, between Borça and Hopa.

Löbl, 1974a: 61 (records)

Iablokoff-Khnzorian, 1985: 141 (records)

DISTRIBUTION. Armenia; Georgia; Turkey; "Caucasus".

Scaphisoma simplex Löbl

Scaphisoma simplex Löbl, 1972b: 88, Figs 21, 22. Holotype male, ZMUB; type locality: Philippines: Luzon, Bangui.

Löbl & Ogawa, 2016b: 1352 (characters)

DISTRIBUTION. Philippines: Luzon.

***Scaphisoma simplexoides* Löbl & Ogawa**

Scaphisoma simplexoides Löbl & Ogawa, 2016b: 1352, Figs 46, 47. Holotype male, MHNG; type locality: Philippines: Luzon, Lagunas, Mt. Makiling, summit road, 600 m.

DISTRIBUTION. Philippines: Luzon.

***Scaphisoma simplicipenis* Löbl**

Scaphisoma simplicipenis Löbl, 1992a: 545, Figs 131, 132. Holotype male, MHNG; type locality: Nepal: Phulcoki, 2550 m, Patan District.

Löbl, 2005: 181 (records)

DISTRIBUTION. Nepal.

***Scaphisoma simulans* Löbl**

Scaphisoma simulans Löbl, 1973b: 324, Figs 23, 24. Holotype male, NHMW; type locality: New Caledonia: Pic du Pin.

Löbl, 1981a: 373 (characters, records)

DISTRIBUTION. New Caledonia.

***Scaphisoma singaporense* Löbl**

Scaphisoma singaporense Löbl, 1986b: 97, Figs 14–16. Holotype male, NHML; type locality: Singapore.

DISTRIBUTION. Singapore.

***Scaphisoma sinuatum* Löbl & Ogawa**

Scaphisoma sinuatum Löbl & Ogawa, 2016b: 1413, Figs 31, 174, 175. Holotype male, MHNG; type locality: Philippines: Luzon, Lagunas Prov., Mt. Makiling, summit road, ca 600 m.

DISTRIBUTION. Philippines: Luzon.

***Scaphisoma skanda* Löbl**

Scaphisoma skanda Löbl, 1979a: 104, Figs 25, 26. Holotype male, MHNG; type locality: India: Tamil Nadu, 39 km east Kodaikanal, 650 m, Palni Hills.

DISTRIBUTION. India: Tamil Nadu.

***Scaphisoma solutum* Löbl**

Scaphisoma solutum Löbl, 1990b: 562, Figs 78–81. Holotype male, MHNG; type locality: Thailand: Khao Sabap Nat. Park, near Phliu Waterfalls, 150–300 m.

DISTRIBUTION. Thailand.

Scaphisoma soror Löbl

Scaphisoma soror Löbl, 1970c: 743, Figs 7, 8. Holotype male, MHNG; type locality: Turkey: Antakya, Madenli.

DISTRIBUTION. Turkey.

Scaphisoma spatulatum Löbl

Scaphisoma spatulatum Löbl, 2015b: 182, Figs 44, 45. Holotype male, SMNS; type locality: Indonesia: Maluku Islands, Halmahera Is., Sidangoli, Batu putih, 100 m.

DISTRIBUTION. Indonesia: Maluku Islands: Halmahera Is.

Scaphisoma spatuloides Löbl & Ogawa

Scaphisoma spatuloides Löbl & Ogawa, 2016b: 1431, Figs 224–227. Holotype male, MHNG; type locality: Philippines: Luzon, Lagunas Prov., Mt. Makiling, above Mad Springs, 400–700 m.

DISTRIBUTION. Philippines: Luzon.

Scaphisoma spiniger Löbl

Scaphisoma spiniger Löbl, 1981d: 161, Fig. 8. Holotype male, MHNG; type locality: East Malaysia: Sabah, Quoin Hill, 750 ft, Tawau.

DISTRIBUTION. East Malaysia: Sabah.

Scaphisoma spinosum Löbl

Scaphisoma spinosum Löbl, 2015b: 180, Figs 42, 43. Holotype male, SMNS; type locality: Indonesia: Maluku Islands, Halmahera Is., Sidangoli, Batu putih, 100 m.

DISTRIBUTION. Indonesia: Maluku Islands: Halmahera Is.

Scaphisoma spissum Löbl

Scaphisoma spissum Löbl, 1990b: 579, Figs 111–113. Holotype male, MHNG; type locality: Thailand: Doi Suthep, 1400 m, Chiang Mai.

DISTRIBUTION. Thailand.

Scaphisoma spurium Löbl

Scaphisoma spurium Löbl, 1971c: 973, Figs 40, 41. Holotype male, MHNG; type locality: Sri Lanka: ca 5 km NW Bibile.

Löbl, 1986c: 346 (records)

Löbl, 1992a: 534 (records)

DISTRIBUTION. India: Uttarakhand; Sri Lanka.

***Scaphisoma stephani* Leschen & Löbl**

Scaphisoma stephani Leschen & Löbl, 1990: 287, Figs 2, 9, 26–34. Holotype male, SEMC; type locality: USA: Bull Slough Wildlife Management area, Mayflower, Faulkner Co., Arkansas.

DISTRIBUTION. United States: Arkansas, Georgia, Oklahoma, Virginia.

***Scaphisoma stictum* Löbl**

Scaphisoma stictum Löbl, 1977e: 53, Figs 23, 80, 81. Holotype male, MVMA; type locality: Australia: Mt. Tomah, New South Wales.

DISTRIBUTION. Australia: Victoria, New South Wales, Queensland.

***Scaphisoma stigmatipenne* Heller**

Scaphosoma stigmatipenne Heller, 1917: 46. Lectotype male, SMTD; type locality: Philippines: Luzon, Mt. Makiling.

Scutoscaphosoma luteoapicale Pic, 1926b: 2 [secondary junior homonym of *Scaphisoma luteoapicale* Pic, 1923]. Lectotype male, MNHN; type locality: Philippines: Luzon, Los Banos.

Löbl, 1970b: 127, Fig. 3 (lectotype fixations by inference for *S. stigmatipenne* and *S. luteoapicale*, synonymy, characters)

Löbl, 1981d: 163, Fig. 16 (characters)

Löbl & Ogawa, 2016b: 1414, Figs 32, 176 (characters, records)

DISTRIBUTION. Philippines: Leyte, Luzon, Mindanao.

***Scaphisoma striolatum* Vinson**

Scaphosoma striolatum Vinson, 1943: 195, Figs 5, 15. Syntypes, NHML; type locality: Mauritius: Macabé Forest.

DISTRIBUTION. Mauritius.

***Scaphisoma styloides* Löbl**

Scaphisoma styloides Löbl, 2000: 621, Figs 18–20. Holotype male, MHNG; type locality: China: Hubei, Shennonglia Nat. Res., 2000–2200 m.

DISTRIBUTION. China: Hubei.

***Scaphisoma subalpinum subalpinum* Reitter**

Scaphisoma subalpinum Reitter, 1880d: 44. Lectotype male, HNHM; type locality: Czech Republic: Beskyde Mts.

Reitter, 1885: 362 (characters)

Kuthy, 1897: 86 (records)

Ganglbauer, 1899: 344 (characters)

Roubal, 1930: 298 (records)

Scheerpeltz & Höfler, 1948: 149 (fungal hosts, habitat)
Horion, 1949: 255 (records)
Palm, 1951: 145 (habitat)
Benick, 1952: 47 (fungal hosts)
Lundblad, 1952: 28, Figs 1A, 2A, 3A (characters)
Tamanini, 1955: 13, Figs 21–26 (characters, records)
Löbl, 1965g: 732 (records)
Palm, 1966: 44, Fig. B (characters)
Koch, 1968: 85 (records)
Kofler, 1968: 41, Figs 1E, 2E (characters, records)
Tamanini, 1969a: 486 (records)
Tamanini, 1969b: 364, Fig. 4F (records, fungal hosts)
Kofler, 1970: 58 (records)
Löbl, 1970a: 13, Figs 13, 20 (characters, records)
Löbl, 1970c: 758, Figs 33, 34 (lectotype fixed by inference, characters, records)
Tamanini, 1970: 12, 16, Figs 1B, 3F, 10A (characters, records, fungal hosts)
Franz, 1970: 267 (records)
Freude, 1971: 346, Fig. 3: 6 (characters)
Strand, 1975: 10 (records)
Holzschuh, 1977: 31 (records)
Burakowski et al., 1978: 235 (records)
Angelini, 1986: 55 (records)
Angelini, 1987: 22 (records)
Schawaller, 1990: 235 (records)
Merkl, 1996: 261 (records)
Nikitsky et al., 1996: 25 (fungal hosts)
Krasutskij, 1996a: 97 (records, fungal hosts)
Krasutskij, 1997a: 307 (habitat, fungal hosts)
Krasutskij, 1997b: 774 (fungal hosts, habitat)
Van Meer, 1999: 8 (records)
Byk, 2001: 354 (records)
Kapp, 2001: 177 (records, fungal hosts)
Nikitsky & Schigel, 2004: 9, 12 (fungal hosts)
Tsinkevich, 2004: 19 (records, fungal hosts)
Krasutskij, 2005: 38 (fungal hosts)
Mateleshko, 2005: 128 (fungal hosts)
Borowski, 2006: 54–58 (records, fungal hosts)
Nikitsky et al., 2008: 120 (records, fungal hosts)
Vinogradov et al., 2010: 16 (records, fungal hosts)
Zamotajlov & Nikitsky, 2010: 105 (records, fungal hosts)

Schigel, 2011: 325, 329, 331, 332, 337 (fungal hosts)
Löbl & Ogawa, 2016c: 39 (records)
Garpebring, 2016: 2 (records)
Solodovnikov, 2016: 55 (records)
Nikitsky et al., 2016: 138 (records, fungal hosts)
Semenov, 2017: 202 (records)
Vlasov & Nikitsky, 2017: 7 (records, fungal hosts)

DISTRIBUTION. Albania; Austria; Belarus; Bosnia and Herzegovina; Bulgaria; Croatia; Czech Republic; Estonia; Finland; France; Germany; Greece; Hungary; Italy; Latvia; Lithuania; Macedonia; Norway; Poland; Romania; Russia: European, Siberia; Serbia; Slovakia; Slovenia; Sweden; Switzerland; Ukraine.

***Scaphisoma subalpinum ussuricum* Pic**

Scaphosoma ussuricum Pic, 1921b: 1. Syntypes, ?NMPC, not traced; type locality: Russian Far East: Vladivostok.
Löbl, 1967b: 107, Fig. 8 (characters, records)
Löbl, 1970c: 760, Fig. 35 (characters, records)
Iablokoff-Khnzorian, 1985: 142 (records)
Li, 2015: 39, Fig. 4 (as valid species, characters, records)

DISTRIBUTION. China: Heilongjiang; Russia: Far East.

***Scaphisoma subanun* Löbl & Ogawa**

Scaphisoma subanun Löbl & Ogawa, 2016b: 1380, Figs 33, 108, 109. Holotype male, SMTD; type locality: Philippines: Mindanao, South Coabato, Pt. Banga.

DISTRIBUTION. Philippines: Mindanao.

***Scaphisoma subconvexum* Pic**

Scphosoma subconvexum Pic, 1926b: 1. Lectotype male, MNHN; type locality: Philippines: Luzon [Los Banos].
Löbl, 1972b: 106, Figs 57, 58 (lectotype fixed by inference, characters)
Löbl & Ogawa, 2016b: 1353, Fig. 34 (records)

DISTRIBUTION. Philippines: Luzon.

***Scaphisoma subelongatulum* Löbl**

Scaphosoma subelongatum Pic, 1946: 83. Syntypes, MNHN; type locality: Kenya: Mt. Elgon.

Scaphisoma subelongatulum Löbl, 1997: xii; replacement name for *Scaphosoma subelongatum* Pic, 1946 (secondary homonym of *Pseudoscaphosoma subelonatum* Pic, 1915, see under *Scaphisoma testaceomaculatum* (Pic, 1915)).

DISTRIBUTION. Kenya.

***Scaphisoma subfasciatum* Pic**

Scaphosoma subfasciatum Pic, 1926b: 2. Lectotype male, MNHN; type locality: Philippines: Luzon, Los Banos.

Löbl, 1972b: 105, Figs 55, 56 (lectotype fixed by inference, characters)

Löbl & Ogawa, 2016b: 1382 (characters)

DISTRIBUTION. Philippines: Luzon.

***Scaphisoma subgracile* Löbl & Ogawa**

Scaphisoma subgracile Löbl & Ogawa, 2016b: 1353, Figs 35, 48, 49. Holotype male, MHNG; type locality: Philippines: Palawan, Central, Olangoan 18 km NE San Rafael.

DISTRIBUTION. Philippines: Palawan.

***Scaphisoma sublimbatum* Löbl**

Scaphisoma sublimbatum Löbl, 1977e: 57, Figs 25, 86, 87. Holotype male, SAMA; type locality: Australia: Queensland, Cairns District.

DISTRIBUTION. Australia: Queensland.

***Scaphisoma submaculatum* Pic**

Scaphosoma submaculatum Pic, 1920f: 24. Lectotype male, MNHN; type locality: Indonesia: Kalimantan [Riam Kanan, Martapura].

Löbl, 2015c: 139, Figs 6–8 (lectotype designation, characters)

DISTRIBUTION. Indonesia: Kalimantan.

***Scaphisoma subplanatum* Löbl & Ogawa**

Scaphisoma subplanatum Löbl & Ogawa, 2016b: 1354, Figs 50, 51. Holotype male, FMNH; type locality: Philippines: Mindanao, Davao Province, Meran, east slope Mt. Apo, 6000 ft.

DISTRIBUTION. Philippines: Mindanao.

***Scaphisoma subpunctulum* Löbl & Ogawa**

Scaphisoma subpunctulum Löbl & Ogawa, 2016b: 1355, Fig. 52. Holotype male, FMNH; type locality: Philippines: Mindanao, Cotabato Province, Upi, Burungkot.

DISTRIBUTION. Philippines: Mindanao.

***Scaphisoma subtile* Löbl**

Scaphisoma subtile Löbl, 2000: 620, Figs 14, 15. Holotype male, MHNG; type locality: China: Sichuan, south Xichang, Lunji, 2300–2500 m.

DISTRIBUTION. China: Sichuan.

Scaphisoma suknense Löbl

Scaphisoma suknense Löbl, 1986a: 151, Figs 12, 13. Holotype male, MHNG; type locality: India: West Bengal, Sukna, 200 m.

DISTRIBUTION. India: West Bengal (Darjeeling District).

Scaphisoma sulcatum Löbl

Scaphisoma sulcatum Löbl, 1975a: 393, Figs 34, 35. Holotype male, MHNG; type locality: Papua New Guinea: Quelilag.

DISTRIBUTION. Papua New Guinea.

Scaphisoma sumatranum Löbl

Scaphisoma sumatranum Löbl, 1975b: 277, Figs 10, 11. Holotype male, MHNG; type locality: Indonesia: Sumatra, Gunung Singgalan, 1800 m.

DISTRIBUTION. Indonesia: Sumatra.

Scaphisoma surigaosum (Pic)

S*cutoscaphosoma luteoapicale* Pic var. *surigaosum* Pic, 1926b: 3. Lectotype male, MNHN; type locality: Philippines: Surigao.

Löbl, 1970b: 128, Fig. 4 (lectotype fixed by inference, raised to species status, characters)

Löbl, 1981d: 163, Fig. 14 (characters)

Löbl & Ogawa, 2016b: 1414, Figs 36, 177 (characters, records)

DISTRIBUTION. Philippines: Luzon, Palawan, Mindanao.

Scaphisoma surya Löbl

Scaphisoma surya Löbl, 1986b: 172, Figs 40–42. Holotype male, MHNG; type locality: India: Meghalaya, between Mawsynram and Balat, 16 km from Mawsynram, 1000 m, Khasi Hills.

DISTRIBUTION. India: Meghalaya.

Scaphisoma suspiciosum Löbl

Scaphisoma suspiciosum Löbl, 2000: 646, Figs 70–73. Holotype male, MHNG; type locality: China: Yunnan, Xishiuangbanna.

DISTRIBUTION. China: Yunnan.

Scaphisoma suthepense Löbl

Scaphisoma suthepense Löbl, 1990b: 550, Figs 58, 59. Holotype male, MHNG; type locality: Thailand: Doi Suthep, 1180 m, Chiang Mai.

DISTRIBUTION. Thailand.

Scaphisoma suturale LeConte

Scaphisoma suturale LeConte, 1860: 323. Lectotype male, MCZC; type locality: USA: Middle and Southern States.
Casey, 1893: 527 (characters, records)
Blatchley, 1910: 496 (characters, records)
Weiss & West, 1922: 199 (fungal hosts)
Leonard, 1928: 314 (records)
Brimley, 1938: 149 (records)
Leschen et al., 1990: 285, Figs 6, 21, 22 (lectotype designation, characters, records, fungal hosts)
Downie & Arnett, 1996: 366 (records)
Sikes, 2004: 105 (distribution)

DISTRIBUTION. Canada: Quebec; United States: Arkansas, Connecticut, Indiana, Missouri, New York, North Carolina, Rhode Island.

Scaphisoma swapna Löbl

Scaphisoma swapna Löbl, 1979a: 105, Figs 27, 28. Holotype male, MHNG; type locality: India: Tamil Nadu, 36 km east Kodaikanal, 850 m, Palni Hills.

DISTRIBUTION. India: Tamil Nadu.

Scaphisoma tagalog Löbl & Ogawa

Scaphisoma tagalog Löbl & Ogawa, 2016b: 1396, Figs 138, 139. Holotype male, FMNH; type locality: Philippines: Mindanao, Cotabato Prov., Burungkot, Upi, 1500 ft.

DISTRIBUTION. Philippines: Mindanao.

Scaphisoma taiwanum Löbl

Scaphisoma taiwanum Löbl, 1980a: 106, Figs 20, 21. Holotype male, MHNG; type locality: Taiwan: Fenchihu, 1400 m.
Löbl, 1982a: 327 (records)
Löbl, 1993: 38 (records)
Hoshina, 2008: 60 (records)

DISTRIBUTION. Japan: Honshu, Ryukyus; Russia: Far East; Taiwan.

Scaphisoma tamaninii Löbl

Scaphosoma tamaninii Löbl, 1965h: 3, Figs 2, 3. Holotype male, HNHM; type locality: Japan: Yamanashi Pref., Lakeside of Shoji.
Löbl, 1970c: 773, Figs 52, 53 (characters, records)

DISTRIBUTION. Japan: Hokkaido, Honshu, Kyushu, Russia: Far East.

Scaphisoma tannaense Löbl

Scaphisoma tannaense Löbl, 1978b: 109, Figs 1, 2. Holotype male, NHML; type locality: Vanuatu: Tanna.

DISTRIBUTION. Vanuatu.

Scaphisoma tarsale Löbl

Scaphisoma tarsale Löbl, 2015c: 141, Figs 9–12. Holotype male, NMPC; type locality: Indonesia: Kalimantan, ca. 55 km west of Balikpapan, PF Fajar Surya Swadaya area, 100 m.

DISTRIBUTION. Indonesia: Kalimantan.

Scaphisoma taylori Löbl

Scaphisoma taylori Löbl, 1975b: 280, Figs 14, 15. Holotype male, MHNG; type locality: East Malaysia: Sabah, Quion Hill, Tawau.

DISTRIBUTION. East Malaysia: Sabah.

Scaphisoma telnovi Löbl & Ogawa

Scaphisoma telnovi Löbl & Ogawa, 2017: 375, Figs 1, Pl. 82. Holotype male, NMEC; type locality: Indonesia: Western New Guinea, Doberaki Peninsula, Arfak Mts, Anggi Gigi Lake, 1°17'10"S, 133° 54'18"E.

DISTRIBUTION. Indonesia: Western New Guinea.

Scaphisoma teres Löbl

Scaphisoma teres Löbl, 1977e: 26, Figs 15, 32, 33. Holotype male, MVMA; type locality: Australia: Cairns.
Newton, 1984: 318 (fungal hosts)

DISTRIBUTION. Australia: New South Wales, Queensland.

Scaphisoma terminatum Melsheimer

Scaphisoma terminatum Melsheimer, 1846: 104. Lectotype male, MCZC; type locality: USA: Pennsylvania.

Scaphisoma evanescens Casey, 1893: 528. Lectotype male, USNM; type locality: USA: Iowa.

Baeocera punctipennis Blatchley, 1910: 494. Lectotype male, PURC; type locality: USA: Marion, Wells Co., Indiana.

Baeocera blatchleyi Achard, 1915c: 292; replacement name for *Baeocera punctipennis* Blatchley, 1910: 494 (nec *Baeocera punctipennis* Matthews, 1888).
LeConte, 1860: 323 (characters, records)
Casey, 1893: 527 (characters, records)
Bowditch, 1896: 5 (records)
Blatchley, 1910: 494 (records)

Fall, 1910: 118 (characters)
Leonard, 1928: 314 (records)
Blatchley, 1930: 35 (lectotype designation for *B. punctipennis* Blatchley)
Cornell, 1967: 2 (transfer of *B. blatchleyi* to *Scaphisoma*)
Leschen et al., 1990: 289, Figs 10, 35, 36 (lectotype designations for *S. terminatum*, *S. evanescens* and *B. punctipennis* [invalid for *B. punctipennis*], synonymy of *S. evanescens* and *S. blatchleyi* with *S. terminatum*, characters, records, fungal hosts)
Downie & Arnett, 1996: 366 (as *S. blatchleyi*, characters)

DISTRIBUTION. United States: Arkansas, Florida, Indiana, Iowa, Kansas, Kentucky, Missouri, New Hampshire, New York, Ohio, Oklahoma, Texas.

***Scaphisoma testaceiventre* Pic**

Scaphosoma testaceiventre Pic, 1928e: 76. Syntypes, MNHN/NMPC; type locality: Brazil: Sao Paulo.

DISTRIBUTION. Brazil.

***Scaphisoma testaceomaculatum testaceomaculatum* (Pic)**

Pseudoscaphosoma testaceomaculatum Pic, 1915b: 31. Syntypes, MNHN; type locality: Indonesia: Java, Pengalengon.

Pseudoscaphosoma subelongatum Pic, 1915b: 31. Syntypes, MNHN; type locality: Indonesia: Java.
Achard, 1920d: 131 (synonymy of *P. subelongatum* with *P. testaceomaculatum*)
Achard, 1921b: 86 (records)
Löbl, 1975b: 270 (transfer to *Scaphisoma*)

DISTRIBUTION. Indonesia: Java.

***Scaphisoma testaceomaculatum conjunctum* (Pic)**

Pseudoscaphosoma testaceomaculatum var. *conjunctum* Pic, 1920e: 97. Syntypes, MCSN; type locality: Indonesia: Sumatra, Si Rambe [near Balige].

DISTRIBUTION. Indonesia: Sumatra.

***Scaphisoma tetrastictum* Champion**

Scaphosoma tetrastictum Champion, 1927: 275. Holotype male, NHML; type locality: India: River Sarda Gorge, Kumaon.
Löbl, 1979a: 108 (records)
Löbl, 1980a: 116 (records)
Löbl, 1981f: 109 (characters, records)
Löbl, 1986a: 190 (records)
Löbl, 1986c: 347 (records)
Löbl, 1990a: 121 (records)

Löbl, 1990b: 573 (records)
Löbl, 1992a: 535, 587, Fig. 34 (characters, records)
Mazumder & al., 2001: 59 (fungal hosts)
Mazumder & al., 2008: 45 (fungal hosts, seasonal activity)

DISTRIBUTION. India: Assam, Himachal Pradesh, Kerala, Meghalaya, Tamil Nadu, Uttarakhand; Myanmar; Nepal; Taiwan; Thailand; Vietnam.

Scaphisoma thoracicum **Matthews**

Scaphisoma thoracicum Matthews, 1888: 174. Syntypes, NHML; type locality: Mexico: Jacale, Volcan de Orizaba.

DISTRIBUTION. Mexico.

Scaphisoma tonkineum **Pic**

Scaphosoma tonkineum Pic, 1922a: 1. Lectotype male, MNHN; type locality: Vietnam: Tonkin [Hoa Binh].
Löbl, 1976c: 223, Figs 3, 4 (characters)
Löbl, 1992a: 539 (lectotype designation, characters, records)

DISTRIBUTION. India: Uttarakhand (Kumaon); Nepal; Vietnam.

Scaphisoma tortile **Löbl**

Scaphisoma tortile Löbl, 1984c: 1004, Figs 12, 13. Holotype male, MHNG; type locality: Myanmar: Wetwun, 800 m, 23 km east of Maymyo, Shan State.
Löbl, 1990b: 583 (records)

DISTRIBUTION. Myanmar; Thailand.

Scaphisoma toxopeusi **Löbl**

Scaphisoma toxopeusi Löbl, 1976a: 10, Figs 5, 6. Holotype male, NBCL; type locality: Indonesia: Maluku Islands, Buru, Station 8.

DISTRIBUTION. Indonesia: Maluku Islands: Buru.

Scaphisoma transforme **Löbl**

Scaphisoma transforme Löbl, 1984c: 1001, Figs 9–10. Holotype male, MHNG; type locality: Myanmar: Kalaw, 1300 m, Shan State.

DISTRIBUTION. Myanmar.

Scaphisoma transparens **Löbl**

Scaphisoma transparens Löbl, 1973b: 310, Figs 3, 4. Holotype male, NHMW; type locality: New Caledonia: Table d'Union, near Col d'Amieu.
Löbl, 1981a: 363 (characters, records)

DISTRIBUTION. New Caledonia.

Scaphisoma transversale **Löbl & Ogawa**

Scaphisoma transversale Löbl & Ogawa, 2016b: 1415, Figs 37, 178, 179. Holotype male, MHNG; type locality: Philippines: Luzon, Lagunas Prov., Mt. Makiling summit road, 600 m.

DISTRIBUTION. Philippines: Luzon.

Scaphisoma tricolor **Heller**

Scaphosoma tricolor Heller, 1917: 46. Lectotype male, SMTD; type locality: Philippines: Mt. Makiling, Los Banos.

Scaphosoma latum Pic, 1920b: 3. Lectotype male, MNHN; type locality: Philippines: Los Banos.

Löbl, 1970b: 126 (lectotype designations by inference for *S. tricolor* and *S. latum*, synonymy of *S. latum* with *S. tricolor*)

Löbl, 1971a: 250, Fig. 5 (characters, records)

Löbl, 1981b: 77 (characters, records)

Löbl, 1982b: 105 (records)

Morimoto, 1985: 256 (records)

Ogawa, 2015: 85, Figs 4-14e, f (doubtful identity, as cf *tricolor*; characters, records)

Löbl & Ogawa, 2016b: 1434, Figs 38, 39, 234–237 (characters, records)

DISTRIBUTION. Japan: ?Ogasawara, ?Okinawa; Philippines: Luzon, Palawan, Balabac, Leyte, Masbate, Mindoro, Dinagat, Bucas, Mindanao.

Scaphisoma tricoloratum **Löbl & Ogawa**

Scaphisoma tricoloratum Löbl & Ogawa, 2016b: 1435, Figs 238, 239. Holotype male, MHNG; type locality: Philippines: Mindanao, Baracatan, 1500 m.

DISTRIBUTION. Philippines: Mindanao.

Scaphisoma tricolorinotum **Löbl & Ogawa**

Scaphisoma tricolorinotum Löbl & Ogawa, 2016b: 1436, Figs 40, 240, 241. Holotype male, MHNG; type locality: Philippines: Mindanao, Baracatan, 1500 m.

DISTRIBUTION. Philippines: Mindanao.

Scaphisoma tricoloripenne **Löbl & Ogawa**

Scaphisoma tricoloripenne Löbl & Ogawa, 2016b: 1437, Figs 41, 242, 243. Holotype male, SMND; type locality: Philippines: Luzon, Baguio.

DISTRIBUTION. Philippines: Luzon.

Scaphisoma tricoloroides Löbl

Scaphosoma tricolor Pic, 1923c: 269. Lectotype male, MNHN; type locality: Vietnam: Tonkin, Lac Tho.

Scaphisoma tricoloroides Löbl, 1975b: 271; replacement name for *Scaphosoma tricolor* Pic, 1923 (nec *Scaphisoma tricolor* Heller, 1917).

Löbl, 1975b: 271 (lectotype fixed by inference for *S. tricolor* Pic)

Löbl, 1981b: 75 (characters)

Löbl, 1982f: 5 (records)

DISTRIBUTION. Thailand; Vietnam.

Scaphisoma tridentatum Löbl

Scaphisoma tridentatum Löbl, 1975a: 408, Figs 64–66. Holotype male, HNHM; type locality: Papua New Guinea: Friedrich-Wilhelms-Hafen [Madang], Astrolabe Bay.

Löbl, 1977i: 818 (records)

Löbl, 1981b: 74 (characters, records)

DISTRIBUTION. Micronesia: Caroline Islands; Papua New Guinea; Samoa.

Scaphisoma tridens Löbl & Ogawa

Scaphisoma tridens Löbl & Ogawa, 2016b: 1438, Figs 244–247. Holotype male, MHNG; type locality: Philippines: Luzon, Lagunas, Mt. Makiling, above Mad Springs, 400–700 m.

DISTRIBUTION. Philippines: Luzon.

Scaphisoma trifurcatum Löbl & Ogawa

Scaphisoma trifurcatum Löbl & Ogawa, 2016b: 1439, Figs 42, 248–250. Holotype male, EUMJ; type locality: Philippines: Negros Oriental, Yagumyum, Mt. Talinis, ca. 1400-1
600 m.

DISTRIBUTION. Philippines: Negros.

Scaphisoma trilobatum Löbl

Scaphisoma trilobatum Löbl, 1981f: 108, Figs 6, 7. Holotype male, HNHM; type locality: Vietnam: Ninh Binh, Cuc Phuong.

DISTRIBUTION. Vietnam.

Scaphisoma trilobum Löbl & Ogawa

Scaphisoma trilobum Löbl & Ogawa, 2016b: 1440, Figs 43, 44, 251–254. Holotype male, MHNG; type locality: Philippines: Luzon, Mountain Prov., near Sagada, Latan cave.

DISTRIBUTION. Philippines: Luzon.

Scaphisoma trimaculatum **Löbl & Ogawa**

Scaphisoma trimaculatum Löbl & Ogawa, 2016b: 1441, Figs 255, 256. Holotype male, EUMJ; type locality: Philippines: Luzon, Mountain Prov., Sagada, Banga-an, 1650 m.

DISTRIBUTION. Philippines: Luzon.

Scaphisoma triste **Löbl**

Scaphisoma triste Löbl, 1975a: 396, Figs 40, 41. Holotype male, MHNG; type locality: Papua New Guinea: near Sattelberg [Sattelburg], Huon Golf.

DISTRIBUTION. Papua New Guinea.

Scaphisoma tropicum tropicum **Kirsch**

Scaphisoma tropicum Kirsch, 1873: 136. Syntypes, MNHN; type locality: Peru.

DISTRIBUTION. Peru.

Scaphisoma tropicum andreinii **Pic**

Scaphosoma tropicum var. *andreinii* Pic, 1920e: 96. Syntypes, MCSN; type locality: Brazil: Santos.

DISTRIBUTION. Brazil.

Scaphisoma turkomanorum **Reitter**

Scaphisoma turkomanorum Reitter, 1887: 507. Lectotype male, HNHM; type locality: Turkmenistan: Hodscha Kala.
Löbl, 1964c: Fig. 3 (characters)
Löbl, 1967b: 109, Figs 4, 9, 19, 13, 14 (lectotype designation, characters, records)
Löbl, 1970c: 790, Figs 74, 75 (characters, records)
Iablokoff-Khnzorian, 1985: 142 (records)

DISTRIBUTION. Tajikistan; Turkmenistan; Uzbekistan.

Scaphisoma unicolor **Achard**

Scaphosoma unicolor Achard, 1923: 113. Lectotype female, NMPC; type locality: Japan: Kyoto.
Miwa & Mitono, 1943: 540 (characters)
Nakane, 1955c: 56 (characters)
Löbl, 1970c: 772, Figs 50, 51 (lectotype fixed by inference, characters, records)
Hisamatsu, 1977: 194 (records)
Löbl, 1980a: 110, Fig. 26 (characters, records)
Morimoto, 1985: Pl. 45, Fig. 30 (characters)
Löbl, 1990b: 549 (records)
Löbl, 1992a: 533 (records)

Löbl, 1993: 38 (records)
Löbl, 2000: 611 (records)
DISTRIBUTION. China: Yunnan; Japan: Honshu, Kyushu, Shikoku; Nepal; Russia: Far East; Taiwan; Thailand.

Scaphisoma unifasciatum **Pic**
Scaphosoma unifasciatum Pic, 1956b: 72. Lectotype female, HNHM; type locality: Papua New Guinea: Sattelberg [Sattelburg], Huon Golf.
Löbl, 1975a: 404, Figs 52–54 (lectotype fixed by inference, characters)
DISTRIBUTION. Papua New Guinea.

Scaphisoma uniforme **Löbl**
Scaphisoma uniforme Löbl, 1986a: 157, Fig. 20. Holotype male, MHNG; type locality: India: between Algarah and Labha, 7 km from Algarah, 1900 m, Darjeeling District.
Löbl, 1992a: 532 (records)
Löbl, 2000: 611 (characters, records)
DISTRIBUTION. China: Yunnan; India: West Bengal (Darjeeling District); Nepal.

Scaphisoma unimaculatum **Löbl**
Scaphisoma unimaculatum Löbl, 1975a: 405, Figs 55–57. Holotype male, NHML; type locality: Papua New Guinea: Kokoda, 1200 ft.
DISTRIBUTION. Papua New Guinea.

Scaphisoma vagans **Löbl**
Scaphisoma vagans Löbl, 1990b: 543, Figs 47, 48. Holotype male, MHNG; type locality: Thailand: Khao Sabah Nat. Park, near Phliu Waterfalls, 150–300 m.
DISTRIBUTION. Thailand.

Scaphisoma vagenotatum **Pic**
Scaphosoma vagenotatum Pic, 1926b: 2. Lectotype female, MNHN; type locality: Philippines: Mindanao, Dansalan.
Löbl, 1972b: 106 (lectotype designation, characters)
Löbl & Ogawa, 2016b: 1443 (characters)
DISTRIBUTION. Philippines: Mindanao.

Scaphisoma valens **Löbl**
Scaphisoma valens Löbl, 1990b: 573; Figs 97–99. Holotype male, MHNG; type locality: Thailand: Doi Inthanon, 1650 m, Chiang Mai.
DISTRIBUTION. Thailand.

Scaphisoma validum Löbl

Scaphisoma validum Löbl, 1973b: 317, Figs 13, 14. Holotype male, NHMW; type locality: New Caledonia: Roche d'Ouaième near Hienghène, 500–700 m.
Löbl, 1981a: 366 (characters, records)

DISTRIBUTION. New Caledonia.

Scaphisoma variabile Löbl

Scaphisoma variabile Löbl, 1981a: 375, Figs 32–38. Holotype male, NZAC; type locality: New Caledonia: Col des Roussettes, 400 m.

DISTRIBUTION. New Caledonia.

Scaphisoma varians Löbl

Scaphisoma varians Löbl, 1992a: 556. Holotype male, MHNG; type locality: Nepal: south Mangsingma, 2200 m, Sankhuwasabha District.
Löbl, 1986a: 205, Figs 98–101 (as *Scaphisoma* cf. *bedeli*, characters)

DISTRIBUTION. India: Sikkim, West Bengal (Darjeeling District); Nepal.

Scaphisoma varium Löbl

Scaphisoma varium Löbl, 1986a: 189, Figs 70, 71. Holotype male, MHNG; type locality: Bhutan: Paro.
Löbl, 1992a: 536, Fig. 31 (records)

DISTRIBUTION. Bhutan; India: West Bengal (Darjeeling District); Nepal.

Scaphisoma velox Löbl

Scaphisoma velox Löbl, 1990b: 560, Figs 74–77. Holotype male, MHNG; type locality: Thailand: Kaeng Krachan Nat. Park, 30 km from Headquarters, 450 m.

DISTRIBUTION. Thailand.

Scaphisoma vernicatum Achard

Scaphosoma vernicatum Achard, 1920i: 241. Syntypes, NMPC; type locality: Madagascar: Tananarive.

DISTRIBUTION. Madagascar.

Scaphisoma versicolor Fierros-López

Scaphisoma versicolor Fierros-López, 2006a: 64, Figs 1d, 2d, 3d, 4j-l, 6d. Holotype male, CZUG; type locality: Mexico: Jalisco, Casimiro Castillo, Arroyo Tacubaya, 600 m.

DISTRIBUTION. Mexico.

***Scaphisoma vestigator* Löbl**

Scaphisoma vestigator Löbl, 2000: 651, Figs 82–85. Holotype male, MHNG; type locality: China. Sichuan, Mt. Emei, 1700 m.

DISTRIBUTION. China: Sichuan.

***Scaphisoma vexator* Löbl**

Scaphisoma vexator Löbl, 2000: 634, Figs 40, 42. Holotype male, MHNG; type locality: China: Yunnan, Gaoligong Mts., 1500–2500 m.

DISTRIBUTION. China: Yunnan.

***Scaphisoma vicinum* Pic**

Scaphosoma vicinum Pic, 1930a: 88. Syntypes, MRAC/MNHN; type locality: Democratic Republic of the Congo: Biruwe, Buhunde.

DISTRIBUTION. Democratic Republic of the Congo.

***Scaphisoma viduum* Löbl**

Scaphisoma viduum Löbl, 1973b: 326, Figs 27, 28. Holotype male, NHMW; type locality: New Caledonia: Mt. Koghis.

Löbl, 1981a: 373 (characters, records)

DISTRIBUTION. New Caledonia.

***Scaphisoma vietum* Löbl**

Scaphisoma vietum Löbl, 1976c: 222, Figs 1, 2. Holotype male, MHNG; type locality: Vietnam: Saigon.

DISTRIBUTION. Vietnam.

***Scaphisoma villiersi* Pic**

Scaphosoma villiersi Pic, 1942: 1. Syntypes, MNHN; type locality: Cameroon [SE Mt. Cameroon, 1800–2000 m].

DISTRIBUTION. Cameroon.

***Scaphisoma viti* Löbl**

Scaphisoma viti Löbl, 1986e: 387, Figs 1, 2. Holotype male, MHNG; type locality: Pakistan: forest south of Islamabad.

Löbl, 1992a: 531 (records)

DISTRIBUTION. Nepal; Pakistan.

***Scaphisoma volitatum* Löbl**

Scaphisoma volitatum Löbl, 2000. 650, Figs 77–80. Holotype male, MHNG; type locality: China: Yunnan, Xishuangbanna.

DISTRIBUTION. China: Yunnan.

***Scaphisoma werneri* Löbl & Ogawa**

Scaphisoma werneri Löbl & Ogawa, 2016b: 1403, Figs 45, 153, 154. Holotype male, FMNH; type locality: Philippines: Mindanao, Cotabato Prov., Burungkot, Upi, 1500 ft.

DISTRIBUTION. Philippines: Mindanao.

***Scaphisoma wolong* Löbl**

Scaphisoma wolong Löbl, 2000: 621, Figs 16, 17. Holotype male, MHNG; type locality: China: Sichuan, Wolong Nat. Res., 1700 m.

DISTRIBUTION. China: Sichuan.

***Scaphisoma yapense* Löbl**

Scaphisoma yapense Löbl, 1981b: 76, Figs 7, 8. Holotype male, BPBM; type locality: Micronesia: Caroline Islands Tomil District, Yap.

DISTRIBUTION. Micronesia: Caroline Islands.

***Scaphisoma zimmermani* Löbl**

Scaphisoma zimmermani Löbl, 1977i: 821, Figs 5, 6. Holotype male, BPBM; type locality: Fiji: Ketira, Moala Is.

Löbl, 1980c: 392, Figs 25, 26 (characters, records)

DISTRIBUTION. Fiji.

***Scaphisoma*, nomen dubium**

Sphaeridium pulicarium Rossi, 1792: 21. Syntypes, not traced ?ZMUB; type locality: Italy: Etruria [Tuscany].

Gyllenhal, 1808: 187 (in synonymy with *S. agaricinum*, characters)

Scaphobaeocera Csiki

Scaphobaeocera Csiki, 1909: 341. Type species: *Scaphobaeocera papuana* Csiki, 1909; by monotypy. Gender: feminine.

Nesotoxidium Scott, 1922: 22. Type species: *Nesotoxidium typicum* Scott, 1922; by original designation. Gender: neuter.

Baeotoxidium Löbl, 1971c: 990. Type species: *Baeotoxidium lanka* Löbl, 1971; by original designation. Gender: neuter.

Löbl, 1969c: 347 (as *Nesotoxidium*, characters)

Löbl, 1969c: 350 (synonymy of *Nesotoxidium* with *Scaphobaeocera*)

Löbl, 1971c: 986, 990 (keys to Sri Lankan species)

Löbl, 1975a: 414 (key to New Guinean species)

Löbl, 1977e: 58 (key to Australian species)

Löbl, 1979a: 112 (key to South Indian species)

Löbl, 1980a: 117 (key to Taiwanese species)
Löbl, 1981b: 78 (key to Micronesian species)
Löbl, 1981c: 230 (key to Japanese species)
Newton, 1984: 319 (slime mould hosts)
Löbl, 1984a: 81, 82 (keys to North Indian and Bhutanese species)
Löbl, 1986d: 469 (key to Australian species)
Lawrence, 1989: 12 (feeding)
Newton & Stephenson, 1990: 204 (slime mould hosts)
Löbl, 1990b: 592 (characters, key to Thai species)
Löbl, 1992a: 562 (*Baeotoxidium*, characters, key to species)
Löbl, 1992a: 564 (characters, key to Himalayan species)
Löbl, 1999: 735 (key to Chinese species)
Betz et al., 2003: 203, Fig. 13 (characters)
Leschen & Löbl, 2005: 25 (synonymy of *Baeotoxidium* with *Scaphobaeocera*)
Hoshina et al., 2009: 326 (key to Korean species)
Löbl, 2011e: 695 (key to Philippine species)
Ogawa, 2015: 108 (characters, Sulawesi species)
Löbl, 2015a: 115 (key to Lesser Sunda species)
Löbl, 2017: 33 (New Guinean species)

Scaphobaeocera aberrans **Löbl**

Scaphobaeocera aberrans Löbl, 1984a: 92, Figs 56–58. Holotype male, MHNG; type locality: India: West Bengal, between Ghoom and Lopchu, 2000 m.

DISTRIBUTION. India: West Bengal (Darjeeling District).

Scaphobaeocera abnormalis **Löbl**

Scaphobaeocera abnormalis Löbl, 1981c: 240, Figs 18–20. Holotype male, MHNG; type locality: Japan: Mt. Ishizuchi Nat. Park, 1000 m, Ehime Pref.

DISTRIBUTION. Japan: Shikoku.

Scaphobaeocera alticola **Löbl**

Scaphobaeocera alticola Löbl, 1990b: 597, Figs 144, 145. Holotype male, MHNG; type locality: Thailand: Doi Inthanon, 2500 m, Chiang Mai.

DISTRIBUTION. Thailand.

Scaphobaeocera amicalis **Löbl**

Scaphobaeocera amicalis Löbl, 2003a: 74, Figs 26–28. Holotype male, MHNG; type locality: China: Jianggashan, 700–900 m, Jiangxi.

DISTRIBUTION. China: Jiangxi.

Scaphobaeocera antennalis Löbl

Scaphobaeocera antennalis Löbl, 1975a: 416, Figs 75, 76. Holotype male, HNHM; type locality: Papua New Guinea: Sattelberg [Sattelburg], Huon Golf.

DISTRIBUTION. Papua New Guinea.

Scaphobaeocera australiensis Löbl

Scaphobaeocera australiensis Löbl, 1977e: 60, Figs 92, 93. Holotype male, SAMA; type locality: Australia: Queensland, near Maipoton north of Brisbane.

DISTRIBUTION. Australia: Queensland.

Scaphobaeocera baliensis Löbl

Scaphobaeocera baliensis Löbl, 2015a: 115, Figs 112, 113. Holotype male, MHNG; type locality: Indonesia: Bali, Mt. Batukaru near Luhur Temple.

DISTRIBUTION. Indonesia: Bali.

Scaphobaeocera balkei Löbl

Scaphobaeocer balkei Löbl, 2017: 35, Figs 1–3. Holotype male, NMEC; type locality: Indonesia: Western New Guinea Nabire area, road Nabire-Ilaga, 03°29'517"S 135°43'913"E, 750 m.

DISTRIBUTION. Indonesia: Western New Guinea.

Scaphobaeocera bengalensis (Löbl)

Baeotoxidium bengalense Löbl, 1984a: 81, Figs 32–34. Holotype male, MHNG; type locality: India: between Ghoom and Lopchu, 2000 m, Darjeeling District.

Löbl, 1992a: 563 (records)

Leschen & Löbl, 2005: 26 (transfer to *Scaphobaeocera*)

DISTRIBUTION. India: West Bengal (Darjeeling District); Nepal.

Scaphobaeocera brevipennis (Pic)

Toxidium brevipenne Pic, 1947a: 10. Syntype female, MNHN; type locality: Ivory Coast: Park National de Banco.

Löbl & Leschen, 2010: 88 (transfer to *Scaphobaeocera*, characters)

DISTRIBUTION. Ivory Coast.

Scaphobaeocera bulbosa Löbl

Scaphobaeocera bulbosa Löbl, 2011e: 698, Figs 1, 2. Holotype male, MHNG; type locality: Luzon: Lagunas Prov., Mt. Makiling, 400 m.

Löbl, 2015a: 116 (records)

DISTRIBUTION. Indonesia: Bali; Philippines: Luzon.

Scaphobaeocera burckhardti **Löbl**

Scaphobaeocera burckhardti Löbl, 1990b: 606, Figs 159, 160. Holotype male, MHNG; type locality: Thailand: Khao Sabap Nat. Park, above Phliu Waterfalls, 150–300 m.

DISTRIBUTION. Thailand.

Scaphobaeocera cacumina **(Vinson)**

Nesotoxidium cacuminum Vinson, 1943: 200, Fig. 20. Syntypes, NHML/MIMM; type locality: Mauritus: Mt. Cocotte.
Löbl, 1969c: 350 (transfer to *Scaphobaeocera*)

DISTRIBUTION. Mauritus.

Scaphobaeocera cognata **Löbl**

Scaphobaeocera cognata Löbl, 1984a: 89, Figs 47, 48. Holotype male, MHNG; type locality: India: Meghalaya, Mawphlang, 1800 m, Khasi Hills.
Löbl, 1986c: 350 (records)
Löbl, 1992a: 566 (records)
Löbl, 1999: 736 (records)

DISTRIBUTION. China: Shaanxi, Sichuan, Yunnan; India: Meghalaya, Uttarakhand (Garhwal); Nepal.

Scaphobaeocera complicans **Löbl**

Scaphobaeocera complicans Löbl, 2011e: 698, Figs 3–6. Holotype male, MHNG; type locality: Philippines: Luzon, Lagunas Prov., Mt. Makiling, summit road ca 600 m.

DISTRIBUTION. Philippines: Luzon.

Scaphobaeocera confusa **Löbl**

Scaphobaeocera confusa Löbl, 1986d: 466, Figs 3, 4. Holotype male, ANIC; type locality: Australia: Queensland: Fraser Is.

DISTRIBUTION. Australia: Queensland.

Scaphobaeocera curvipes **Löbl**

Scaphobaeocera curvipes Löbl, 1977b: 61, Figs 94, 95. Holotype male, NHML; type locality: Australia: New South Wales, Illawarra.

DISTRIBUTION. Australia: New South Wales.

Scaphobaeocera cyrta **Löbl**

Scaphobaeocera cyrta Löbl, 1980a: 120, Figs 44–46. Holotype male, MHNG; type locality: Taiwan: Fenchihu, 1400 m.

DISTRIBUTION. Taiwan.

Scaphobaeocera data Löbl

Scaphobaeocera data Löbl, 2011e: 700, Figs 7–10. Holotype male, MHNG; type locality: Philippines: Luzon, Mount Data Lodge, 2200–2300 m.

DISTRIBUTION. Philippines: Luzon.

Scaphobaeocera davaoana Löbl

Scaphobaeocera davaoana Löbl, 2011e: 702. Holotype male, FMNH; type locality: Philippines: Mindanao, Davao Prov., east slope Mt. McKinley, 6400 ft.

DISTRIBUTION. Philippines: Mindanao.

Scaphobaeocera delicatula Löbl

Scaphobaeocera delicatula Löbl, 1971c: 986, Figs 56, 57. Holotype male, MHNG; type locality: Sri Lanka: Inginiyagala.

Löbl, 1990b: 597 (records)

DISTRIBUTION. Sri Lanka; Thailand.

Scaphobaeocera difficilis Löbl

Scaphobaeocera difficilis Löbl, 1979a: 113, Figs 35, 36. Holotype male, MHNG; type locality: India: near Aliyar Dam, 300 m, Anaimalai Hills, Tamil Nadu.

Löbl, 1984a: 88 (records)
Löbl, 1986c: 350 (records)
Löbl, 1990b: 599 (records)
Löbl, 1992a: 567 (records)
Löbl, 1999: 736 (records)
Löbl, 2004a: 346 (records)
Löbl, 2005: 180 (records)

DISTRIBUTION. China: Hubei; India: Madhya Pradesh, Meghalaya, Tamil Nadu, Uttarakhand, West Bengal (Darjeeling District); Nepal; Pakistan; Thailand.

Scaphobaeocera discreta Löbl

Scaphobaeocera discreta Löbl, 1984a: 88, Figs 45, 46. Holotype male, MHNG; type locality: India: Manas Wild Life Sanctuary, 200 m, Assam.

Löbl, 1986c: 350 (records)
Löbl, 1990b: 600 (records)

DISTRIBUTION. India: Assam, Meghalaya, Uttarakhand (Kumaon); Thailand.

Scaphobaeocera dispar Löbl

Scaphobaeocera dispar Löbl, 1980a: 118, Figs 42, 43. Holotype male, MHNG; type locality: Taiwan: Fenchihu, 1400 m.

DISTRIBUTION. Taiwan.

***Scaphobaeocera dorsalis* Löbl**

Scaphobaeocera dorsalis Löbl, 1980a: 118, Figs 40, 41. Holotype male, HNHM; type locality: Taiwan: Pilam.
Löbl, 1984a: 91, Figs 49, 50 (characters, records)
Löbl, 1990b: 597 (records)
Löbl, 1992a: 566 (records)
Löbl, 1999: 736 (records)
Hoshina et al., 2009: 326, Figs 1a, 2a, 3a, 3d (characters, records)
Hoshina, 2011: 196 (records)

DISTRIBUTION. China: Sichuan, Yunnan; India: Assam, Meghalaya, West Bengal (Darjeeling District); Japan: Ryukyus: Yonaguni-jima; South Korea; Nepal; Taiwan; Thailand.

***Scaphobaeocera elegans* (Löbl)**

Baeotoxidium elegans Löbl, 1971c: 993, Fig. 66. Holotype male, MZLU; type locality: Sri Lanka: Deerwood Kuruwita, 10 km NNW Ratnapura.
Leschen & Löbl, 2005: 26 (transfer to *Scaphobaeocera*)

DISTRIBUTION. Sri Lanka.

***Scaphobaeocera episternalis* Löbl**

Scaphobaeocera episternalis Löbl, 2011e: 704, Figs 14–16. Holotype male, SMNS; type locality: Philippines: Mindanao, Davao Prov., 25 km west of New Batan.

DISTRIBUTION. Philippines: Mindanao.

***Scaphobaeocera escensa* Löbl**

Scaphobaeocera escensa Löbl, 2011e: 704, Figs 17–19. Holotype male, SMNS; type locality: Philippines: Mindanao, 30 km NW of Maramag, Bagongsilang, 1700 m.

DISTRIBUTION. Philippines: Mindanao.

***Scaphobaeocera excisa* Löbl**

Scaphobaeocera excisa Löbl, 2011e: 706, Figs 20–23. Holotype male, MHNG; Philippines: Luzon, Banguio Prov., Mt. Santo Thomas, ca 1850 m.

DISTRIBUTION. Philippines: Luzon.

***Scaphobaeocera formosana* (Miwa & Mitono)**

Toxidium formosanum Miwa & Mitono, 1943: 547, 555: Fig. H. Lectotype female, TARI; type locality: Taiwan: Akau.
Löbl, 1980a: 117 (transfer to *Scaphobaeocera*, characters, records)
Löbl, 2011b: 206 (lectotype designation)

DISTRIBUTION. Taiwan.

Scaphobaeocera franzi Löbl

Scaphobaeocera franzi Löbl, 1977e: 63, Figs 96, 97. Holotype male, SAMA; type locality: near Maipoton north Brisbane.

DISTRIBUTION. Australia: Queensland.

Note. Löbl, 1973b: 329 reported this species as *Scaphobaeocera* sp. from New Caledonia, based on a mislabelled specimen.

Scaphobaeocera fratercula Löbl

Scaphobaeocera fratercula Löbl, 1984a: 85. Holotype male, MHNG; type locality: India: Meghalaya: Mawphlang, Khasi Hills.

DISTRIBUTION. India: Meghalaya.

Scaphobaeocera fujiana Löbl

Scaphobaeocera fujiana Löbl, 2003a: 74, Figs 22–25. Holotype male, ZIBC; type locality: China: Fujian, Wuyi Shan, Giligiao-Guadun road, 1000–1300 m.

DISTRIBUTION. China: Fujian.

Scaphobaeocera gagata (Löbl)

Baeotoxidium gagatum Löbl, 1971c: 993, Figs 64, 65. Holotype male, MHNG; type locality: Sri Lanka: Pidurutalagala, 1950 m, near Nuwara Eliya.

Leschen & Löbl, 2005: 26 (transfer to *Scaphobaeocera*)

DISTRIBUTION. Sri Lanka.

Scaphobaeocera gemina Löbl

Scaphobaeocera gemina Löbl, 2017: 36, Figfs 4–7. Holotype male, MHNG; type locality: Papua New Guinea: Morobe District, Biaru Road, Mt. Kolorong, 2200 m.

DISTRIBUTION. Papua New Guinea.

Scaphobaeocera gracilis Löbl

Scaphobaeocera gracilis Löbl, 1981c: 235, Figs 9–11. Holotype male, MHNG; type locality: Japan: Mt. Kasuga, Nara Pref.

DISTRIBUTION. Japan: Honshu.

Scaphobaeocera hamata Löbl

Scaphobaeocera hamata Löbl, 2011e: 707, Figs 24–26. Holotype male, MHNG; Philippines: Luzon, Laguna Prov., Mt. Makiling.

DISTRIBUTION. Philippines: Luzon.

Scaphobaeocera hisamatsui Hoshina

Scaphobaeocera hisamatsui Hoshina, 2008a: 141, Figs 1–7. Holotype male, MNHA; type locality: Japan: Sonai, Iriomote Is.

DISTRIBUTION. Japan: Ryukyus: Iriomote, Ishigaki.

Scaphobaeocera incisa Löbl

Scaphobaeocera incisa Löbl, 1990b: 595, Figs 141–143. Holotype male, MHNG; type locality: Thailand: Chiang Mai, road to Wab Pang An, 000 m, 50 km NE Chiang Mai.

Löbl, 1999: 735 (records)

DISTRIBUTION. China: Yunnan; Thailand.

Scaphobaeocera indica (Löbl)

Baeotoxidium indicum Löbl, 1979a: 112, Figs 33, 34. Holotype male, MHNG; type locality: India: Kerala, Valara Falls, 46 km SW Munnar, 450–500 m.

Leschen & Löbl, 2005: 25 (transfer to *Scaphobaeocera*)

DISTRIBUTION. India: Kerala.

Scaphobaeocera inexpectata Löbl

Scaphobaeocera inexpectata Löbl, 1981c: 237, Figs 12–14. Holotype male, MHNG; type locality: Japan: below Usui Pass, 850 m, Gunma Pref.

Löbl, 1993: 39 (records)

Löbl, 2003a: 63 (records)

Hoshina et al., 2009: 327, Figs 1b, 2b, 3b, 3e (characters, records)

DISTRIBUTION. China: Anhui, Jiangxi; Japan: Honshu; Russia: Far East; South Korea.

Scaphobaeocera instriata (Pic)

Scaphosoma instriatum Pic, 1955: 51. Lectotype male, MRAC; type locality: Rwanda: forest Rugege, 2400 m, Tshuruyaga.

Löbl, 1987d: 846 (lectotype designation, transfer to *Scaphobaeocera*)

DISTRIBUTION. Rwanda.

Scaphobaeocera integra (Reitter)

Toxidium integrum Reitter, 1908: 34. Holotype female, NHMW; type locality: Tanzania: Amani.

Löbl & Leschen, 2010: 88 (transfer to *Scaphobaeocera*, characters)

DISTRIBUTION. Tanzania.

***Scaphobaeocera japonica* (Reitter)**

Toxidium japonicum Reitter, 1880a: 49. Lectotype female, MNHN; type locality: Japan.

Achard, 1923: 119 (as *Toxidium*, characters)
Miwa & Mitono, 1943: 547 (as *Toxidium*, characters)
Nakane, 1955c: 57 (as *Toxidium*, characters)
Nakane, 1963b: 79 80 (as *Toxidium*, characters)
Löbl, 1969c: 348, Figs 2, 4, 5 (lectotype fixed by inference, transfer to *Nesotoxidium*, characters, records)
Löbl, 1969c: 350 (transfer to *Scaphobaeocera*)
Hisamatsu, 1977: 194 (as *Toxidium*, records)
Löbl, 1981c: 230, Figs 1, 2 (characters, records)
Löbl, 1992a: 621, Figs 181, 183 (characters)
Löbl, 2003a: 63 (records)

DISTRIBUTION. China: Sichuan; Japan: Honshu, Kyushu.

***Scaphobaeocera kraepelini* (Pic)**

Toxidium kraepelini Pic, 1933a: 72. Lectotype MNHN; type locality: Indonesia: Java, Buitenzorg [Bogor].

Löbl, 1984a: 82 (transfer to *Scaphobaeocera*)
Löbl, 2015a: 117, Figs 116, 117 (lectotype designation, characters)

DISTRIBUTION. Indonesia: Java.

***Scaphobaeocera laevis* Löbl**

Scaphobaeocera laevis Löbl, 1990b: 594, Figs 138–140. Holotype male, MHNG; type locality: Thailand: Doi Suthep, south slope, 1450 m, Chian Mai.

DISTRIBUTION. Thailand.

***Scaphobaeocera lamellifera* Löbl**

Scaphobaeocera lamellifera Löbl, 1984a: 85. Holotype male, MHNG; type locality: India: Meghalaya, Songsak, Garo Hills.

Löbl, 1999: 735 (records)

DISTRIBUTION. China: Yunnan; India: Meghalaya.

***Scaphobaeocera lanka* (Löbl)**

Baeotoxidium lanka Löbl, 1971c: 991, Figs 62, 63. Holotype male, MHNG; type locality: Sri Lanka: above Talatuoya, 850 m.

Löbl, 1992a: 621, Figs 180, 191 (characters)
Leschen & Löbl, 2005: 26 (transfer to *Scaphobaeocera*)

DISTRIBUTION. Sri Lanka.

Scaphobaeocera lombokensis **Löbl**

Scaphobaeocera lombokensis Löbl, 2015a: 116, Figs 114, 115. Holotype male, MHNG; type locality: Indonesia: Lombok, Pusuk Pass, 300 m.

DISTRIBUTION. Indonesia: Lombok.

Scaphobaeocera lycocorax **Löbl**

Scaphobaeocera lycocorax Löbl, 2017: 37, Figs 8–12. Holotype male, MHNG; type locality: Papua New Guinea: Morobe District, Bulldog Road., Mt. Naiko, 2750 m.

DISTRIBUTION. Papua New Guinea.

Scaphobaeocera maculata **Löbl**

Scaphobaeocera maculata Löbl, 1990b: 596. Holotype female, MHNG; type locality: Thailand: Khao Sabah Nat. Park, near Phliu Waterfalls, 150–300 m.

DISTRIBUTION. Thailand.

Scaphobaeocera minuta **(Achard)**

Toxidium minutum Achard, 1920f: 364. Lectotype female, NMPC; type locality: India: Darjeeling District, Kurseong.

Löbl, 1971c: 989 (lectotype fixed by inference, transfer to *Scaphobaeocera*)

Löbl, 1984a: 86, Figs 43, 44 (characters, records)

Löbl, 1990b: 599 (records)

Löbl, 1992a: 565 (records)

DISTRIBUTION. India: Meghalaya, West Bengal (Darjeeling District); Nepal; Thailand.

Scaphobaeocera minutissima **(Löbl)**

Nesotoxidium minutissimum Löbl, 1969c: 349. Holotype male, MHNG; type locality: Philippines: Luzon, Mt. Makiling.

Löbl, 1969c: 350 (transfer to *Scaphobaeocera*)

DISTRIBUTION. Philippines: Luzon.

Scaphobaeocera molesta **Löbl**

Scaphobaeocera molesta Löbl, 1999: 738, Figs 66, 67. Holotype male, MHNG; type locality: China: Mengyang Nat. Res., ca 500 m, Yunnan.

DISTRIBUTION. China: Yunnan.

Scaphobaeocera monticola **Löbl**

Scaphobaeocera monticola Löbl, 2011e: 708, Figs 27–30. Holotype male, MHNG; type locality: Philippines: Luzon, Banguio Prov., Mt. Santo Thomas, ca 2150 m.

DISTRIBUTION. Philippines: Luzon.

Scaphobaeocera montivagans **Löbl**

Scaphobaeocera montivagans Löbl, 2011e: 710. Holotype male, MHNG; type locality: Philippines: Luzon, Ifugao Prov., Mt. Pangao, 2350 m, near Data.

DISTRIBUTION. Philippines: Luzon.

Scaphobaeocera mussardi **Löbl**

Scaphobaeocera mussardi Löbl, 1971c: 989, Figs 60, 61. Holotype male, MHNG; type locality: Sri Lanka: Weragamtota.

Löbl, 1979a: 113 (records)
Löbl, 1984a: 86 (characters, records)
Löbl, 1992a: 567 (records)

DISTRIBUTION. Bhutan; India: Kerala, Meghalaya, Tamil Nadu, West Bengal (Darjeeling District); Nepal; Sri Lanka.

Scaphobaeocera nobilis **Löbl**

Scaphobaeocera nobilis Löbl, 1984a: 91, Figs 51, 52. Holotype male, NHMB; type locality: Bhutan: Balu Jhura, 6 km east Phuntsholing, 200 m.

Löbl, 1990b: 599 (records)
Löbl, 1999: 736 (records)
Löbl, 2003a: 63 (records)

DISTRIBUTION. Bhutan; China: Fujian, Yunnan; Thailand.

Scaphobaeocera notata **Löbl**

Scaphobaeocera notata Löbl, 2017: 38, Figs 13–15. Holotype male, MHNG; type locality: Papua New Guinea: Morobe District, Mt. Mission, Bitoi Road, 1350 m.

DISTRIBUTION. Papua New Guinea.

Scaphobaeocera nuda **Löbl**

Scaphobaeocera nuda Löbl, 1979a: 117. Holotype female, MHNG; type locality: India: Valara Falls, 46 km SW Munnar, 450–500 m, Kerala.

Löbl, 1984a: 84, Figs 35–38 (characters, records)
Löbl, 1990b: 594 (records)
Löbl, 1992a: 565 (records)

DISTRIBUTION. India: Kerala, Meghalaya, West Bengal; Nepal; Thailand.

Scaphobaeocera obducta **Löbl**

Scaphobaeocera obducta Löbl, 1990b: 604, Figs 154, 155. Holotype male, MHNG; type locality: Thailand: Khao Yai Nat. Park, near Headquarters, 750–850 m.

DISTRIBUTION. Thailand.

Scaphobaeocera oberthueri (Reitter)

Toxidium oberthüri Reitter, 1881: 141. Syntype, MNHN; type locality: Ethiopia ["Abyssinien"].

Toxidium reitteri Oberthür, 1883: 16. Syntype, MNHN; type locality: Ethiopia ["Abyssinien"].

Löbl & Leschen, 2010: 87 (transfer to *Scaphobaeocera*, synonymy of *T. reitteri* with *T. oberthueri*, characters)

DISTRIBUTION. Ethiopia.

Scaphobaeocera ornata (Pic)

Toxidium ornatum Pic, 1956b: 73. Holotype male, HNHM; type locality: Papua New Guinea: Stephansort, Astrolabe Bay.

Löbl, 1975a: 416, Figs 73, 74 (transfer to *Scaphobaeocera*, characters)

DISTRIBUTION. Papua New Guinea.

Scaphobaeocera orousseti Löbl

Scaphobaeocera orousseti Löbl, 2011e: 710, Figs 31–34. Holotype male, MHNG; type locality: Philippines: Luzon, Sagada, above resurgence of Ambasing.

DISTRIBUTION. Philippines: Luzon.

Scaphobaeocera palawana Löbl

Scaphobaeocera palawana Löbl, 2011e: 711, Figs 35, 36. Holotype male, MHNG; type locality: Philippines: Olangoan, 18 km NE San Rafael, Palawan.

DISTRIBUTION. Philippines: Palawan.

Scaphobaeocera papuana Csiki

Scaphobaeocera papuana Csiki, 1909: 342. Lectotype male, HNHM; type locality: Papua New Guinea: Friedrich-Wilhems-Hafen [Madang], Astrolabe Bay.

Löbl, 1975a: 414, Figs 71, 72 (lectotype fixed by inference, characters)

Löbl, 2017: 38 (records)

DISTRIBUTION. Papua New Guinea.

Scaphobaeocera pecki Löbl

Scaphobaeocera pecki Löbl, 1981c: 238, Figs 15–17. Holotype male, MHNG; type locality: Japan: Omogo Valley, Mt. Ichizushi Nat. Park, 700 m, Ehime Pref.

Löbl, 1997: 146 (records)

Löbl, 2004: 503 (records)

Hoshina et al., 2009: 327, Figs 1c, 2c, 3c, 3f (characters, records)

DISTRIBUTION. Japan: Honshu, Shikoku; South Korea.

Scaphobaeocera piceoapicalis Löbl

Scaphobaeocera piceoapicalis Löbl, 1977e: 61, Figs 92, 93. Holotype male, MHNG; type locality: Australia: New South Wales, Acacia Plateau.

DISTRIBUTION. Australia: New South Wales.

Scaphobaeocera ponapensis Löbl

Scaphobaeocera ponapensis Löbl, 1981b: 79, Fig. 11. Holotype male, BPBM; type locality: Micronesia: Caroline Islands, Awakpa, Uh District, Ponape.

DISTRIBUTION. Micronesia: Caroline Islands.

Scaphobaeocera pseudotenella Löbl

Scaphobaeocera pseudotenella Löbl, 2011e: 713, Figs 37, 38. Holotype male, MHNG; Philippines: Luzon, Mt. Makiling, summit road, 600 m.

Löbl, 2015a: 116 (records)

DISTRIBUTION. Philippines: Luzon; Indonesia: Bali.

Scaphobaeocera pseudovalida Löbl

Scaphobaeocera pseudovalida Löbl, 1999: 739, Figs 68, 69. Holotype male, MHNG; type locality: China: Mengyang Nat. Res., ca 500 m, Yunnan.

DISTRIBUTION. China: Yunnan.

Scaphobaeocera ptiliformis Löbl

Scaphobaeocera ptiliformis Löbl, 1975a: 417, Figs 77, 78. Holotype male, HNHM; type locality: Papua New Guinea: Bussu river, Lae.

DISTRIBUTION. Papua New Guinea.

Scaphobaeocera pubiventris Löbl

Scaphobaeocera pubiventris Löbl, 2011e: 714, Figs 39, 40. Holotype male, MHNG; Philippines: Luzon, Mountain Prov., north and northeast of Sagada.

DISTRIBUTION. Philippines: Luzon.

Scaphobaeocera punctata Löbl

Scaphobaeocera punctata Löbl, 2017: 39, Figs 16–18. Holotype male, HNHM; type locality: Papua New Guinea: Lae.

DISTRIBUTION. Papua New Guinea.

Scaphobaeocera queenslandica Löbl

Scaphobaeocera queenslandica Löbl, 1986d: 468, Fig. 5. Holotype male, ANIC; type locality: Australia: Queensland, Palm Park.

DISTRIBUTION. Australia: Queensland.

***Scaphobaeocera querceti* Löbl**

Scaphobaeocera querceti Löbl, 1984a: 94, Figs 59–61. Holotype male, MHNG; type locality: India: Shillong Peak, 1850–1950 m, Khasi Hills.
Löbl, 1986c: 349 (records)

DISTRIBUTION. India: Meghalaya, ?Himachal Pradesh.

***Scaphobaeocera remota* Löbl**

Scaphobaeocera remota Löbl, 1981b: 78, Figs 9, 10. Holotype male, FMNH; type locality: Micronesia: Caroline Islands, Peleliu, near coast, Palau.
Löbl, 2017: 40 (records)

DISTRIBUTION. Indonesia: Western New Guinea; Micronesia: Caroline Islands.

***Scaphobaeocera robustula* Löbl**

Scaphobaeocera robustula Löbl, 1990b: 606, Figs 161, 162. Holotype male, MHNG; type locality: Thailand: Doi Suthep, north slope, 1450 m, Chiang Mai.

DISTRIBUTION. Thailand.

***Scaphobaeocera sabapensis* Löbl**

Scaphobaeocera sabapensis Löbl, 1990b: 598, Figs 146, 147. Holotype male, MHNG; type locality: Thailand: Khao Sabap Nat. Park, 150–300 m.
Löbl, 2011e: 715 (records)

DISTRIBUTION. Philippines; Thailand.

***Scaphobaeocera schouteni* Löbl**

Scaphobaeocera schouteni Löbl, 1980b: 223, Figs 3, 4. Holotype male, MHNG; type locality: Papua New Guinea: New Ireland, Limbin, Lelet Plateau, 1100 m.

DISTRIBUTION. Papua New Guinea: New Ireland.

***Scaphobaeocera serpentis* Löbl**

Scaphobaeocera serpentis Löbl, 2011e: 715, Figs 41–43. Holotype male, MHNG; type locality: Philippines: Luzon, Lagunas Prov., Mt. Makiling.

DISTRIBUTION. Philippines: Luzon.

***Scaphobaeocera siamensis* (Löbl)**

Baeotoxidium siamense Löbl, 1990b: 592, Figs 135–136. Holotype male, MHNG; type locality: Thailand: 33 km NE Chiang Mai, via Chiang Rai, ca 500 m.
Leschen & Löbl, 2005: 26 (transfer to *Scaphobaeocera*)

DISTRIBUTION. Thailand.

Scaphobaeocera simplex Löbl

Scaphobaeocera simplex Löbl, 1999: 738, Figs 64, 65. Holotype male, MHNG; type locality: China: Wolong Nat. Res., 1500 m, Sichuan.

DISTRIBUTION. China: Sichuan.

Scaphobaeocera smetanai Löbl

Scaphobaeocera smetanai Löbl, 1981c: 233, Figs 4–6. Holotype male, MHNG; type locality: Japan: Mt. Ishizuchi, 1550 m, Ehime Pref.
Morimoto, 1985: Pl. 45, Fig. 35 (characters)

DISTRIBUTION. Japan: Honshu, Shikoku.

Scaphobaeocera soror Löbl

Scaphobaeocera soror Löbl, 1979a: 114, Figs 37, 38. Holotype male, MHNG; type locality: India: 16 km east Kodaikanal, 1400 m, Palni Hills, Tamil Nadu.

DISTRIBUTION. India: Tamil Nadu.

Scaphobaeocera spinigera Löbl

Scaphobaeocera spinigera Löbl, 1979a: 116, Figs 39, 40. Holotype male, MHNG; type locality: India: Biligiri Rangan Hills, 10 km NE Dhimbam, 1200 m, Tamil Nadu.
Löbl, 1984a: 91 (records)
Löbl, 1986c: 349 (records)
Löbl, 1990b: 602 (records)
Löbl, 1992a: 566 (records)
Löbl, 1993a: 63 (records)
Löbl, 1999: 736 (records)

DISTRIBUTION. China: Fujian, Hong Kong, Sichuan; India: Assam, Kerala, Meghalaya, Tamil Nadu, West Bengal (Darjeeling District); Nepal; Pakistan; Thailand.

Scaphobaeocera spira Löbl

Scaphobaeocera spira Löbl, 1990b: 602, Figs 152, 153. Holotype male, MHNG; type locality: Thailand: Khao Yai Nat. Park, near Headquarters, 750–850 m.
Löbl, 1992a: 567 (records)
Löbl, 1999: 736 (records)

DISTRIBUTION. China: Yunnan; Nepal; Thailand.

Scaphobaeocera stephensoni Löbl

Scaphobaeocera stephensoni Löbl, 1988a: 375, Figs 3, 4. Holotype male, FMNH; type locality: India: Himachal Pradesh, Dhangiara, Junee Valley, 1800 m.

Newton & Stephenson, 1990: 212 (slime mould hosts)
Löbl, 1992a: 566 (records)
DISTRIBUTION. India: Himachal Pradesh; Nepal.

Scaphobaeocera stipes **Löbl**
Scaphobaeocera stipes Löbl, 1971c: 988, Figs 58, 59. Holotype male, MHNG; type locality: Sri Lanka: Inginiyagala.
DISTRIBUTION. Sri Lanka.

Scaphobaeocera sunadai **Hoshina & Sugaya**
Scaphobaeocera sunadai Hoshina & Sugaya, 2003: 36, Figs 1 [misleading], 3, 5, 7, 9–11. Holotype male, MNHA; type locality: Japan: Ie-Rindô, 300 m, Okinawa Is.
DISTRIBUTION. Japan: Ryukyus: Okinawa.

Scaphobaeocera tenella **Löbl**
Scaphobaeocera tenella Löbl, 1990b: 601, Figs 150, 151. Holotype male, MHNG; type locality: Thailand: Khao Yai Nat. Park, near Headquarters, 750–850 m.
Löbl, 1992a: 566 (records)
DISTRIBUTION. India: Meghalaya; Nepal; Thailand.

Scaphobaeocera tibialis **Löbl**
Scaphobaeocera tibialis Löbl, 1984a: 94, Figs 62–64. Holotype male, MHNG; type locality: India: West Bengal, betwenn Ghoom and Lopchu, 2000 m, 13 km east Ghoom.
DISTRIBUTION. India: West Bengal (Darjeeling District).

Scaphobaeocera timida **Löbl**
Scaphobaeocera timida Löbl, 1984a: 92, Figs 53–55. Holotype male, NHMB; type locality: Bhutan: 70 km from Phuntsholing, via Thimbu.
Löbl, 1986c: 349 (records)
Newton & Stephenson, 1990: 204 (slime mould hosts)
Löbl, 1992a: 567 (records)
Löbl, 2003a: 63 (records)
Löbl, 2005: 180 (records)
Löbl, 2011d: 183 (records)
DISTRIBUTION. Bhutan; China: Jiangxi; India: Himachal Pradesh, Uttarakhand (Kumaon); Nepal.

Scaphobaeocera typica **(Scott)**

Nesotoxidium typicum Scott, 1922: 229, Pl. 22, Figs 28–30. Syntypes, NHML; type localities: Seychelles: near Pot-à-eau, Silhouettes; Cascade Estate, forest behind Trois Frères, Mahé; Code-de-Mer forest, Vallée de Mai, Praslin.
Löbl, 1969c: 350 (transfer to *Scaphobaeocera*)

DISTRIBUTION. Seychelles.

Scaphobaeocera uncata **Löbl**

Scaphobaeocera uncata Löbl, 1990b: 600, Figs 148, 149. Holotype male, MHNG; type locality: Thailand: W. Tak.

DISTRIBUTION. Thailand.

Scaphobaeocera valida **Löbl**

Scaphobaeocera valida Löbl, 1990b: 604, Figs 156–158. Holotype male, MHNG; type locality: Thailand: Khao Yai Nat. Park, east Heo Suwat Waterfalls, 900 m.

DISTRIBUTION. Thailand.

Scaphobaeocera variabilis **Löbl**

Scaphobaeocera variabilis Löbl, 1981c: 234, Figs 7, 8. Holotype male, MHNG; type locality: Japan: below Usui Bypass, 700 m, Gunma Pref.

DISTRIBUTION. Japan: Honshu, Shikoku.

Scaphobaeocera watrousi **Löbl**

Scaphobaeocera watrousi Löbl, 2011e: 716, Figs 44–46. Holotype male, MHNG; type locality: Philippines: Luzon, Lagunas Prov., Mt. Makiling.

DISTRIBUTION. Philippines: Luzon.

Scaphobaeocera werneri **Löbl**

Scaphobaeocera werneri Löbl, 2011e: 718, Figs 47–51. Holotype male, FMNH; type locality: Philippines: Mindanao, Davao Prov., east slope Mt. McKinley, 3200 ft.

DISTRIBUTION. Philippines: Mindanao.

Scaphobaeocera yeti **(Löbl)**

Baeotoxidium yeti Löbl, 1992a: 563, Fig. 166. Holotype male, MHNG; type locality: Nepal: Malemci, 2800 m, Sindhupalcok District.
Leschen & Löbl, 2005: 26 (transfer to *Scaphobaeocera*)

DISTRIBUTION. Nepal.

Scaphobaeocera zdenae **Löbl**

Scaphobaeocera zdenae Löbl, 1992a: 568. Holotype female, MHNG; type locality: Nepal: NE Kuwapani, 2500 m, Sankhuwasabha District.

DISTRIBUTION. Nepal.

Scaphoxium

Scaphoxium Löbl, 1979a: 118. Type species: *Toxidium madurense* Pic, 1920, by original designation. Gender: neuter.
Löbl, 1971c: 995 (as *Toxidium*, key to Sri Lankan species)
Löbl, 1977f: 829 (as *Toxidium*, key to Fijian species)
Löbl, 1979a: 120 (key to South Indian species)
Löbl, 1980c: 396 (characters, key to Fijian species)
Löbl, 1984a: 100 (key to Northeast Indian species)
Löbl, 1990b: 611 (key to Thai species)
Löbl, 1992a: 569 (characters, key to Asian species)
Löbl, 2002c: 473 (key to South Pacific species)
Hoshina, 2009: 1 (key to Ryukyu species)
Löbl, 2010b: 128 (key to African and Malagasy species)
Ogawa, 2015: 114 (characters, Sulawesi species)

Scaphoxium alesi **Löbl**

Scaphoxium alesi Löbl, 2011a: 310, Figs 16, 17. Holotype male, MHNG; Philippines: Luzon, Mt. Makiling, 4 km SE Los Banos.

DISTRIBUTION. Philippines: Luzon.

Scaphoxium assamense **Löbl**

Scaphoxium assamense Löbl, 1984a: 103, Figs 80–82. Holotype male, MHNG; type locality: India: Manas Wild Life Sanctuary, 200 m.

DISTRIBUTION. India: Assam.

Scaphoxium avidum **Löbl**

Scaphoxium avidum Löbl, 1990b: 612, Figs 171–173. Holotype male, MHNG; type locality: Thailand: Chiang Mai, road to Wab Pang An, 900 m, 50 km NE Chiang Mai.

DISTRIBUTION. Thailand.

Scaphoxium bilobum **Löbl**

Scaphoxium bilobum Löbl, 2015a: 117, Figs 118–120. Holotype male, MHNG; type locality: Indonesia: Lombok, Mt. Rinjani near Waterfalls, ca 400 m.

DISTRIBUTION. Indonesia: Lombok.

***Scaphoxium biroi* (Pic)**

Toxidium biroi Pic, 1956b: 73. Holotype female, HNHM; type locality: Papua New Guinea: Sattelberg [Sattelburg], Huon Golf.
Löbl, 1975a: 418 (as *Toxidium*, characters)
Löbl, 1979a: 120 (transfer to *Scaphoxium*)
DISTRIBUTION. Papua New Guinea.

***Scaphoxium cuspidatum* (Löbl)**

Toxidium cuspidatum Löbl, 1977e: 68, Fig. 102. Holotype male, SAMA; type locality: Australia: Queensland, Cairns District.
Löbl, 1979a: 120 (transfer to *Scaphoxium*)
DISTRIBUTION. Australia: Queensland.

***Scaphoxium eximium* Löbl**

Scaphoxium eximium Löbl, 1986c: 365 Figs 22–24. Holotype male, MHNG; type locality: India: Uttarakhand, Bhim Tal, 1500 m, Kumaon.
Löbl, 1992a: 570 (records)
DISTRIBUTION. India: Uttarakhand (Kumaon); Nepal.

***Scaphoxium gibbosum* (Champion)**

Toxidium gibbosum Champion, 1927: 273. Lectotype female, NHML; type locality: India: Haldwani District, Kumaon.
Löbl, 1979a: 120 (lectotype designation, transfer to *Scaphoxium*)
DISTRIBUTION. India: Uttarakhand (Kumaon).

***Scaphoxium grande* Löbl**

Scaphoxium grande Löbl, 1986b: 101, Figs 19–21. Holotype male, MHNG; type locality: West Malaysia: Cameron Highlands.
Löbl, 2012b: 183 (records)
DISTRIBUTION. West Malaysia.

***Scaphoxium hartmanni* Löbl**

Scaphoxium hartmanni Löbl, 2001: 183, Figs 9–11. Holotype male, NMEC; type locality: Nepal: Sankhuwasabha District, SW Chichila, 2040 m, Kosi.
DISTRIBUTION. Nepal.

***Scaphoxium heissi* Löbl**

Scaphoxium heissi Löbl, 2010b: 126, Figs 17–19. Holotype male, MHNG; type locality: Madagascar: Diego Suarez, Mt. d'Ambre.
DISTRIBUTION. Madagascar.

Scaphoxium impedicum **Löbl**

Scaphoxium impedicum Löbl, 2002c: 472, Figs 7–9. Holotype male, MHNG; type locality: Papua New Guinea: Morobe District, Mt. Kaindi, 1350 m.

DISTRIBUTION. Papua New Guinea.

Scaphoxium intermedium **Löbl**

Scaphoxium intermedium Löbl, 1984a: 101, Figs 75–79. Holotype male, MHNG; type locality: India: Songsak, 400 m, Garo Hills, Meghalaya.

Löbl, 1986c: 350 (records)

Löbl, 1990b: 612 (records)

Löbl, 1999: 740 (records)

Löbl, 2003a: 63 (records)

DISTRIBUTION. China: Anhui, Yunnan; India: Meghalaya, Uttarakhand (Kumaon), West Bengal (Darjeeling District); Thailand.

Scaphoxium japonicum **Löbl**

Scaphoxium japonicum Löbl, 1981c: 242, Figs 21–23. Holotype male, MHNG; type locality: Japan: Mt. Kasuga, Nara Pref.

DISTRIBUTION. Japan: Honshu, Shikoku.

Scaphoxium kenyanum **Löbl**

Scaphoxium kenyanum Löbl, 2010b: 123, Figs 11–13. Holotype male, MHNG; type locality: Kenya: Shimla Hills Nat. Park, 400 m, Makadara Forest.

DISTRIBUTION. Kenya.

Scaphoxium keralense **Löbl**

Scaphoxium keralense Löbl, 1979a: 123, Figs 57–59. Holotype male, MHNG; type locality: India: Kerala, Kaikatty, 900 m, Nelliampathi Hills.

DISTRIBUTION. India: Kerala.

Scaphoxium kunigamiense **Hoshina & Sugaya**

Scaphoxium kunigamiense Hoshina & Sugaya, 2003: 39, Figs 2, 4, 6, 8, 12–14. Holotype male, MNHA; type locality: Japan: Mt. Yonahadake, Kunigami Village, Okinawa Is.

Note. Fig. 2 is misleading.

DISTRIBUTION. Japan: Okinawa.

Scaphoxium leleupi **(Pic)**

Toxidium leleupi Pic, 1954b: 38. Holotype male, MRAC; type locality: Democratic Republic of the Congo: Kundelungu, Katanga.

Löbl, 2010b: 120, Figs 1–3 (transfer to *Scaphoxium*, characters)

DISTRIBUTION. Democratic Republic of the Congo.

***Scaphoxium lemairei* Löbl**

Scaphoxium lemairei Löbl, 1980b: 223, Figs 5–7. Holotype male, MHNG; type locality: Papua New Guinea: New Ireland, near Limbin, Lelet Plateau, ca 1200 m.
Löbl, 1997: 150 (misspelled *lemarei*)

DISTRIBUTION. Papua New Guinea: New Ireland.

***Scaphoxium madurense* (Pic)**

Toxidium madurense Pic, 1920f: 24. Lectotype male, MNHN; type locality: India: Tamil Nadu, Shembaganur.
Löbl, 1979a: 120, Figs 43–50 (lectotype designation, transfer to *Scaphoxium*, characters, records)

DISTRIBUTION. India: Tamil Nadu.

***Scaphoxium mahnerti* Löbl**

Scaphoxium mahnerti Löbl, 2010b: 125, Figs 14–16. Holotype male, MHNG; type locality: Ivory Coast: Man, Issoneu, 6 km west of Sanguiné.

DISTRIBUTION. Ivory Coast.

***Scaphoxium malekulense* (Löbl)**

Toxidium malekulense Löbl, 1977i: 827. Type material not designated.
Löbl, 1978b: 110, Figs 3–5 (subsequently issued description). Holotype male, BPBM; type locality: Vanuatu, Malekula, Malua Bay.
Löbl, 1979a: 120 (transfer to *Scaphoxium*)
Löbl, 1980c: 396, Figs 38, 39 (characters, records)

DISTRIBUTION. Fiji; Vanuatu.

***Scaphoxium oblitum* (Löbl)**

Toxidium oblitum Löbl, 1971c: 996, Figs 69, 70. Holotype male, MHNG; type locality: Sri Lanka: Hatton, 1400 m.
Löbl, 1979a: 120 (transfer to *Scaphoxium*)

DISTRIBUTION. Sri Lanka.

***Scaphoxium occidentale* Löbl**

Scaphoxium occidentale Löbl, 2010b: 122, Figs 8–10. Holotype male, MHNG; type locality: Ivory Coast: Forêt de Yapo near Yapo Gare.

DISTRIBUTION. Ivory Coast; Ghana.

***Scaphoxium oxyurum* (Löbl)**

Toxidium oxyurum Löbl, 1977e: 66, Fig. 101. Holotype male, QMBA; type locality: Australia: Queensland, The Boulders, via Babinda.
Löbl, 1979a: 120 (transfer to *Scaphoxium*)

DISTRIBUTION. Australia: Queensland.

Scaphoxium papuanum Löbl

Scaphoxium papuanum Löbl, 2002c: 470, Figs 4–6. Holotype male, MHNG; type locality: Papua New Guinea: Morobe District, Biaru Road, Mt. Kolorong, 2000 m.

DISTRIBUTION. Papua New Guinea.

Scaphoxium pigneratum Löbl

Scaphoxium pigneratum Löbl, 2002c: 469, Figs 1–3. Holotype male, MHNG; type locality: Papua New Guinea: Morobe District, at Wau Ecology Institute, 1200 m.

DISTRIBUTION. Papua New Guinea.

Scaphoxium praeustum (Reitter)

Toxidium praeustum Reitter, 1908: 33. Holotype male, NHMW; type locality: Tanzania: Amani.

Toxidium evanescens Reitter, 1908: 34. Holotype male, NHMW; type locality: Tanzania: Amani.

Löbl, 2010b: 121, Figs 4–7 (transfer to *Scaphoxium*, synonymy of *T. evanescens* with *S. praeustum*, characters)

DISTRIBUTION. Tanzania.

Scaphoxium praslinense (Scott)

Toxidium praslinense Scott, 1922: 227, Pl.21, Figs 24b, 25b, 26b, 27b. Syntypes, NHML/CUMZ; type locality: Seychelles: Coco-de-Mer forest in Vallée de Mai, Côtes d'Or Estate, Praslin.

Löbl & Leschen, 2010: 89 (transfer to *Scaphoxium*, characters)

DISTRIBUTION. Seychelles.

Scaphoxium prospector Löbl

Scaphoxium prospector Löbl, 2010b: 128, Figs 20–22. Holotype male, MHNG; type locality: Madagascar: Maromizah rainforest south Périnet.

DISTRIBUTION. Madagascar.

Scaphoxium puetzi Löbl

Scaphoxium puetzi Löbl, 2001: 183, Figs 12–14. Holotype male, PCAP; type locality: China: West Sichuan, Xiaoxiang Ling, side valley from Nanay near Caluo, 11 km south Shimian.

DISTRIBUTION. China: Sichuan.

Scaphoxium reductum Löbl

Scaphoxium reductum Löbl, 1979a: 123, Figs 54–56. Holotype male, MHNG; type locality: India: Valara Falls, 46 km SW Munnar, 450–500 m, Kerala.

DISTRIBUTION. India: Kerala, Tamil Nadu.

Scaphoxium saigoi Hoshima

Scaphoxium saigoi Hoshima, 2009: 2, Figs 1–7. Holotype male, MNHA; type locality: Japan: Mt. Kôchi, Amami-Ôshima Is.

Ogawa & Hoshina, 2012: 264, Figs 1d-f (characters, records)

DISTRIBUTION. Japan: Amami-Ôshima, Shikoku.

Scaphoxium seychellense (Scott)

Toxidium seychellense Scott, 1922: 226, Pl. 21, Figs 23, 24a, 25a, 26a, 27a. Syntypes, NHML/CUMZ/NMPC; type localities: Seychelles: high forest near Pot-à-eau and above Mare aux Cochons, Silhouette; high forest of Morne blanc, forest above Cascade, Mahé.

Löbl & Leschen, 2010: 89 (transfer to *Scaphoxium*, characters)

DISTRIBUTION. Seychelles.

Scaphoxium simulans (Löbl)

Toxidium simulans Löbl, 1971c: 997, Figs 71, 72. Holotype male, MHNG; type locality: Sri Lanka: Kandy, Udawattekele Sanctuary, ca 600 m.

Löbl, 1979a: 120 (transfer to *Scaphoxium*)

DISTRIBUTION. Sri Lanka.

Scaphoxium singlanum Löbl

Scaphoxium singlanum Löbl, 1984a: 101, Figs 72–74. Holotype male, MHNG; type locality: India: West Bengal, Darjeeling District, Singla, 300 m.

Löbl, 1990b: 611 (characters, records)

Löbl, 2003a: 64 (records)

DISTRIBUTION. China: Anhui; India: Meghalaya, West Bengal (Darjeeling District); Thailand.

Scaphoxium sparsum Löbl

Scaphoxium sparsum Löbl, 1979a: 121, Figs 51–53. Holotype male, MHNG; type locality: 6 km east Coonoor, 1400 m, Nilgiri Hills, Tamil Nadu.

Löbl, 1984a: 100 (records)

Löbl, 1986c: 350 (records)

Löbl, 1990b: 612 (records)
Löbl, 1992a: 570 (records)

DISTRIBUTION. India: Assam, Meghalaya, Tamil Nadu, Uttarakhand (Garhwal), West Bengal (Darjeeling District); Nepal; Thailand.

Scaphoxium taiwanum **Löbl**

Scaphoxium taiwanum Löbl, 1980a: 121, Figs 47–49. Holotype male, MHNG; type locality: Taiwan: Fenchihu, 1400 m.
Löbl, 1986c: 351 (records)
Löbl, 1990b: 612 (records)
Löbl, 1992a: 570 (records)
Löbl, 1999: 740 (doubtfully identified, from Sichuan, China)
Hoshina, 2008: 60 (records)

DISTRIBUTION. India: Uttarakhand (Kumaon); Japan: Ryukyus (Yaeyama); Nepal; Taiwan; Thailand.

Scaphoxium taylori **Löbl**

Scaphoxium taylori Löbl, 1981e: 101, Figs 1–3. Holotype male, MHNG; type locality: East Malaysia: Sarawak, Semengoh Forest Reserve, 11 mi SW Kuching.
Löbl, 2011a: 312, Figs 18–20 (characters, records)

DISTRIBUTION. East Malaysia: Sarawak; Philippines: Luzon, Mindanao, Palawan.

Scaphoxium topali **Löbl**

Scaphoxium topali Löbl, 1981e: 102, Figs 4–6. Holotype male, HNHM; type locality: Vietnam: Ninh Binh, Cuc Phuong.

DISTRIBUTION. Vietnam.

Scaphoxium ventrale **(Löbl)**

Toxidium ventrale Löbl, 1977i: 828, Fig. 16. Holotype male, BPBM; type locality: Fiji: Nandarivatu, 3700 ft, Viti Levu.
Löbl, 1979a: 120 (transfer to *Scaphoxium*)
Löbl, 1980c: 396, Figs 35–37 (characters, records)

DISTRIBUTION. Fiji.

Scaphoxium vitianum **(Löbl)**

Toxidium vitianum Löbl, 1977i: 828, Fig. 15. Holotype male, BPBM; type locality: Fiji: Mit. Korombamba, 1200 ft, Viti Levu.
Löbl, 1979a: 120 (transfer to *Scaphoxium*)
Löbl, 1980c: 396, Figs 40, 41 (characters, records)

DISTRIBUTION. Fiji.

Scaphoxium zebra **(Löbl)**

Toxidium zebra Löbl, 1971c: 995, Figs 67, 68. Holotype male, MHNG; type locality: Sri Lanka: Hakgala, 1800 m.

Löbl, 1979a: 120 (transfer to *Scaphoxium*)

DISTRIBUTION. Sri Lanka.

Sphaeroscapha Leschen & Löbl

Sphaeroscapha Leschen & Löbl, 2005: 26. Type species: *Pseudobironium globosum* Löbl, 1981; by original designation. Gender: feminine.

Sphaeroscapha globosa **(Löbl)**

Pseudobironium globosum Löbl, 1981a: 349, Figs 1, 2. Holotype male, NZAC; type locality: New Caledonia: Col d'Amieu, 550 m.

Leschen & Löbl, 2005: 26, Figs 19–22 (transfer to *Sphaeroscapha*)

DISTRIBUTION. New Caledonia.

Sphaeroscapha punctata **Löbl**

Sphaeroscapha punctata Löbl, 2010a: 37, Figs 1–3. Holotype male, MHNG; type locality: New Caledonia: Col d'Amieu.

DISTRIBUTION. New Caledonia.

Spinoscapha Leschen & Löbl

Spinoscapha Leschen & Löbl, 2005: 28. Type species: *Baeocera rufa* Broun, 1881; by original designation. Gender: feminine.

Spinoscapha rufa **(Broun)**

Baeocera rufa Broun, 1881a: 665. Syntypes, NHML; type locality: New Zealand: Whangarei Harbour, Parua.

Baeocera armata Broun, 1881b: 891. Holotype, NHML; type locality: New Zealand: Woodhill Kaipara Railway.

Löbl & Leschen, 2003b: 22, 51, Figs 6, 26, 27, 108, 110, 111, Map 15 (as *Brachynopus*, synonymy of *B. armata* with *S. rufa*, characters, fungal hosts)

Leschen & Löbl, 2005: 29, Figs 23–27, 61 (transfer to *Spinoscapha*, characters)

DISTRIBUTION. New Zealand: North Island.

Termitoscaphium Löbl

Termitoscaphium Löbl, 1982c: 31. Type species: *Termitoscaphium kistneri* Löbl, 1982; by original designation. Gender: neuter.

Termitoscaphium kistneri **Löbl**

Termitoscaphium kistneri Löbl, 1982c: 33, Figs 6–16. Holotype male, MZBI; type locality: Indonesia: Sulawesi, Lake Lindu area, ca 950 m.
Ogawa, 2015: 77 (characters)
Ogawa & Maeto, 2015: Figs 2A-C.
Note. Ex fungus garden of *Odontotermes takensis* Ahmad.
DISTRIBUTION. Indonesia: Sulawesi.

Toxidium LeConte
(31 species)

Toxidium LeConte, 1860: 324. Type species: *Toxidium gammaroides* LeConte, 1860, by monotypy. Gender: neuter.
Casey, 1893: 521, 522 (characters, key to north American species)
Matthew, 1888: 178 (characters, key to Central American species)
Blatchley, 1910: 495, Fig. 176 (characters, natural history)
Miwa & Mitono, 1943: 546 (key to Japanese and Taiwanese species)
Löbl, 1969c: 344, Figs 1, 3, 6 (characters)
Löbl, 1984a: 9 (species groups, key to *T. aberrans* group)
Newton, 1984: 319 (fungal hosts)
Löbl, 1990b: 607 (key to Thai species)
Löbl, 1992a: 571 (characters, key to Asian species)
Leschen, 1993: 73 (larval mandible)
Leschen, 1994: 4, Figs A, B (larval behavior)
Márquez Luna & Navarrete Heredia, 1995: 35 (association with debris of *Atta mexicana* (F. Smith)
Downie & Arnett, 1996: 368 (key to North American species
Newton et al., 2000: 376 (habitat, notes)
Löbl & Leschen, 2010: 78 (characters, key to Malagasy species)

Toxidium aberrans **Achard**

Toxidium aberrans Achard, 1923: 119. Lectotype female, NHML; type locality: Japan: Nikko.
Miwa & Mitono, 1943: 547 (characters)
Nakane, 1955c: 57 (characters)
Nakane, 1963b: 80 (characters)
Löbl, 1969c: 346 (lectotype fixed by inference, characters, records)
Morimoto, 1985: Pl. 45, Fig. 34 (characters)
DISTRIBUTION. Japan: Amami-ôshima, Honshu, Shikoku.

***Toxidium acuminatum* Pic**

Toxidium acuminatum Pic, 1920f: 24. Syntypes, MNHN; type locality: Brazil.

DISTRIBUTION. Brazil.

***Toxidium bifasciatum* Matthews**

Toxidium bifasciatum Matthews, 1888: 179, Pl. 4, Fig. 14. Syntypes, NHML; type localities: Guatemala: Capetillo; San Géronimo.
Fierros-López, 2006b: 41 (characters, records)

DISTRIBUTION. Costa Rica; Guatemala.

***Toxidium cavicola* Löbl & Faille**

Toxidium cavicola Löbl & Faille, 2017: 345. Holotype male, MNHN; type locality: Madagascar: Namoroka, Grotte Canyon S16°24.676' E045°19.645.

DISTRIBUTION. Madagascar.

***Toxidium compressum* Zimmerman**

Toxidium compressum Zimmerman, 1869: 251. Syntypes, MCZH; type locality: USA: Louisiana.
LeConte, 1869: 251 (characters)
Casey, 1893: 522 (characters, records)
Blatchley, 1910: 495, Fig. 176 (characters, natural history)
Kirk, 1969: 30 (records)
Campbell, 1991: 125 (records)
Leschen, 1994: 4 (larval behavior)
Downie & Arnett, 1996: 368 (records)

DISTRIBUTION. Canada: Ontario; United States: Florida, Illinois, Indiana, Louisiana, Kansas, Michigan, Nebraska, North Carolina, South Carolina, Texas.

***Toxidium curtilineatum* Champion**

Toxidium curtilineatum Champion, 1927: 273. Syntypes, NHML; type locality: India: Haldwani Division, Kumaon.
Löbl, 1984a: 97, Figs 65, 66 (characters, records)
Löbl, 1986c: 350 (records)
Löbl, 1992a: 572, Figs 168, 169 (characters, records)

DISTRIBUTION. India: Meghalaya, Uttarakhand; Nepal.

***Toxidium diffidens* Löbl**

Toxidium diffidens Löbl, 1984a: 97, Fig. 67. Holotype male, MHNG; type locality: India: Meghalaya, Songsak, 400 m, Khasi Hills.

DISTRIBUTION. India: Meghalaya.

Toxidium donckieri Pic

Toxidium donckieri Pic, 1928a: 3. Syntypes, MNHN; type locality: Argentina: Tucuman.

DISTRIBUTION. Argentina.

Toxidium gammaroides LeConte

Toxidium gammaroides LeConte, 1860: 324. Syntypes, MCZC; type locality: USA: Southern and Western States.

Matthews, 1888: 180, Pl. 4, Fig. 15 (characters, records)
Casey, 1893: 522 (characters, records)
Scott, 1922: Pl. 21, Fig. 27c (characters)
Leonard, 1928: 314 (records)
Brimley, 1938: 149 (records)
Kirk, 1969: 30 (records)
Leschen, 1988a: 231 (fungal hosts)
Campbell, 1991: 125 (records)
Löbl, 1992a: 622, Figs 1804, 190 (characters)
Downie & Arnett, 1996: 368 (records)
Peck & Thomas, 1998: 46 (distribution)
Webster et al., 2012: 248 (records, habitat, fungal hosts)

DISTRIBUTION. Canada: New Brunswick, Ontario, Quebec; Guatemala; Mexico; United States: Florida, Georgia, Lousiana, Maine, Maryland, New Hampshire, New York, North Carolina, Rhode Island, North Carolina, South Carolina, Virginia.

Toxidium incompletum Löbl

Toxidium incompletum Löbl, 1990b: 609, Figs 169, 170. Holotype male, BPBM; type locality: Thailand: Chiang Dao, 450 m.

DISTRIBUTION. Thailand.

Toxidium janaki Löbl & Leschen, 2010

Toxidium janaki Löbl & Leschen, 2010: 82, Figs 14–17. Holotype male, MHNG; type locality: East Madagascar: Vohitrosa forest 30 km ESE of Betroka, 1600–1650 m, 3 km NE summit, East Madagascar.

DISTRIBUTION. Madagascar.

Toxidium lunatum Löbl

Toxidium lunatum Löbl, 2012b: 182, Fig. 15. Holotype male, MHNG; type locality: West Malaysia: Gn. Beremban, Cameron Highlands.

DISTRIBUTION. West Malaysia.

Toxidium nigrum **Pic**

Toxidium nigrum Pic, 1951a: 6. Syntypes, ?MNHN, not traced; type locality: Madagascar.

DISTRIBUTION. Madagascar.

Toxidium ornatipenne **Achard**

Toxidium ornatipenne Achard, 1922c: 43. Syntypes, MNHN; type locality: Bolivia: Cochabamba.

DISTRIBUTION. Bolivia.

Toxidium ovatum **Matthews**

Toxidium ovatum Matthews, 1888: 180. Holotype, NHML; type locality: Guatemala: Totonicapam, 10.000 ft.

DISTRIBUTION. Guatemala.

Toxidium parvum **Matthew**

Toxidium parvum Matthews, 1888: 181. Syntypes, NHML; type locality: Guatemala: Very Paz, Balheu.

DISTRIBUTION. Guatemala.

Toxidium pubistylis **Löbl**

Toxidium pubistylis Löbl, 1990b: 607, Figs 163–165. Holotype male, MHNG; type locality: Thailand: Kha Yai Nat. Park, east Heo Suwat Waterfalls, 800–900 m.

DISTRIBUTION. Thailand.

Toxidium punctatum **Matthews**

Toxidium punctatum Matthews, 1888: 179. Syntypes, NHML; type locality: Mexico: Toluca.

DISTRIBUTION. Mexico.

Toxidium pygidiale **Pic**

Toxidium pygidiale Pic, 1923b: 18. Syntypes, MNHN; type locality: Vietnam: Tonkin.

DISTRIBUTION. Vietnam.

Toxidium robustum **Pic**

Toxidium robustum Pic, 1930b: 58. Lectotype female, NMPC; type locality: Myanmar: Tenasserim.

Löbl, 1984a: 97, Figs 70, 71 (lectotype designation, characters)

Löbl, 1990b: 607 (records)
DISTRIBUTION. Myanmar; Thailand.

Toxidium rougemonti **Löbl & Leschen**
Toxidium rougemonti Löbl & Leschen, 2010: 85, Figs 18–20. Holotype male, MHNG; type locality: Madagascar: Andasibe-Mantadia NP, East Madagascar.
DISTRIBUTION. Madagascar.

Toxidium rufonotatum **Pic**
Toxidium rufonotatum Pic, 1915c: 35. Syntypes, MNHN; type locality: Madagascar.
Löbl & Leschen, 2010: 81, Figs 11–13 (characters, records)
DISTRIBUTION. Madagascar.

Toxidium sikorai **Pic**
Toxidium sikorai Pic, 1915c: 35. Syntypes, MNHN; type locality: Madagascar: Annanarive.
Löbl & Leschen, 2010: 80 (characters, records)
DISTRIBUTION. Madagascar.

Toxidium spectabile **Löbl**
Toxidium spectabile Löbl, 1992a: 573. Holotype female, MHNG; type locality: Nepal: above Yamputhin, left bank of Kabeli Khola, 1800–2000 m, Taplejung District.
DISTRIBUTION. Nepal.

Toxidium sternale **Achard**
Toxidium sternale Achard, 1922c: 43. Syntypes, NMPC; type locality: Bolivia: Cochabamba.
DISTRIBUTION. Bolivia.

Toxidium storeyi **Löbl & Leschen, 2010**
Toxidium storeyi Löbl & Leschen, 2010: 78, Figs 7–10. Holotype male, QMBA; type locality: Australia: Topaz, Hughes Road, 650 m, Queensland
DISTRIBUTION. Australia: Queensland.

Toxidium styligerum **Löbl**
Toxidium styligerum Löbl, 1990b: 609, Figs 166–168. Holotype male, MHNG; type locality: Thailand: Doi Suthep, 1600 m, Chiang Mai.
DISTRIBUTION. Thailand.

***Toxidium usambarense* (Reitter)**

Baeocera usambarensis Reitter, 1908: 33. Holotype male, NHMW; type locality: Tanzania: Amani.

Löbl, 1987d: 859, Figs 22, 23 (transfer to *Toxidium*, characters)

DISTRIBUTION. Tanzania.

***Toxidium vagans* Löbl**

Toxidium vagans Löbl, 1984a: 98, Figs 68, 69. Holotype male, MHNG; type locality: India: West Bengal, Darjeeling District, Algarah, 1800 m.

Löbl, 1992a: 621, Fig. 178, 185 (characters)

DISTRIBUTION. India: West Bengal (Darjeeling District).

***Toxidium variegatum* Achard**

Toxidium variegatum Achard, 1922c: 44. Syntypes, NMPC; type locality: Bolivia: Cochabamba.

DISTRIBUTION. Bolivia.

***Toxidium villosum* Löbl**

Toxidium villosum Löbl, 1999: 741, Figs 70–72. Holotype male, MHNG; type locality: China: Wolong Nat. Res., 1500 m, Sichuan.

Löbl, 2011b: 206 (records)

DISTRIBUTION. China: Sichuan; Taiwan.

Tritoxidium Leschen & Löbl

Tritoxidium Leschen & Löbl, 2005: 30. Type species: *Toxidium indicum* Achard, 1915; by original designation. Gender: neuter.

***Tritoxidium indicum* (Achard)**

Toxidium indicum Achard, 1915a: 561. Lectotype male, NMPC; type locality: India: Tamil Nadu, Chambaganor [Shembaganur].

Scaphosoma pygidiale Pic, 1916c: 19. Lectotype male, MNHN; type locality: India: Tamil Nadu, Chambaganor [Shembaganur].

Löbl, 1971c: 996 (lectotype designations for *T. indicum* and *S. pygidiale*, synonymy of *S. pygidiale*)

Löbl, 1979a: 118 (as *Toxidium*, characters, records)

Leschen & Löbl, 2005: 32, Figs 28–33 (transfer to *Tritoxidium*, characters)

DISTRIBUTION. India: Kerala, Tamil Nadu.

Vickibella Leschen & Löbl

Vickibella Leschen & Löbl, 2005: 32. Type species: *Scaphisoma apicella* Broun, 1880, by original designation. Gender: feminine.

Vickibella apicella (Broun)

Scaphisoma apicella Broun, 1880: 160. Lectotype male, NHML; type locality: New Zealand: Whangarei Heads.

Löbl & Leschen, 2003b: 20, 49, Figs 7, 28, 29, 74–76, Map. 13 (lectotype designation, transfer to *Brachynopus*, characters, records)

Leschen & Löbl, 2005: 32, Figs 34–37 (transfer to *Vickibella*, characters)

DISTRIBUTION. New Zealand: North Island.

Vituratella Reitter

Vituratella Reitter, 1908: 35. Type species: *Vituratella eichelbaumi* Reitter, 1908; by monotypy. Gender: feminine.

Antongilium Pic, 1920f: 22. Type species: *Antongilium nitidum* Pic, 1920; by monotypy. Gender: neuter.

Mysthrix Champion, 1927: 278. Type species: *Mysthrix termitophilum* Champion, 1927; by original designation. Gender: neuter.

[*Mystrix* Löbl, 1992a: 577 misspelling].

Termitoxidium Pic, 1928b: 38. Type species: *Termitoxidium longicolle* Pic, 1928; by monotypy. Gender: neuter.

Trichoscaphella Reitter, 1908: 34. Type species: *Trichoscaphella suturisulcata* Reitter, 1908; by monotypy. Gender: feminine.

Achard, 1924b: 30 (synonymy of *Antongilium* with *Vituratella*)

Pic, 1925a: 194 (ressurection of *Antongilium*)

Löbl, 1992a: 577 (as *Mysthrix*, characters)

Leschen & Löbl, 2005: 34, 44, Fig. 50 (synonymy of *Antongilium*, *Mysthrix*, *Termitoxidium*, *Trichoscaphella*, characters, termitophily)

Ogawa, 2015: 78 (characters)

Vituratella angustata (Pic)

Antongilium angustatum Pic, 1948b: 71. Holotype, MNHN; type locality: Ivory Coast: Banco.

Leschen & Löbl, 2005: 36 (transfer to *Vituratella*)

DISTRIBUTION. Ivory Coast.

Vituratella atra (Pic)

Antongilium atrum Pic, 1928b: 40. Holotype, MRAC; type locality: Democratic Republic of the Congo: Tsheta.

Leschen & Löbl, 2005: 36 (transfer to *Vituratella*)
DISTRIBUTION. Democratic Republic of the Congo.

Vituratella bipartita **(Pic)**
Antongilium bipartitum Pic, 1955: 54. Syntypes, MRAC; type locality: Rwanda: east Muhavura, 2100 m, terr. Ruhengeri.
Leschen & Löbl, 2005: 36 (transfer to *Vituratella*)
DISTRIBUTION. Congo; Rwanda.

Vituratella bisbinotata **(Pic)**
Antongilium bisbinotatum Pic, 1954b: 36. Holotype, MRAC; type locality: Democratic Republic of the Congo: Haute-Luvulu, 2650 m, terr. Unira, Kivu.
Leschen & Löbl, 2005: 36 (transfer to *Vituratella*)
DISTRIBUTION. Democratic Republic of the Congo.

Vituratella burgeoni burgeoni **(Pic)**
Antongilium burgeoni Pic, 1928b: 41. Syntypes, MRAC/MNHN; type locality: Democratic Republic of the Congo: Haute-Uelé, Abimva.
Antongilium burgeoni var. *semibrunneum* Pic, 1930a: 87. Syntypes, MRAC; type locality: Democratic Republic of the Congo: Blukwa, Nizi.
Antongilium burgeoni var. *casseti* Pic, 1948b: 71. Holotype, MNHN; type locality: Ivory Coast: Haute Cavally.
Pic, 1955: 53 (implicit synonymy of varieties *semibrunneum* and *casseti* with *burgeoni*, records)
Leschen & Löbl, 2005: 36 (transfer to *Vituratella*)
DISTRIBUTION. Democratic Republic of the Congo; Rwanda.

Vituratella burgeoni vicina **(Pic)**
Antongilium burgeoni forma *vicinum* Pic, 1928b: 41. Holotype, MRAC; type locality: Democratic Republic of the Congo: Kikionga.
Leschen & Löbl, 2005: 36 (transfer of nominal subspecies to *Vituratella*)
DISTRIBUTION. Democratic Republic of the Congo; Ivory Coast; Rwanda.

Vituratella cinctipennis cinctipennis **(Pic)**
Antongilium cinctipenne Pic, 1954b: 38. Holotype, MRAC; type localities: Democratic Republic of the Congo: Bumba; Ituri: Yebo Moto.
Leschen & Löbl, 2005: 36 (transfer to *Vituratella*)
DISTRIBUTION. Democratic Republic of the Congo.

Vituratella cinctipennis boutakoffi **(Pic)**

Antongilium cinctipenne var. *boutakoffi* Pic, 1954a: 38. Holotype, MRAC; type locality: Democratic Republic of the Congo: Mabwita, Kivu.

DISTRIBUTION. Democratic Republic of the Congo.

Vituratella collarti **(Pic)**

Antongilium collarti Pic, 1928b: 39. Syntypes, MRAC/MNHN; type localities: Democratic Republic of the Congo: Tschibo N'Goy.
Leschen & Löbl, 2005: 36 (transfer to *Vituratella*)

DISTRIBUTION. Democratic Republic of the Congo.

Vituratella eichelbaumi **Reitter**

Vituratella eichelbaumi Reitter, 1908: 35. Holotype, NHMW; type locality: Tanzania: Mt. Bomole, Amani.
Leschen & Löbl, 2005: 63 (invalid lectotype designation)

DISTRIBUTION. Tanzania.

Vituratella elongatior elongatior **(Pic)**

Antongilium elongatior Pic, 1955: 54. Syntypes, MRAC/MNHN; type localities: Rwanda: Gitarama, 1850 m, terr. Nyxanza; Rwanda: Shangugu, 1500 m; Rwanda, Kayove, 2000 m, terr. Kisenyi.
Leschen & Löbl, 2005: 36 (transfer to *Vituratella*)

DISTRIBUTION. Rwanda.

Vituratella elongatior obscura **(Pic)**

Antongilium elongatior var. *obscurum* Pic, 1955: 54. Holotype, MRAC; type locality: Rwanda: Shangugu, 1500 m.

DISTRIBUTION. Rwanda.

Vituratella humeralis **(Pic)**

Antongilium humerale Pic, 1931b: 420. Syntypes, MNHN; type locality: Angola: Saint Amaro.
Leschen & Löbl, 2005: 36 (transfer to *Vituratella*)

DISTRIBUTION. Angola.

Vituratella inapicalis **(Pic)**

Antongilium inapicale Pic, 1928b: 41. Syntypes, MNHN; type locality: Congo.
Leschen & Löbl, 2005: 36 (transfer to *Vituratella*)

DISTRIBUTION. Democratic Republic of the Congo.

***Vituratella innotata* (Pic)**

Antongilium innotatum Pic, 1946: 82. Syntypes, MNHN; type locality: Kenya: Mont Elgon [Suam fishing hut].

Leschen & Löbl, 2005 (transfer to *Vituratella*)

DISTRIBUTION. Kenya.

***Vituratella kistneri* (Löbl)**

Mysthrix kistneri Löbl, 1979b: 321, Figs 1, 2. Holotype male, FMNH; type locality: Indonesia: Sumatra, Lankat Reserve near Bukit Lawang.

Leschen & Löbl, 2005: 37 (transfer to *Vituratella*)

DISTRIBUTION. Indonesia: Sumatra.

***Vituratella lateralis* (Pic)**

Antongilium laterale Pic, 1946: 83. Syntypes, MNHN; type locality: Ethiopia: Omo, Bourillé.

Leschen & Löbl, 2005: 36 (transfer to *Vituratella*)

DISTRIBUTION. Ethiopia.

***Vituratella longicollis* (Pic)**

Termitoxidium longicolle Pic, 1928b: 39. Syntypes, ISNB/MRAC/MNHN; type localities: Democratic Republic of the Congo: N'Goy, Tschiobo; Bondo Mabe, Arebi.

Pic, 1955: 53 (records)

Leschen & Löbl, 2005: 37 (transfer to *Vituratella*)

DISTRIBUTION. Democratic Republic of the Congo, Rwanda.

***Vituratella luteonotata* (Pic)**

Antongilium luteonotatum Pic, 1955: 54. Holotype, MRAC; type locality: Burundi: Nyamasumu, 1500 m, east Usumbura.

Leschen & Löbl, 2005: 36 (transfer to *Vituratella*)

DISTRIBUTION. Burundi.

***Vituratella luteosignata* (Pic)**

Antongilium luteosignatum Pic, 1954b: 37. Syntypes, MRAC/MNHN; type localities: Democratic Republic of the Congo: Mongbwalu, Kilo; Equateur: Bokuma.

Leschen & Löbl, 2005: 37 (transfer to *Vituratella*)

DISTRIBUTION. Democratic Republic of the Congo.

Vituratella minuta **(Pic)**
Antongilium minutum Pic, 1928b: 40. Holotype, MRAC; type locality: Democratic Republic of the Congo: Tschibo N'Goy.
Leschen & Löbl, 2005: 37 (transfer to *Vituratella*)
DISTRIBUTION. Democratic Republic of the Congo.

Vituratella nigrosignata **(Pic)**
Antongilium nigrosignatum Pic, 1954b: 37. Holotype, MRAC; type locality: Democratic Republic of the Congo, Kasenyi, Kibali-Ituri.
Leschen & Löbl, 2005: 37 (transfer to *Vituratella*)
DISTRIBUTION. Democratic Republic of the Congo.

Vituratella nitida **(Pic)**
Antongilium nitidum Pic, 1920f: 23. Syntypes, MNHN; type locality: Madagascar: south Antongil Bay.
Leschen & Löbl, 2005: 37 (transfer to *Vituratella*)
DISTRIBUTION. Madagascar.

Vituratella noeli **(Pic)**
Antongilium noeli Pic, 1948b: 71. Holotype, MNHN; type locality: Cameroon: Yaoundé.
Leschen & Löbl, 2005: 37 (transfer to *Vituratella*)
DISTRIBUTION. Cameroon.

Vituratella notata **(Pic)**
Antongilium notatum Pic, 1948b: 71. Holotype, MNHN; type locality: Ivory Coast: Banco.
Leschen & Löbl, 2005: 37 (transfer to *Vituratella*)
DISTRIBUTION. Ivory Coast.

Vituratella notatipennis **(Pic)**
Antongilium notatipenne Pic, 1933b: 120. Holotype, MZUF; type locality: Eritrea, Adi Caiè.
Leschen & Löbl, 2005: 37 (transfer to *Vituratella*)
DISTRIBUTION. Eritrea.

Vituratella pauliani **(Pic)**
Antongilium pauliani Pic, 1948b: 71. Holotype, MNHN; type locality: Ivory Coast: Banco.
Leschen & Löbl, 2005: 37 (transfer to *Vituratella*)
DISTRIBUTION. Ivory Coast.

Viturattella perrieri **Achard**

Viturattella perrieri Achard, 1920i: 242. Syntypes, NMPC/MNHN; type locality: Madagascar: Antongil Bay.

DISTRIBUTION. Madagascar.

Viturattella punctata **(Pic)**

Antongilium punctatum Pic, 1928b: 41. Holotype, MRAC; type locality: Democratic Republic of the Congo: Buende Suindi.

Leschen & Löbl, 2005: 37 (transfer to *Viturattella*)

DISTRIBUTION. Democratic Republic of the Congo.

Viturattella subelongata subelongata **(Pic)**

Antongilium subelongatum Pic, 1928b: 40. Holotype, MRAC; type locality: Democratic Republic of the Congo: Elisabethville [Lubumbashi].

Pic, 1933b: 120 (records)

Leschen & Löbl, 2005: 37 (transfer to *Viturattella*)

DISTRIBUTION. Eritrea; Democratic Republic of the Congo.

Viturattella subelongata lata **(Pic)**

Antongilium subelongatum var. *latum* Pic, 1928b: 40. Holotype, MRAC; type locality: Democratic Republic of the Congo, Kiniati-Zobe.

DISTRIBUTION. Democratic Republic of the Congo.

Viturattella suturisulcata **(Reitter)**

Trichoscaphella suturisulcata Reitter, 1908: 34. Holotype, NHMW; type locality: Tanzania: Amani.

Leschen & Löbl, 2005: 63 (invalid lectotype designation, transfer to *Viturattella*)

DISTRIBUTION. Tanzania.

Viturattella termitophila **(Champion)**

Mysthrix termitophilum Champion, 1927: 278, Pls 5, 6. Syntypes, NHML; type locality: India: Haldwani Division, Kumaon.

Löbl, 1992a: 577 (records)

Rougemenot, 1996: 12 (records)

Löbl, 1999: 734 (records)

Leschen & Löbl, 2005: 37 (transfer to *Viturattella*)

Ogawa, 2015: 78, Figs 4-13a-f (characters, records)

Ogawa & Maeto, 2015: 302, Figs 1A-F (characters, records)

DISTRIBUTION. China: Hong Kong; India: Uttarakhand (Kumaon); Indonesia: Sulawesi; Nepal.

Vituratella testaceiventris **(Pic)**

Antongilium testaceiventre Pic, 1954b: 37. Holotype, MRAC; type locality: Democratic Republic of the Congo: Bokuma.
Leschen & Löbl, 2005: 37 (transfer to *Vituratella*)

DISTRIBUTION. Democratic Republic of the Congo.

Vituratella vittipennis **(Pic)**

Antongilium vittipenne Pic, 1937: 207. Holotype, ZMUH; type locality: South Africa: KwaZulu-Natal, Pietermaritzburg: Fort Napier.
Leschen & Löbl, 2005: 37 (transfer to *Vituratella*)

DISTRIBUTION. South Africa: Natal.

Xotidium Löbl

Xotidium Löbl, 1992a: 573; type species: *Xotidium uniforme* Löbl, 1992; by original designation. Gender: neuter
Löbl, 1971c: 994 (in *Toxidium*, key to Sri Lankan species)
Löbl, 2015a: 119 (key to species)
Ogawa, 2015: 111 (characters, Sulawesi species)
Ogawa & Löbl, 2016a: 156, 168 (characters, key to species)

Xotidium bolmarum **Löbl**

Xotidium bolmarum Löbl, 2015a: 119, Figs 121–124. Holotype male, MHNG; type locality: Indonesia: Lombok, Senaro, north slope of Rinjani, 1100 m.
Ogawa & Löbl, 2016a: 156, Figs 3a, 6a, 7g, 9a (characters)

DISTRIBUTION. Indonesia: Lombok.

Xotidium flagellum **Ogawa & Löbl**

Xotidium flagellum Ogawa & Löbl, 2016a: 159, Figs 5b, 6h, 8d. Holotype male, MHNG; type locality: East Malayasia: Sabah, Mt. Kinabalu N.P., summit trail Pontok Lowii, 2300–2400 m.

DISTRIBUTION. East Malaysia: Sabah.

Xotidium heissi **Ogawa & Löbl**

Xotidium heissi Ogawa & Löbl, 2016a: 160, Figs 5a, 6i, 8a. Holotype male, MHNG; type locality: Brunei: Temburong, Kuala Belalong 60–300 m.

DISTRIBUTION. Brunei; East Malaysia: Sabah.

Xotidium mauritianum **(Vinson)**

Toxidium mauritianum Vinson, 1943: 201, Fig. 21. Holotype, NHML; type locality: Mauritius: The Pouce.

Leschen & Löbl, 2005: 63 (transfer to *Xotidium*)
Ogawa & Löbl, 2016a: 157, Figs 4b, 6b, 9g (characters)
DISTRIBUTION. Mauritius.

Xotidium meridionale **Ogawa & Löbl**
Xotidium meridionale Ogawa & Löbl, 2016a: 163, Figs 5d, 6j, 8e, f. Holotype male, MZBI: Indonesia: S. Sulawesi, Mt. Lompobatang, Malino, ca 1700 m.
DISTRIBUTION. Indonesia: Sulawesi.

Xotidium montanum **(Löbl)**
Toxidium montanum Löbl, 1971c: 1000, Figs 76, 77. Holotype male, MHNG; type locality: Sri Lanka: Pidurutalagala, 1950 m, near Nuwara Eliya.
Löbl, 1992a: 574 (transfer to *Xotidium*)
Ogawa & Löbl, 2016a: 157, Figs 3c, d, 6c, 7e, 9b (characters, records)
DISTRIBUTION. Sri Lanka.

Xotidium notatum **(Löbl)**
Toxidium notatum Löbl, 1977e: 65, Figs 98–100. Holotype male, NHML; type locality: Australia: New South Wales, Acacia Plateau.
Löbl, 1992a: 574 (transfer to *Xotidium*)
Ogawa & Löbl, 2016a: 157, Figs 4c, d, 6d, 7c, 9f (characters)
DISTRIBUTION. Australia: New South Wales.

Xotidium pygmaeum **(Löbl)**
Toxidium pygmaeum Löbl, 1971c: 998, Figs 73–75. Holotype male, MHNG; type locality: Sri Lanka: Kandy.
Löbl, 1992a: 574 (transfer to *Xotidium*)
Ogawa & Löbl, 2016a: 159, Figs 4a, 6e, 7a, 9d (characters)
DISTRIBUTION. Sri Lanka.

Xotidium smetanai **Ogawa & Löbl**
Xotidium smetanai Ogawa & Löbl, 2016a: 166, Figs 2a-e, 5c, 6k, 8b. Holotype male, MHNG; type locality: East Malayasia: Sabah, Mt. Kinabalu N.P., HQ Silau-Silau, trail 1550 m.
DISTRIBUTION. East Malaysia: Sabah.

Xotidium tarantulatum **Ogawa & Löbl**
Xotidium tarantulatum Ogawa & Löbl, 2016a: 166, Figs 2f-i, 5e, 6g, 8d, 9e. Holotype male, MZBI; type locality: Indonesia: N. Sulawesi, Mt. Tilongkabila, ca 1300 m.
DISTRIBUTION. Indonesia: Sulawesi.

Xotidium tubuliferum **Löbl**

Xotidium tubuliferum Löbl, 2011e: 720, Figs 52–55. Holotype male, FMNH; type locality: Philippines: Leyte, Tarragona.
Ogawa & Löbl, 2016a: 159, Fig. 7f (characters)

DISTRIBUTION. Philippines: Leyte.

Xotidium uniforme **Löbl**

Xotidium uniforme Löbl, 1992a: 575, 621, Figs 179, 189. Holotype male, MHNG; type locality: Nepal: Induwa Khola Valley, 2100 m, Sankhuwasabha District.
Ogawa & Löbl, 2016a: 159, Figs 1, 3b, 5f, 7d, 9c (characters)

DISTRIBUTION. India: Himachal Pradesh, Uttarakhand (Kumaon); Nepal.

Zinda Löbl

Zinda Löbl, 1981a: 354; type species: *Zinda teres* Löbl, 1981; by original designation. Gender: feminine.
Löbl, 1981a: 354 (key to species)

Zinda abdita **Löbl**

Zinda abdita Löbl, 1981a: 356, Figs 18, 19. Holotype male, NZAC; type locality: New Caledonia: Table d'Union, 800 m, near Col d'Amieu.

DISTRIBUTION. New Caledonia.

Zinda teres **Löbl**

Zinda teres Löbl, 1981a: 356, Figs 6–17. Holotype male, NZAC; type locality: New Caledonia: Col d'Amieu, 550 m.

DISTRIBUTION. New Caledonia.

Appendix: Fungus and Myxomycetes Hosts

[host-names are given as originally published]

Cyparium concolor **Fabricius**

Leschen, 1988a: 231: *Tremellodendron pallidum* (as *C. flavipes* LeConte)

Leschen, 1988b: 15: *Agaricus* sp., *Amanita* sp., *Boletus* sp., *Laccaria laccata, L. ochropurpurea, Marasmius* sp., *Meripilus giganteus, Nematoloma* sp., *Oudemansiella radiata, Pluteus cervinus, Psathyrella* sp., *Stroblomyces floccopus, Tremelodendron pallidum,* Trichalomataceae.

Cyparium sibiricum **Solsky**

Kompantsev & Pototskaya, 1987: 94: *Pleurotus* sp.

Cyparium sallei **Matthews**

Newton, 1984: 317: Agaricales.

Cyparium terminale **Matthews**

Lawrence & Newton, 1980: 137: *Clavaria* sp.

Newton, 1984: 317: ?*Clavaria*, ?*Pleurotus*

Newton, 1991: 338: ?*Clavaria.*

Cyparium yapalli **Fierroz-López**

Fierroz-López, 2002: 10: *Agaricus* sp.

Cyparium **sp.**

Newton, 1984: 317:?*Clavaria, Pterula.*

Persescaphium pari **Löbl & Ogawa**

Löbl & Ogawa, 2016c: 40: *Hericium coralloides.*

Ascaphium sulcipenne **Lewis**

Leschen & Löbl, 1995: 450: *Trametes versicolor.*

Ascaphium tibiale **Lewis**

Leschen & Löbl, 1995: 450: *Trametes versicolor.*

Scaphium castanipes **Kirby**

Ashe, 1984: 368: *Collybia* sp., *Cortinarius* sp., *Russula* sp.

Newton (pers. comm.): *Cortinarius* sp. (prob. *armillatus*)

Scaphium immaculatum **(Olivier)**
Benick, 1952: 46: *Russula nigricans*.
Burakowski et al., 1978: 232: *Lactarius piperatus* (ex Kolbe, 1927).

Scaphium quadraticolle **Solsky**
Kompantsev & Pototskaya, 1987: 97: Agaricales.

Scaphidium amurense **Solsky**
Kompantsev & Pototskaya, 1987: 91: ?Stereaceae.

Scaphidium atrum **Matthews**
Navarrete-Heredia, 1991: 126: *Poria* sp.
Fierros-López, 2005: 25: *Coriolus versicolor*, *Phanerochaete chryzorhiza*.

Scaphidium exornatum **Oberthür**
Hawkeswood, 1989: 94: *Poria* sp.
Hawkeswood, 1990: 96: *Poria* sp.

Scaphidium geniculatum **Oberthür**
Fierros-López, 2005: 42: *Favolus hexagonalis*.

Scaphidium lescheni **Fierros-López**
Fierros-López, 2005: 50: *Favolus* sp.

Scaphidium loebli **Fierros-López**
Fierros-López, 2005: 55: *Phanerochaetes chryzorhiza*, *Hydnopolyporus palmatus*, *Collybia* sp.

Scaphidium matthewi **Csiki**
Fierros-López, 2005: 64: *Favolus* sp.

Scaphidium nigripes **Chevrolat [as *S. mexicanum*]**
Fierros-López, 1998: 36: *Auricularia delicata*, *Xylaria* sp., *Favolus* sp.
Fierros-López, 2005: 69: *Auricularia delicata*, *Xylaria* sp., *Inonotus* sp.

Scaphidium piceum **Melsheimer**
Leschen, 1988b: 15: *Irpex lacteus*, *Melanoleuca* sp.

Scaphidium punctipenne **Macleay**
Hawkeswood, 1989: 94: *Poria* sp.
Hawkeswood, 1990: 97: *Poria* sp.

Scaphidium quadriguttatum **Say**

Weiss & West, 1920: 5: *Polyporus vesicolor.*

Newton, 1984: 317: *Coriolus versicolor, Polyporus squamosus.*

Leschen, 1988b: 15: *Irpex lacteus, Melanoleuca* sp., *Meripilus giganteus, Schizopora paradoxa.*

Scaphidium quadrimaculatum **Olivier**

Scheerpeltz & Höfler, 1948: 149: *Panus rudis, Trametes gibbosa, Hypholoma lacrymabundum, Pholiota mutabilis, Pluteus cervinus.*

Benick, 1952: 46: *Lenzites betulae, Placoderma betulinum, Polyporus squamosus, Polystictus hirsutus, Polystictus versicolor.*

Rehfous, 1955: 19: *Trametes gibbosa, Polyporus adustus, Coriolus versicolor, Panellus stipticus, Panus flabelliformis, Pholiota mutabilis, Cantharellus cibarius.*

Nuss, 1975: 109: *Laetiporus sulphureus, Phellinus robustus, Piptoporus betulinus, Polyporus squamosus.*

Tamanini, 1969b: 354: *Melanopus squamosus, Trametes unicolor, T. versicolor, T. suaveolens, Leptoporus adustus, Ganoderma applanatum, Fomes fomentarius.*

Nikitsky et al., 1996: 24: *Steccherinum ochraceum, Oxyporus corticola, Oxyporus* sp., *Cerrena unicolor, Fomes fomentarius, Laetiporus sulphureus, Trametes* sp.

Krasutskij, 1996a: 96: *Fomes fomentarius, Lentinus tigrinus.*

Krasutskij, 1997a: 307: *Fomes fomentarius, Lentinus cyathiformis, L. tigrinus.*

Nikitsky et al., 1998: 7: *Phlebia centrifuga*

Kofler, 1998: 651: *Trametes suaveolens.*

Nikitsky & Schigel, 2004: 8: *Cerrena unicolor, Bjerkandera adusta.*

Mateleshko, 2005: 128: *Pleurotus* sp.

Borowski, 2006: 54–58: *Phellinus igniarius, Fomitopsis pinicola, Fomes fomentarius, Daedaelopsis confragosa.*

Nikitsky et al., 2008: 120: *Fomes fomentarius, Laetiporus sulphureus, Oxyporus corticola.*

Zamotajlov & Nikitsky, 2010: 105: *Steccherinum ochraceum, Oxyporus corticola, Oxyporus* sp., *Cerrena unicolor, Fomes fomentarius, Laetiporus sulphureus, Trametes* sp.

Vinogradov et al., 2010: 15: *Fomes fomentarius, Pleurotus pulmonarius.*

Chillo, 2012: 45: *Daedalea* sp.

Torrella Alegue, 2013: 45 *Russula cyanoxantha.*

Nikitsky et al., 2016: 138: *Steccherinum ochraceum, Oxyporus corticola, Oxyporus* sp., *Cerrena unicolor, Fomes fomentarius, Laetiporus sulphureus, Trametes* sp.

Scaphidium reitteri **Lewis**

Hoshina, 2015: 46: *Pleurotus pulmonarius.*

Scaphidium tlatlauhqui **Fierros-López**

Fierros-López, 2005: 108: *Auricularia delicata, Favolus brasiliensis, Oligopoeus floriformis, Hydnopolyporus fimbriatus, Rigtidoporus microporus, Stereum* sp.

Scaphidium uinduri **Fierros-López**

Fierros-López, 2005: 122: *Pleurotus* sp., *Oligoporus floriformis.*

Scaphidium **sp.**

Newton, 1984: 317: *Coriolus tenuis, Earliella corrugata, Ganoderma* sp., Polyporales, Tremellales.

Wheeler, 1991: 42 [ex Navarret-Heredia]: *Amanita pantherina, Armillariella mellea, Tricholoma* sp., *Cortinarius* sp., *Boletus* aff. *edulis.*

Navarrete-Heredia, 1991: 127: *Hydnopolyporus palmatus, Collybia.*

Baeocera abrupta **Löbl & Leschen**

Löbl & Leschen, 2003b: 30: *Ganoderma* sp.

Baeocera actuosa **(Broun)**

Löbl & Leschen, 2003b: 30: *Arcyria incarnata, Ceratiomyxa fructiculosa, Fuligo ?septica, Fuligo*, sp., *Hemitrichia serpula, Lamproderma* sp., *Lycogala+Ceratiomyxa, Lycogala epidendrum, Stemonitis* sp., *Stemonitis fusca, Trichia floriformis, Hypoxylon* sp., *Amanita* (infected with *Hyphomyces* sp., *Auricularia polytrichia, Bjerkandra adusta, Favolaschia pustulosus, Ganoderma applanatum, Ganoderma* sp., *Scizopora paradoxa, Stereum* sp.

Baeocera charybda **(Cornell)**

Lawrence & Newton, 1980: 137: *Comatrichia longa.*

Löbl & Stephan, 1993: 690: *Stemonitis exifera.*

Baeocera elenae **Löbl & Leschen**

Löbl & Leschen, 2003b: 32: undetermined Myxomycetes.

Baeocera gutierrezberaudi **Fierros-López**

Fierros-López, 2010: 206: *Ganoderma australe, Poria* sp., *Trichia* sp.

Baeocera inermis **Löbl**

Newton & Stephenson, 1990: 212: *Trichia decipiens, Trichia favoginea.*

Baeocera nana **Casey**

Cornell, 1967: 15: *Stemonites fusca.*

Lawrence & Newton, 1980: 137: *Stemonitis* sp.

Löbl & Stephan, 1993: 694: *Comatotrichia typhoides, Stemonitis fusca, Badhamia affinis.*

Baeocera microptera **Löbl**

Newton & Stephenson, 1990: 212: *Cribraria intricata, Physarum flavicomum.*

Baeocera picea **Casey**

Lawrence & Newton, 1980: 137: *Didymium*? sp., *Fuligo septica, Stemonitis* sp.

Newton, 1984: 319: *Ceratiomyxa fruticulosa.*

Newton, 1991: 338: *Ceratiomyxa fruticulosa.*

Löbl & Stephan, 1993: 689: *Arcynia incarnata, A. denudata, Tubifera ferruginosa, Stemonitis exifera, Ceratiomyxa fruticulosa, Comatrichia typhoides.*

Baeocera pulchella **Fierros-López**

Fierros-López, 2010: 202: *Poria* sp., *Trichia* sp.

Baeocera **sp.**

Lawrence & Newton, 1980: 138: *Arcyria cinerea, A. incarnata, Fuligo septica, Stemonitis axifera, Stemonitis fusca, Stemonitis splendens, Stemonitis* sp., *Tubifera ferruginosum.*

Newton, 1984: 319: *Lenzites elegans, Hirschioporus pargamenus, Trichaptum trichomallum*, Polyporales sp., Agaricales sp.

Leschen, 1988b: 14: *Arcyria* sp., *Fuligo septica, Phellinus* sp., *Physarium polycephalum, Pleurotus ostreatus, Polyporus brasiliensis, Russula* sp., *Schizopora paradoxa, Trametes cervinus, Tremellodendron pallidum.*

Hammond & Lawrence, 1989: 294: specialists on slime mold.

Wheeler, 1991: 42 [ex Navarrete-Heredia]: *Pleurotus ostreatus.*

Brachynopus latus **Broun**

Löbl & Leschen, 2003b: 22: *Arcyria denudata, Stemonitis, Agrocybe* sp., *Aseroe* sp., *Coltricia cinnamonea, Ganoderma* sp., *Phanerochaeta sordita, Phellinus* sp., *Schizopora* sp., ?*Scytonostroma* sp., *Stereum* sp., resupinate polypore, corticioid sp.

Brachynopus scutellaris **(Redtenbacher)**

Kuschel, 1990: 4: *Corynocarpus laevigatus, Hedycarya arborea, Melicope ternata.*

Löbl & Leschen, 2003b: 24: *Arcyria incarnata, Fuligo septica, Physarum* cf. *leucophaeum, Stemonitis* sp., *Trichia floriformis, Bjerkandra adusta, Ganoderma* sp., *Hyphodontia* sp., *Inonotus nothophagi, Phellinus kamahi, Poria* sp., *Schizopora* sp., *Trametes versicolor, Trametes* (=*Coriolus*) sp.

Scaphisoma agaricinum **(Linnaeus)**

Saalas, 1917: 395: *Fomitopsis ungulata.*

Scheerpeltz & Höfler, 1948: 149: *Ganoderma lucidum, Pluteus cervinus, Leptoporus trabus, Polyporus squamosus, Clitocybe infundibuliformis, Coriolus hirsutus.*

Benick, 1952: 46: *Armillaria mellea, Daedalea quercina, Fomes applanatus, Fomes igniarius, Hydnum (Dryodon) coralloides, Hypholoma fascicularis, Lactarius piperatus, Lenzites betulae, Pholiota mutabilis, Placoderma betulinum, Polyporus adustus, Polyporus brumalis, Polyporus giganteus, Polyporus squamosus, Polystichus velutinus, Polystichus versicolor, Trametes gibbosa, Fomes fomentarius, Paxillus lamellirugus.*

Tamanini, 1969b: 368 *Trametes hirsuta, T. versicolor, Poria medullapanis, Grifola sulphurea*

Tamanini, 1970: 23: *Trametes hirsuta, T. versicolor, Poria medullapanis, Grifola sulphurea.*

Nuss, 1975: 110: *Buglossoporus pulvinus, Daedalea quercina, Ganoderma adspersum, G. applanatum, Inonotus radiatus, Laetiporus sulphureus* (including on chlamydosperm form of *Ceriomyces aurantiacus*), *Onnia triqueter, Phellinus igniarius, Piptoporus betulinus, Polyporus ciliatus, P. squamosus.*

Klimaszewski & Peck, 1987: 543: *Polyporellus squamosus.*

Nikitsky et al., 1996: 24: *Fomitopsis pinicola, Fomes fomentarius, Cerrena unicolor, Inonotus rheades, Trichaptum biforme, Bjerkandera adusta, Inonotus obliquus, Daedalea quercina, Ganoderma applanatum, Piptoporus betulinus, Trametes versicolor, Oxyporus corticola, Phanerochaete* sp., *Clavicorona pyxidata, Exidia glandulosa, Pleurotus pulmonarius.*

Krasutskij, 1996a: 96: *Daedaleopsis confragosa, Fomes fomentarius, F. trogii, Fomitopsis pinicola, Piptoporus betulinus, Trametes versicolor, T. biforme, T. rheades, Phellinus pini, Ganoderma applanatum, Lentinus lepideus, Pleurotus atricapilus, P. pulmonarius.*

Krasutskij, 1996b: 276: *Fomes fomentarius.*

Krasutskij, 1997a: 307: *Fomes fomentarius, Fomitopsis pinicola.*

Kofler, 1998: 645, 646, 648: *Ganoderma applanatum, Heterobasidion anossum, Phellinus ferruginosus, Phillinus igniarius.*

Krasutskij, 2000: 80: *Lentinus cyathiformis.*

Nikitsky & Schigel, 2004: 8–14: *Bjerkandera adusta, Cerrana unicolor, Fomitopsis pinicola, Ganoderma lipsiense, Inocutis rheades, Inonotus obliquus, Phellinus igniarius, Piptoporus betulinus, Rigidoporus corticola, Trichaptum pargamenum.*

Schigel et al., 2004: 41, 42: *Gleoporus dichrous, Polyporus brumalis.*

Mateleshko, 2005: 128: *Pleurotus* sp.

Krasutskij, 2005: 37–42: *Fomes fomentarius, Fomitopsis pinicola, Piptoporus betulinus, Daedaleopsis confragosa, Trametes* sp.

Borowski, 2006: 54–59: *Ganoderma applanatum, Phellinus robustum, P. igniarius, P. pini, Fomitopsis pinicola, Fomes fomentarius, Daedaelopsis confragosa, Piptoporus betulinus.*

Nikitsky et al., 2008: 120: *Fomitopsis pinicola, Fomes fomentarius, Piptoporus betulinus, Pleurotus pulmonarius.*

Krasutskij, 2010: 372: *Trichaptum biforme, Fomitopsis pinicola, Ganoderma applanatum.*

Zamotajlov & Nikitsky, 2010: 105: *Cerrena unicolor, Inonotus rheades, Trichaptum biforme, Fomitopsis pinicola, Fomes fomentarius, Piptoporus betulinus, Pleurotus pulmonarius.*

Vinogradov et al. 2010: 16: *Fomes fomentarius, Laetiporus sulphureus, Ganoderma applanatum, Fomitopsis pinicola, Pleurotus pulmonarius, Pholiota* sp.,

Schigel, 2011: 329, 330, 331, 334, 339, 341: *Fomes fomentarius, Gloeophyllum* sp., *Daedaleopsis seltentrionalis, Gloeoporus dichrous, Antrodia xantha, Piptoporus betulinus, Ceriporia purpurea, Hapalopilus croceus, Ceriporiopsis pseudogilvescens, Spongipellis fissilis, S. spumea.*

Diéguez Fernández, 2014: 241: *Fomitopsis pinicola.*

Nikitsky et al. 2016: 138: *Fomes fomentarius, Fomitopsis pinicola, Cerrena unicolor, Inocutis rheades, Trichaptum biforme, Bjenkandera adusta, Inonotus obliquus, Daedalea quercina, Ganoderma applanatum, Piptoporus betulinus, Trametes versicolor, Oxyporus corticola, Phanerochaete* sp., *Artomyces pyxidatus, Exidia glandulosa, Pleurotus pulmonarius.*

Vlasov & Nikitsky, 2017: 5: *Artomyces pyxidatus, Bjenkandera adusta, Cerrena unicolor, Daedalea quercina, Exidia glandulosa, Fomes fomentarius, Fomitopsis betulina, F. pinicola, Ganoderma applanatum, Inocutis rheades, Inonotus obliquus, Oxyporus corticola, Phanerochaete* sp., *Pleurotus pulmonarius, Trametes versicolor, Trichaptum biforme* [in ref. to Nikitsky et al., 1996]; *Daedalopsis confragosa, Phellinus chrysoloma, Trametes trogii* [in ref. to Krasutskij, 1996]; *Neolentinus cyathiformis, Pleurotus calyptratus* [in ref. to Krasutskij, 2000]; *Phellinus ignarius* [in ref. to Nikitsky & Schigel, 2004]; *Laetiporus sulphureus, Pholiotza flammans, Trametes gibbosa, T. ochracea, Xanthoporia radiata* [in ref. to Tsinkevitch, 2004].

Scaphisoma americanum **(Löbl)**

Ashe, 1984: 361 (as *S. terminata*): *Hericium ramosum.*

Löbl, 1987c: 387: "possibly *Pleurotus* sp.", "fleshy tree fungi".

Leschen, 1988a: 230: *Tremellodendron pallidum.*

Leschen, 1988b: 14: *Armillariella tabescens, Hericium erinaceus, Laetiporus sulphureus, Marasmius* sp., *Meripilus giganteus, Merulius incarnatus, Omphalotus cicarius,* Pleurotus *ostreatus, Schizopora paradoxa.*

Scaphisoma apicefasciatum **Reitter**

Reitter, 1908: 32: *Polyporus.*

Scaphisoma assimile assimile **Erichson**

Benick, 1952: 47: *Clitocybe gilva, Fomes igniarus, Polyporus squamosus, Tremella frondosa.*

Tamanini, 1969b: 366 and 1970: 18: *Fomes fomentarius*, *Grifola sulphurea*: ex Dajoz, 1965, [based on misidentification]

Borowski, 2006: 54, 58: *Phellinus igniarius*, *Fomitopsis pinicola*, *Fomes fomentarius*.

Nikitsky et al., 1996: 25: *Oxyporus latemarginatus*.

Krasutskij, 1996a: 96: *Phellinus igniarius*, *Polyporus squamosus*, *Clitocybe gilva*, *Tremella frondosa*.

Tsinkevich, 2004: 19: *Fomes fomentarius*.

Nikitsky & Schigel, 2004: 13: *Rigidoporus latemarginatus*

Mateleshko, 2005: 128: *Pleurotus* sp.

Vinogradov et al. 2010: 17: *Laetiporus sulphureus*.

Scaphisoma balcanicum **Tamanini**

Tamanini, 1954: 87: *Coriolus versicolor*.

Tamanini, 1970: 7: *Trametes versicolor*.

Nikitsky et al., 1996: 25: *Daedalea quercina*, *Phellinus tremulae*, *Funalia trogii* (=*Trametes trogii*), *Oxyporus* sp.

Kofler, 1998: 643: *Bjerkandera fumosa*

Nikitsky & Schigel, 2004: 8–12: *Daedalea quercina*, *Funalia trogii*, *Phellinus tremulae*.

Borowski, 2006: 54, 58: *Fomitopsis pinicola*, *Fomes fomentarius*.

Nikitsky et al., 2008: 121: *Daedalea quercina*, *Phellinus tremulae*, *Coriolopsis trogii*, *Tyromyces chioneus*.

Vinogradov et al. 2010: 17: *Fomes fomentarius*, *Laetiporus sulphureus*, *Kuehneromyces mutabilis*.

Zamotajlov & Nikitsky, 2010: 105: *Daedalea quercina*, *Phellinus tremulae*, *Funalia trogii* (*Trametes trogii*), *Tyromyces chioneus*.

Nikitsky et al. 2016: 138: *Daedalea quercina*, *Phellinus tremulae*, *Trametes trogii*, *Tyromyces chioneus*.

Vlasov & Nikitsky, 2017: 6: *Cerioperus squamosus*.

Scaphisoma balteatum **Matthews**

Fierros-López, 2006a: 66: *Oligoporus floriformis*

Scaphisoma biplagiatum **Heller**

Löbl & Ogawa, 2016b: 1399: “polypores”.

Scaphisoma boleti **(Panzer)**

Palm, 1951: 145: *Trichoderma lignorum*.

Benick, 1952: 47: *Armillaria mellea*, *Lactarius subdulcis*, *Placoderma betulinum*, *Polyporus squamosus*, *Polystichusvelutinus*, *Piptoporus betulinus*.

Palm, 1959: 224: *Polyporus squamosus*, *P. sulphureus*.

Tamanini, 1969b: 367: *Trametes versicolor, Phellinus ignarius trivialis, Ganoderma applanatum.*
Nuss, 1975: 111: *Ganoderma applanatum, Inonotus radiatus, Laetiporus sulphureus, Phellinus igniarius, Polyporus squamosus.*
Klimaszewski & Peck, 1987: 543: *Polyporellus squamosus.*
Nikitsky et al., 1996: 24: *Clavicorona pyxidata; Armillaria mellea, Fomitopsis pinicola, Piptoporus betulinus.*
Kofler, 1998: 644: *Fomitopsis pinicola.*
Tsinkevich, 2004: 19: *Fomes fomentarius.*
Nikitsky & Schigel, 2004: 8–12: *Bjerkandera adusta, Fomitopsis pinicola, Piptoporus betulinus.*
Borowski, 2006: 53–59: *Ganoderma applanatum, Phellinus igniarius, Fomitopsis pinicola, Fomes fomentarius, Daedaelopsis confragosa, Piptoporus betulinus.*
Nikitsky et al., 2008: 121: *Artomyces* (= *Clavicorona*) *pyxidatus.*
Zamotajlov & Nikitsky, 2010: 105: *Artomyces pyxidatus* (= *Clavicorona pyxidata*).

***Scaphisoma boreale* Lundblad**
Palm, 1959: 224: *Trichoderma lignorum.*
Nikitsky et al., 1996: 25: *Oxyporus corticola, Oxyporus* sp., *Fomes fomentarius, Oxyporus late-marginatus, Funalia trogii, Ganoderma applanatum.*
Nikitsky & Schigel, 2004: 8–13: *Fomes fomentarius, Funalia trogii, Ganoderma lipsiense, Rigidoporus corticola, R. latemarginatus.*
Schigel et al., 2004: 41: *Dichomitus squalens.*
Borowski, 2006: 54, 58, 59: *Phellinus igniarius, Fomitopsis pinicola, Fomes fomentarius, Daedaelopsis confragosa, Piptoporus betulinus.*
Vinogradov et al. 2010: 16: *Ganoderma applanatum.*
Schigel, 2011: 329, 331, 333, 338, 349: *Heterobasidion parviporum, Trametes hirsuta, Gloeophyllum* sp., *Dichomitus campestris, D. squalens, Rigidoporus corticola, Polyporus squamosus, Piptoporus betulinus, Laetiporus sulphureus, Skeletocutis odora, Oligoporus immitis.*
Nikitsky et al. 2016: 138: *Fomes fomentarius, Oxyporus* sp., *Trametes trogii, Ganoderma applanatum.*
Vlasov & Nikitsky, 2017: 6: *Ganoderma latemarginalis, Neolentinus cyathiformis, Oxyporus corticola, O. latemarginatus, Oxyporus* sp., *Trametes trogii.*

***Scaphisoma carolinae* Casey**
Leschen et al., 1990: 292: Basidiomycetes.

***Scaphisoma centripunctulum* Löbl & Ogawa**
Löbl & Ogawa, 2016b: 1390: “polypores”.

***Scaphisoma commune* Löbl**

Hanley, 1996: 38 (as *Scaphisoma castaneum* Motschulsky): *Auriporia aurea, Pleurotus ostreatus, Fomitopsis pinicola, Polyporus alboluteus, Ganoderma* sp.

Hatch, 1957: 282 (as *Scaphisoma castaneum* Motschulsky): *Pleurotus ostreatus.*

***Scaphisoma confusum* Löbl & Ogawa**

Löbl & Ogawa, 2016b: 1419: *Polyporus durus.*

***Scaphisoma convexum* Say**

Leschen, 1988a: 231: *Schizopora paradoxa.*

Leschen, 1988b: 17: *Amanita* sp., *Armilariella melisa, Blerkandra adusta, Irpex lacteus, Laetiporus sulphureus, Marasmius* sp., *M. sillus, Omphalotus oleolaris, Merispilus giganteus, Phaeolus schwenetizii, Phyloporus rhodoxanthus, Pleurotus ostreatus, Strobilomyces floccopus, Trametes cervinus, T. versicolor, Tyromyces* sp., *Xerampelina* sp.

Leschen et al., 1990: 283: *Amanita* sp., *Armillariella mellea, A. tabescens, Bjerkandra adusta, Lactaria* sp., *Marasmius* sp., *Omphalotus oleolaris, Phoeleus schweinertzii, Phylloporus rhodoxanthus, Pleuropus ostreatus, Schizopora paradoxa, Spongipellis pachypon, Strobilomyces flocopus, Trametes cervinus, T. versicolor, Tyromyces* sp., *Xerampelina* sp.

***Scaphisoma cortesaguilari* Fierros-López**

Fierros-López, 2006a: 59: *Sirobasidium sanguineum.*

***Scaphisoma funereum* Löbl**

Löbl & Leschen, 2003b: 40: *Amanita muscaria.*

***Scaphisoma furcigerum* Löbl & Ogawa**

Löbl & Ogawa, 2016b: 1423: "polypores".

***Scaphisoma hamatum* Löbl & Ogawa**

Löbl & Ogawa, 2016b: 1406: "polypores".

***Scaphisoma hanseni* Löbl & Leschen**

Löbl & Leschen, 2003b: 42: *Auricularia polytricha, Ganoderma* sp., *Ganoderma australe, Phellinus gilvus, Phellinus kamahi, Phellinus punctatus* (=*Fuscoporia dryophila*), *Phellinus* sp., *Tyromyces* sp.

***Scaphisoma ilonggo* Löbl & Ogawa**

Löbl & Ogawa, 2016b: 1358: *Fomes applanatus.*

Scaphisoma impunctatum **Reitter**

Leschen et al., 1990: 287: *Bjerkandra adusta, Lactarius subvellereus, Laetiporus sulphureus, Meripilus giganteus, Perenniporia compacta, Phellinus* sp., *Schizopora paradoxa, Trametes cervinus, T. versicolor, Tremellodendron pallidum.*

Scaphisoma inexspectatum **Löbl & Ogawa**

Löbl & Ogawa, 2016b: 1408: "polypores".

Scaphisoma inopinatum **Löbl**

Tamanini, 1969b: 376: *Trametes* sp., *Stereum hirsutum.*

Tamanini, 1970: 22: *Trametes* sp., *Stereum hirsutum.*

Nikitsky et al., 1996: 24: *Steccherium ochraceum, Fomitopsis pinicola.*

Krasutskij, 1996a: 97: *Daedaleopsis confragosa, Fomes fomentarius, F. trogii, Gloeophyllum sepiarium, Trametes T. ochracea, T. versicolor, T. biforme, Inocutis rheades, Phellinus pini, Ganoderma applanatum, Lentinus lepideus, Pleurotus atricapilus, P. calyptratus, P. pulmonarius.*

Krasutskij, 1997a: 307: *Fomes fomentarius, Fomitopsis pinicola, Ganoderma applanatum, Trametes trogii.*

Krasutskij, 1997b: 774: *Fomes fomentarius.*

Kofler, 1998: 646: *Phellinus ferruginosus.*

Krasutskij, 2000: 80: *Lentinus lepidus.*

Nikitsky & Schigel, 2004: 9: *Fomitopsis pinicola.*

Krasutskij, 2005: 37, 38: *Fomes fomentarius, Fomitopsis pinicola, Daedaleopsis confragosa, Trametes* sp.

Borowski, 2006: 54, 58: *Fomitopsis pinicola, Fomes fomentarius.*

Štourač & Rébl, 2009: 121: *Fomes fomentarius.*

Krasutskij, 2010: 374: *Fomes fomentarius, Ganoderma applanatum, Daedaleopsis confragosa, Trametes versicolor, T. trogii, Lentinus lepideus, Pleurotus calyptratus, P. pulmonarius.*

Schigel, 2011: 332, 341: *Bjerkandera adusta, Rhodonia pacenta.*

Wojas, 2016: 140: *Laetiporus sulphureus.*

Süda, 2016: 65: *Fomes fomentarius.*

Vlasov & Nikitsky, 2017: 6: *Fomes fomentarius, Fomitopsis betulina, F. pinicola.*

Scaphisoma italicum **Tamanini**

Tamanini, 1955: 17: *Polyporus velutinus, Coriolus* sp.

Tamanini, 1970: 24: *Trametes pubescens, T. unicolor, T. versicolor.*

Scaphisoma jaliscanum **Fierros-López**

Fierros-López, 2006a: 55: *Favolus brasiliensis, Auricularia delicata, Trametes membranacea, Hydnopolyporus fimbrianus.*

Scaphisoma javanum **Löbl**
Löbl & Ogawa, 2016b: 1427: "polypores".

Scaphisoma kinabaluum **Löbl**
Löbl, 1987a: 97: Clavariaceae.

Scaphisoma limbatum **Erichson**
Benick, 1952: 47: "an Birkenschwämmen".
Kompantsev & Pototskaya, 1987: 99: *Hericium erinaceum*.
Krasutskij, 1996a: 96: *Fomes fomentarius*.
Krasutskij, 1996b: 276: *Fomes fomentarius*.
Nikitsky et al., 1996: 24: *Hericium coralloides, Fomes fomentarius*.
Nikitsky et al., 1998: 7: *Creolophus cirrhatus*.
Tsinkevich, 2004: 19: *Fomes fomentarius*.
Mateleshko, 2005: 128: *Pleurotus* sp.
Borowski, 2006: 54–58: *Fomitopsis pinicola, Fomes fomentarius*.
Vlasov & Nikitsky, 2017: 4: *Hericium coralloides, H. cirrhatum, Fomes fomentarius*.

Scaphisoma loebli **Tamanini**
Tamanini, 1969b: 375: *Trametes versicolor, T. gallica, T. betulina, T. hirsuta, Trametes* sp., *Polyporellus squamosus, Onysomices odoratus, Fomes fomentarius*.
Tamanini, 1970: 24: *Fomes fomentarius, Onysomices odoratus, Polyporellus squamosus, Stereum hirsutum, Trametes versicolor, T. gallica, T. betulina, T. hirsuta, T. unicolor*.

Scaphisoma minutipenis **Löbl & Ogawa**
Löbl & Ogawa, 2016b: 1374: "polypores".

Scaphisoma mirum **Löbl & Ogawa**
Löbl & Ogawa, 2016b: 1375: *Fomes applanatus*.

Scaphisoma nanellum **Löbl & Ogawa**
Löbl & Ogawa, 2016b: 1360: "polypore fungi".

Scaphisoma nigrofasciatum **Pic**
Deepthi et al., 2004: *Pleurotus florida*; as pest on cultivated mushrooms.
Singh & Sharma, 2016: 213: unspecified *Agaricus bisporus, Pleurotus* sp. and *Volvariella* sp. as pest on cultivated mushrooms.

Scaphisoma obenbergeri **Löbl**
Kofler, 1998: 651: *Trametes pubescens*.
Borowski, 2006: 54, 58, 60: *Phellinus igniarius, Fomitopsis pinicola, Fomes fomentariu, Laetiporus sulphureus*.

Scaphisoma ochropenne **Löbl & Ogawa**
Löbl & Ogawa, 2016b: 1410: *Fomes applanatus, Polyporus durus.*

Scaphisoma opochti **Fierros-López**
Fierros-López, 2006a: 61: *Hydnopolyporus palmatus.*

Scaphisoma pulchrum **Löbl & Ogawa**
Löbl & Ogawa, 2016b: 1432: *Fomes applanatus.*

Scaphisoma pusillum **LeConte**
Leschen et al., 1990: 292: Basidiomycetes.

Scaphisoma punctulatum **LeConte**
Leschen et al., 1990: 292: *Schizopora paradoxa.*

Scaphisoma repandum **Casey**
Weiss & West, 1920: 5: *Polyporus gilvus.*
Webster et al., 2012: 246: *Polyporus varius.*

Scaphisoma stephani **Leschen & Löbl**
Leschen et al., 1990: 287, 289: *Armillaris tabescens, Bjerkandra adusta, Laetiporus sulphureus, Meripilus giganteus, Merulius incarnatus, Perenniporia compacta, Phellinus* sp., *Plerotus ostreatus, Polyporus brasiliensis, Schizopora paradoxa, Thelophoraceae* sp., *Trametes cervinus, T. versicolor.*

Scaphisoma stigmatipenne **Heller**
Löbl & Ogawa, 2016b: 1414: "polypores".

Scaphisoma subalpinum **Reitter**
Scheerpeltz & Höfler, 1948: 149: *Ganoderma lucidum.*
Benick, 1952: 47: *Daedalia quercina, Polyporus squamosus, Fomes fomentarius.*
Tamanini, 1969b: 364: *Fomes fomentarius.*
Nikitsky et al., 1996: 25: *Fomes fomentarius, Ganoderma applanatum, Bjerkandera adusta, Fomitopsis pinicola; Daedalea quercina, Polyporus squamosus.*
Krasutskij, 1996a: 97: *Bjerkandera adusta, Fomes fomentarius, Fomitopsis pinicola, Inocutis rheades, Pleurotus pulmonarius.*
Krasutskij, 1997a: 307: *Fomes fomentarius, Fomitopsis pinicola.*
Krasutskij, 1997b: 774: *Fomes fomentarius.*
Kapp, 2001: 177: *Fomes fomentarius.*
Nikitsky & Schigel, 2004: 8–12: *Daedalea quercina, Fomes fomentarius, Ganoderma lipsiense, Inocutis rheades, Polyporus squamosus.*

Tsinkevich, 2004: 19: *Laetiporus sulphureus.*

Mateleshko, 2005: 128: *Pleurotus* sp.

Krasutskij, 2005: 38: *Fomitopsis pinicola.*

Borowski, 2006: 55, 58: *Phellinus igniarius, Fomitopsis pinicola, Fomes fomentarius.*

Nikitsky et al., 2008: 121: *Polyporus squamosus, Fomitopsis pinicola, Fomes fomentarius, Ganoderma applanatum.*

Zamotajlov & Nikitsky, 2010: 105: *Polyporus squamosus, Fomitopsis pinicola, Fomes fomentarius, Ganoderma applanatum.*

Vinogradov et al., 2010: 16: *Fomes fomentarius.*

Schigel, 2011: 325, 329, 331, 332, 339: *Phellinus igniarius, Fomes fomentarius, Trametes hirsuta, Gloeoporus dichrous, Piptoporus betulinus, Protomerulius caryae, Piptoporus betulinus.*

Nikitsky et al. 2016: 138: *Fomes fomentarius, Ganoderma applanatum.*

Vlasov & Nikitsky, 2017: 7: *Fomes fomentarius, Fomitopsis* sp.

***Scaphisoma subplanatum* Löbl & Ogawa**

Löbl & Ogawa, 2016b: 1354: *Fomes applanatus.*

***Scaphisoma surigaosum* (Pic)**

Löbl & Ogawa, 2016b: 1414: “polypores”.

***Scaphisoma suturale* LeConte**

Weiss & West, 1922: 199: *Clavaria* sp.

Leschen et al., 1990: 285: *Irpex lacteus.*

***Scaphisoma teres* Löbl**

Newton, 1984: 318: *Ganoderma* sp., *Bjerkandera adusta.*

Newton (pers. comm.): *Ganoderma applanatum.*

***Scaphisoma terminatum* Melsheimer**

Blatchley, 1910: 494: *Clitocybe illudens* [under *Baeocera punctipennis* Blatchley].

Weiss & West, 1920: 5: *Clitocybe illudens* [under *Baeocera punctipennis* Blatchley].

Leschen et al., 1990: 291: *Auricularia auricula, Laetiporus sulphureus, Meripilus giganteus, Merulius incarnatus, Naematoloma* sp., *Phellinus gilvus, Polyporus brasiliensis, Schizophyllum commune, Schizopora paradoxa, Trametes versicolor, Tyromyces* sp.

***Scaphisoma tetrastictum* Champion, 1927**

Mazumder et al., 2001: 59: *Pleurotus ostreatus*; pest on cultivated mushrooms.

Mazumder et al., 2008: 45: *Pleurotus ostreatus*; pest on cultivated mushrooms, seazonal activity.

***Scaphisoma tricolorinotum* Löbl & Ogawa**
Löbl & Ogawa, 2016b: 1436: "polypores".

***Scaphisoma versicolor* Fierros-López**
Fierros-López, 2006a: 64: *Rigidoporus microporus, Oligoropus floriformis, Trametes membranacea, Hydnopoyporus fimbriatus, Favolus brasiliensis, Ceiba* sp., *Pleurotus* sp.

***Scaphisoma werneri* Löbl & Ogawa**
Löbl & Ogawa, 2016b: 1403: "polypores".

***Scaphisoma* sp.**
Newton, 1984: 318: Tremellares, *Arcyria* sp., *Stemonitis* sp., *Earliella* sp., *Ganoderma* sp., *Rigidophorus* sp., *Clavaria* sp., *Bjerkandera* sp., *Coriolopsis* sp., *Coriolus* sp., *Hexagona* sp.
Leschen, 1988b: 16: *Armilariella tabescens, Blerkandra adusta, Irpex lacteus, Lactarius sulphureus, Meripilus giganteus, Merulius incarnatus, Mycena* sp., *Perennipora compacta, Phanerocaete chryzorhizon, Phellinus gilvus, Pleurotus ostreatum, Polyporus brasiliensis, Schizophilum commune, Schizopora paradoxa, Sparassis crispa,* Thelophoraceae, *Trametes cervinus, T. versicolor, Tramellodendron pallidum.*
Kuschel, 1990: 43: *Fuscoporia dryophila, Amanita muscaria.*
Newton, 1991: 388: *Clavaria coronata.*
Wheeler, 1991: 42 [ex Navarret-Heredia]: *Tricholoma* sp.

***Scaphobaeocera stephensoni* Löbl**
Newton & Stephenson, 1990: 212: *Dictydium cancellatum, Physarum flavicomum, Stemonitis ?splendens, Trichia decipiens.*

***Scaphobaeocera timida* Löbl**
Newton & Stephenson, 1990: 212: *Arcyria cinerea, Cribraria intricata, Dictydium cancellatum, Hemitrichia calyculata, Lycogala epidendrum, Stemonitis hyperopta, Stemonitis smithii.*

***Scaphobaeocera* sp.**
Hammond & Lawrence, 1989: 294: "slime mold".
Newton (per. comm.): *Arcyria* sp., *Stemonitis* sp.

***Spinoscapha rufa* (Broun)**
Löbl & Leschen, 2003: 23: undetermined corticioid.

Toxidium gammaroides **LeConte**

Leschen, 1988b: 17: *Albatrellus* sp., *Armillariella mellea, Crepidotus* sp., *Coniophoraceae, Irpex lacteus, Polyporaceae, Schizopora paradoxa, Strobilomyces floccopus, Trametes cervinus, T. versicolor,* Tricholomataceae, *Tricholomopsis plataphylla, Tremellodendron pallidum, Tyromyces* sp.

Webster et al., 2012: 248: *Pholiota* sp.

Toxidium **sp.**

Newton, 1984: 319: *Ganoderma applanatum, Coriolus* sp., *Lenzites elegans, Coriolopsis capelatus, Daedalea farinaceae, Nigroporus vinosus, Pogonomyces hydnoides, Rigidoporus* sp., *Phellinus pachyphloeus.*

Note

Members of *Scaphisoma* have been often misidentified in the past, in particular *S. inopinatus* Löbl was not distinguished from *S. agaricinum* (Linnaeus), *S. boreale* Lundblad was confused with *S. boleti* (Panzer) and *S. assimile assimile* Erichson, and specimens of *S. balcanicum* Tamanini and *S. obenbergeri* Löbl were identified as *S. subalpinum* Reitter. I have not re-examined voucher specimens of Scaphidiinae on which the fungal host data are based, except those of *S. agaricinum* (Linnaeus) and *S. boleti* (Panzer) in Rehfous (1955) who confused these two species. His data are unreliable and not to be used.

References

Achard, J. 1914. Un scaphidiide nouveau de Birmanie (Col.). *Bulletin de la Société entomologique de France* 1914: 394–396.

Achard, J. 1915a. Descriptions d'espèces nouvelles de Scaphidiides. *Annales de la Société entomologique de France* 83 [1914]: 555–562.

Achard, J. 1915b. Descriptions de deux espèces nouvelles de Scaphidiidae (Col.). *Bulletin de la Société entomologique de France* 1915: 290–291.

Achard, J. 1915c. Synonymie de quelques Scaphidiidae (Col.). *Bulletin de la Société entomologique de France* 1915: 291–292.

Achard, J. 1916. Descriptions de trois nouveaux *Scaphidium* d'Australie (Col. Scaphidiidae). *Bulletin de la Société entomologique de France* 1916: 87–89.

Achard, J. 1920a. Description d'un nouveau *Scaphosoma* du Yunnan (Col. Scaphidiidae). *Annales de la Société entomologique de France* 88 [1919]: 328.

Achard, J. 1920b. Descriptions d'espèces nouvelles de *Scaphidium* (Coléoptères Scaphidiidae) de la région indo-malaise. *Bulletin du Muséum national d'Histoire naturelle* 26: 125–128.

Achard, J. 1920c. Les scaphidiides de la Péninsule de Malacca. *Annales de la Société entomologique de Belgique* 60: 47–58.

Achard, J. 1920d. Notes sur les Scaphidiidae de la faune Indo-Malaise. *Annales de la Société entomologique de Belgique* 60: 123–136.

Achard, J. 1920e. Descriptions de Scaphidiidae (Col.) inédits de la République Argentine. *Bulletin de la Société entomologique de France* 1919: 350–352.

Achard, J. 1920f. Descriptions de nouveaux Scaphidiidae (Col.) du Sikkim. *Bulletin de la Société entomologique de France* 1919: 362–365.

Achard, J. 1920g. Description d'un nouveau genre et d'une nouvelle espèce de Scaphidiidae (Col.). *Bulletin de la Société entomologique de France* 1920: 207–208.

Achard, J. 1920h. Synopsis des *Scaphidium* (Col. Scaphidiidae) de l'Indo-Chine et du Yunnan. *Bulletin de la Société entomologique de France* 1920: 209–212.

Achard, J. 1920i. Diagnoses des espèces nouvelles de Scaphidiidae (Col.). *Bulletin de la Société entomologique de France* 1920: 239–242.

Achard, J. 1920j. Descriptions de nouvelles espèces de Scaphidiidae (Col.). *Bulletin de la Société entomologique de France* 1920: 263–265.

Achard, J. 1920k. Identification du *Scaphidium concolor* F. (Col. Scaphidiidae). *Bulletin de la Société entomologique de France* 1920: 307.

Achard, J. 1920l. Description d'espèces nouvelles du genre *Heteroscapha* (Col. Scaphidiidae). *Insecta, Rennes* 10: 5–9.

Achard, J. 1921a. Une nouvelle espèce du genre *Ascaphium* Lewis (Col. Scaphidiidae). *Bulletin de la Société entomologique de France* (1921): 93.

Achard, J. 1921b. Notes sur les Scaphidiidae du Musée de Leyde. *Zoologische Mededeelingen* 6: 84–91.

Achard, J. 1922a. Essai de groupement des espèces du genre *Scaphidium* Ol. (Col. Scaphidiidae). *Fragments entomologiques*: 10–13.

Achard, J. 1922b. Synopsis des espèces du genre *Hemiscaphium* Achard (Col. Scaphidiidae). *Fragments entomologiques*: 30–35.

Achard, J. 1922c. Descriptions de scaphidides nouveaux (Col. Scaphidiidae). *Fragments entomologiques*: 35–45.

Achard, J. 1922d. Observations sur quelques *Scaphidium* Ol. (Col. Scaphidiidae). *Bulletin de la Société entomologique de France* 1922: 260–263.

Achard, J. 1922e. Description d'un nouveau *Scaphidium* de l'Afrique Equatoriale. *Bulletin du Muséum national d'Histoire naturelle* 28: 489.

Achard, J. 1923. Revision des Scaphidiidae de la faune japonaise. *Fragments entomologiques*: 94–120.

Achard, J. 1924a. Nouvelles espèces de *Scaphidiolum* de la faune indomalaise (Col. Scaphidiidae). *Bulletin de la Société entomologique de France* 1924: 150–153.

Achard, J. 1924b. Essai d'une subdivision nouvelle de la famille des Scaphidiidae. *Annales de la Société entomologique de Belgique* 65: 25–31.

Achard, J. 1924c. Catalogue des Scaphidiidae de la faune paléarctique. *Bulletin de la Société entomologique de Belgique* 6: 143–155.

Achard, J. 1924d. Descriptions de trois variétés nouvelles du genre *Scaphidiolum* Achard (Col. Scaphidiidae). *Sborník Entomologického oddělení Národního Musea v Praze* 2: 91.

Agassiz, L. 1846. Nomenclatoris zoologici index universalis continens nomina systematica classium, ordinum, familiarum et generum animalium omnium, tam viventium quam fossilium, secundum ordinem alphabeticum unicum disposita, adjectis homonymiis plantarum, nec non variis adnotationibus et emendationibus. In: L. Agassiz *Nomenclator Zoologicus, continens nomina systematica generum animalium tam viventium quam fossilium, secundum ordinem alphabeticum disposita, adjectis auctoribus, libris, in quibus reperiuntur, anno editionis, etymologia et familiis, ad quas pertinent, in singulis classibus.* Fasc. 12. Soloduri: Jent et Gassmann, viii + 393 pp.

Ahn, K.-J., Y.-B. Cho, Y.-H Kim., I.-S Yoo. & A.F Newton. 2017. Checklist of the Staphylinidae (Coleoptera) in Korea. *Journal of Asia-Pacific Biodiversity* 10: 279–336.

Alexandrovitch, O.R., I.K. Lopatin, A.D. Pisanenko, V.A. Tsinkevitch & S.M. Snitko 1996. *A catalogue of Coleoptera (Insecta) of Belarus*. Minsk, Nauchnoe Izdanie, 103 pp.

Angelini, F. 1986. Coleotterofauna del Massiccio del Pollino (Basilicata-Calabria) (Coleoptera). *Entomologia* (Bari) 21: 37–125.

Angelini, F. 1987. Coleotterofauna del Promotorio del Gargano. *Atti del Museo civico di Storia naturale di Grosseto* 11/12: 5–84.

Angelini, F., P. Audisio, G. Castellini, R. Poggi, D. Vailati, A. Zanetti & S. Zoia 1995. Coleoptera Polyphaga II (Staphylinoidea escl. Staphylinidae). In: A. Minelli, S. Ruffo & S. La Costa (eds). *Checklist delle specie della fauna italiana*, 47. Calderini, Bologna, 39 pp.

Aragona, L. 1830. *De quibusdam coleopteris Italiae novis aut rarioribus tentamen inaugurale quod annuentibus magnifico domino rectore illustrissimo facultatis directore spectabili d. decano ac clarissimis D. D. professoribus auspice J.M. Zendrini historiae naturalis spec. prof. ord. pro doctoris medicinae laurea rite capessenda Aloysius Aragona Ticinensis in celebratiss. I. R. Archigymnasio Ticinensi publicae disquisitioni submittit. Mense Aprilis MDCCCXXX in thesis adnexas disputabitur in Universitatis Aedibus.* Bizzoni, Ticini Regii, 31 pp.

Arnett, R.H. Jr. 1968. *The beetles of the United States (a manual for identification), 2nd printing*. The American Entomological Institute, Ann Arbor (Mich.), xii + 1112 pp.

Ashe, J.S. 1984. Description of the larva and pupa of *Scaphisoma terminatum* Melsh. and the larva of *Scaphium castanipes* Kirby with notes on their natural history (Coleoptera: Scaphidiidae). *The Coleopterists Bulletin* 38: 361–373.

Bacal, S., N. Muinteanu & I. Toderas 2013. Checklist of beetles (Insecta: Coleoptera) of the Republic of Moldova. *Brukenthal Acta Musei* 3: 415–450.

Bellmann, A., J. Esser, W. Lakomy & A. Rose 2003. Bemerkenswerte und neue Käferfunde aus dem Weser-Ems-Gebiet (Coleoptera) (Teil 5). *Abhandlungen des Naturwissenschaftlichen Verein zu Bremen* 45: 445–448.

Benick, L. 1952. Pilzkäfer und Käferpilze. Ökologische und statistische Untersuchungen. *Acta Zoologica Fennica* 70: 1–250.

Betz, O., M.K. Thayer & A.F. Newton Jr. 2003: Comparative morphology and evolutionary pathways of the mouthparts in spore-feeding Staphylinoidea (Coleoptera). *Acta Zoologica* (Stockholm) 84: 179–238.

Blackburn, T. 1891. Further notes on Australian Coleoptera, with descriptions of new genera and species. *Transactions of the Royal Society of South Australia* 14(1): 65–153.

Blackburn, T. 1903. Further notes on Australian Coleoptera, with descriptions of new genera and species. *Transactions of the Royal Society of South Australia* 27: 91–182.

Blackwelder, R.E. 1944. Checklist of the coleopterous insects of Mexico, Central America, the West Indies, and South America. Part 1. *Bulletin of the United States National Museum* 185: i–xii + 1–188.

Blatchley, W.S. 1910. *An illustrated descriptive Catalogue of the Coleoptera or beetles (exclusive of the Rhynchophora) known to occur in Indiana*. Indiana Department of Geology and Natural Resources Bulletin 1, 1385 pp.

Blatchley, W.S. 1930. *A list of species and varieties of Coleoptera described by W.S. Blatchley, with citation to original descriptions, fixation of single types and designation of known snonyms. Blatchleyana*. The Nature Publishing Co., Indianapolis, pp. 35–50.

Boheman, C.H. 1851. *Insecta Caffrariae Annis 1838–1845 a J.A. Wahlberg collecta. Pars I, Fascic. II. Coleoptera (Buprestides, Elaterides, Cebrionites, Rhipicerides, Cyphonides, Lycides, Lampyrides, Telephorides, Merylides, Clerii, Terediles, Ptiniores, Palpatores, Silphales, Histeres, Scaphidilia, Nitidulariae, Cryptophagidae, Byrhii, Dermestini, Parnidae, Hydrophilidae)*. Nordstedtiana, Holmiae, [1] + 299–626 pp., 2 pls.

Borowski, J. 2006. *Chrząszcze (Coleoptera) grzybów nadrzewnych – studium waloryzacyjne*. Wydawnictwo SGHGW, Warszawa, 1–91 + [1] + [1 annex] pp. + 2 maps, 3 pls.

Borowski, J. 2007. Waloryzacja drzewostanów Gór Świętokrzyskich przy wykorzystaniu mycetobiontycznych chrząszczy grzybów nadrzewnych. In: J. Borowski & S. Mazur (eds). *Waloryzacja ekosystemów leśnych Gór Świętokrzyskich metodą zooindykacyjną*: 119–147. Wydawnictwo SGGW, Warsawa.

Borowski, J., A. Byk & D. Łegowski 2005. 400. *Latridium pseudominutus* (Strand) – chrząszcz nowy dla fauny Polski oraz inne interesująszcze chrząszcze (Coleoptera), odlowiene w okolicach Kwisna na Pojezierzu Pomorskim. *Wiadomości entomologiczne* 24(1): 44–45.

Bousquet, Y. 2016. Litteratura Coleopterologica (1758–1900): a guide to selected books related to the taxonomy of Coleoptera with publication dates and notes. *ZooKeys* 583: 1–776.

Böving, A.G. & F.G. Craighead 1931. An illustrated synopsis of the principal larval forms of the order Coleoptera. *Entomologica americana* 11 (N.S.): 1–86, 125 pls.

Bowditch, F.G. 1896. List of Mt. Washington Coleoptera. *Supplement to Psyche* II, pp. 1–11.

Brancsik, K. 1893. Beiträge zur Kenntnis Nossibés und dessen Fauna nach Sendungen und Mitheilungen des Herrn P. Frey. *Jahresheft des naturwissenschaftlichen Vereines des trencséner Comitates* 15–16: 202–258.

Brimley, C.S. 1938. *The insects of North Carolina. Being a list of the insects of North Carolina and their close relatives*. Department of Agriculture, Knoxville, 560 pp.

Britton, E.B. 1970. Coleoptera (Beetles). Pp. 495–621. In: *Insects of Australia. A textbook for studens and research workers*. Melbourne University Press, Carlton, [1 pl.] + xii + 1029 pp.

Broun, T. 1880. *Manual of the New Zealand Coleoptera*. James Hughes, Wellington, iv–xix + 651 pp.

Broun, T. 1881a. *Manual of the New Zealand Coleoptera*. Part II: 653–744 + xxi–xxiii. Colonial Museum & Geological Survey Department, Wellington.

Broun, T. 1881b. *Manual of the New Zealand Coleoptera*. Part II: 817–973. Colonial Museum & Geological Survey Department, Wellington.

Broun, T. 1886. *Manual of the New Zealand Coleoptera*. Parts III & IV: 745–973 + v–xvii; Colonial Museum & Geological Survey Department, Wellington.

Broun, T. 1914. Descriptions of new genera and species of Coleoptera. (Part iii.). *Bulletin of the New Zealand Institute* 1: 143–266.

Burakowski, B., M. Mroczkowski & J. Stefańska 1978. Chrzaszcze Coleoptera Histeroidea i Staphylinoidea procz Staphylinidae. *Catalogus faunae Poloniae. Cześć XXIII, tom 5*. Panstwowe Wydawnictwo Naukowe, Warszawa, 356 pp, 1 map.

Byk, A. 2001. Próba waloryzacji drzewostanów starszych klas wieku Puszczy Białowieskiej na podstawie struktury zgrupowań chrząszczy (Coleoptera) związanych z rozkładającym się drewnem pni martwych drzew stojących i dziupli. Pp. 333–367. In: A. Szujecki (ed.). *Próba szacunkowej waloryzacji lasów Puszczy Białowieskiej metodą zooindykacyjną*. Wydawnictwo SGGW, Warsaw.

Byk, A., T. Mokrzycki, S. Perliński & A. Rutkiewicz 2006. Saproxylic beetles – in the monitoring of anthropogenic transformations of Białowieża Primeval Forest. Pp. 325–397. In A. Szujecki (ed.). *Zooindication-based monitoring of anthropogenic transformations in Białowieża Primeval Forest*. Warsaw Agricultural University Press, Warsaw.

Campbell, J.M. 1991. Family Scaphidiidae shining fungus beetles, pp. 124–125. In: Y. Bousquet (ed.). *Checklist of beetles of Canada and Alaska*. Publication 1861/E. Agriculture Canada, Ottawa, vi + 430 pp.

Casey, T.L. 1893. Coleopterological Notices V. Scaphidiidae. *Annals of the New York Academy of Sciences* 7: 510–533.

Casey, T.L. 1900. Review of the American Corylophidae, Cryptophagidae, Tritomidae and Dermestidae, with other studies. *Journal of the New York Entomological Society* 8: 51–172.

Champion, G.C. 1913. Notes on various Central American Coleoptera, with descriptions of new genera and species. *Transactions of the Royal Entomological Society of London* (1913): 58–169, pls III, IV.

Champion, G.C. 1927. Some Indian Coleoptera (24). *The Entomologist's Monthly Magazine* 63: 267–279, pls V, VI.

Chandra, A. & V.R. Shivaramakrishnan 1986. A new record of *Scaphidium quadrimaculatum* Olivier (Scaphidiidae: Coleoptera) from Indian region with redescription. *Indian Forester* 1986: 512–516.

Chevrolat, L.A.A. 1844. [new taxon]. In: M F.E. Guérin Méneville: *Iconographie du règne animal de G. Cuvier, ou représentation d'après nature de l'une des espèces les plus remarquables et souvent non encore figurées, de chaque genre d'animaux. Avec un texte descriptif mis au courant de la science. Ouvrage pouvant servir d'atlas à tous les traités de zoologie. Insectes*. J.B. Baillière, Paris, 576 pp.

Chillo, D. 2012: 531-*Scaphidium quadrimaculatum* G.A. Olivier, 1790 (Coleoptera Staphylinidae Scaphidiinae). P. 45. In: Segnalazioni faunistiche italiane. *Bollettino de la Societá entomologica italiana* 144: 44–48.

Chûjô, M. 1961. Coleoptera from the Islands Tsushima being situated between Japan and Korea. *Mikado* 1(1): 1–16.

Cornell, J.F. 1967. A taxonomic study of *Eubaeocera* new genus (Coleoptera: Scaphididae) in North America north of Mexico. *The Coleopterists Bulletin* 21: 1–17.

Crowson, R.A. 1950. The classification of the families of British Coleoptera. Superfamily 3: Staphylinoidea. *The Entomologist's Monthly Magazine* 86: 274–288.

Csiki, E. 1904. Neue Käfernamen. *Wiener entomologische Zeitung* 23: 85.

Csiki, E. 1908. Catalogus Scaphidiidarum. *Rovartani lapok* 15: 151–174.

Csiki, E. 1909. Coleoptera nova in Museo nationali hungarica II. *Annales Musei nationalis Hungarici* 7: 340–343, pl. II.

Csiki, E. 1910. *Coleopterorum Catalogus. Pars 13: Scaphidiidae*. W. Junk, Berlin, 21 pp.

Csiki, E. 1924. Scaphidiidae formosanae novae. *Annales Musei nationalis hungarici* 21: 32.

Dajoz, R. 1965. Morphologie et biologie de la larve de *Scaphosoma assimile* Er. (Coléoptères, Scaphidiidae). *Bulletin de la Société linnéenne de Lyon* 34: 105–110.

Dauphin, P. 2004. Présence en Corse de *Scaphisoma flavonotatum* Pic (Coleoptera Staphylinidae Scaphidiinae). *Bulletin de la Société linnéenne de Bordeaux* 32(4): 259–260.

Dauphin, P. 2005. Sur les *Scaphisoma* de la faune de France (Coleoptera Staphylinidae Scaphidiinae). *Bulletin de la Société linnéenne de Bordeaux* 33: 98–108.

Deepthi, S., M. Suharban, D. Geetha & K. Sudharma 2004. Pests infecting oyster mushroom in Kerala and the seasonality of their occurrence. *Mushroom Research* 13(2): 76–81.

Delwaide, M. & Y. Thieren 2010. Liste de coléoptères observés dans l'ancienne carrière sablonneuse de Vance (Province de Luxemburg, Belgique). *Entomologie faunistique* 62 [2009]: 3–10.

Downie, N.M. & R.H. Arnett. 1996. *The beetles of Northeastern North America. Volume 1: Introduction; Suborders Archostemata, Adephaga, and Polyphaga, thru superfamily Cantharoidea*. Sandhill Crane Press, Gainesville, i-xv + 16–880 pp.

Duff, A.G. (ed.). 2012. *Checklist of beetles of British Isles. 2nd edition*. Pemberley Books, Iver, 171 pp.

Duftschmid, C.E. 1825. *Fauna Austriae oder Beschreibung der österreichischen Insekten für angehende Freunde der Entomologie. Dritter Theil*. Akademie Buchhandlung, Linz und Leipzig, 289 pp.

Dury, C. 1911. Some new beetles from North Carolina, with ecological notes (Coleop.). *Entomological News* 22: 273–275.

Elston, A.H. 1921. Australien Coleoptera. – Part II. *Transactions of the Royal Society of South Australia* 45: 143–168.

Emden, F.K. 1942: Larvae of British beetles. III. Keys to families. *The Entomologist's Monthly Magazine* 78: 206–226, 253–272.

Erichson, W.F. 1845. *Naturgeschichte der Insecten Deutschlands. Erste Abteilung. Coleoptera. Dritter Band. Lieferung 1*. Nicolaische Buchhandlung, Berlin, 320 pp.

Evans, A.V. 2014. *Beetles of Eastern North America*. Princeton and Oxfort: Princeton University Press, 560 pp.

Everts, E.J. 1903. *Coleoptera Neerlandica. De schildvleugeliga insecten (Coleoptera). Volum 1.* Martinus Nijhoff, S-Gravenhage, 676 + [1] pp.

Fabricius, J.C. 1775. *Systema Entomologiae, sistens insectorum classes, ordines, genera, species, adiectis synonymis, locis, descriptionibus, observationibus.* Libraria Kortii: Flensburgi et Lipsiae, [32] + 832 pp.

Fabricius, J.C. 1792. *Entomologia systematica emendata et aucta. Secundum classes, ordines, genera, species adjectis synonimis, locis, observationibus, descriptionibus. Tom. I. Pars II.* Christ. Gottl. Proft, Hafniae, 538 pp.

Fabricius, J.C. 1801. *Systema Eleutheratorum secundum ordines, genera, species: adiectis synonymis, locis, observationibus, descriptionibus. Tomus II.* Bibliopolii Academici Novi, Kiliae, 687 pp.

Fairmaire, L. 1886. Descriptions de coléoptères de l'intérieur de la Chine. *Annales de la Société entomologique de France* (6) 6: 303–356.

Fairmaire, L. 1897. Matériaux pour la faune coléoptérique de la région malgache. 4e note. *Annales de la Société entomologique de Belgique* 41: 363–406.

Fairmaire, L. 1898a. Matériaux pour la faune coléoptérique de la région malgache, 5e note. *Annales de la Société entomologique de Belgique* 42: 222–260.

Fairmaire, L. 1898b. Matériaux pour la faune coléoptérique de la région malgache, 6e note. *Annales de la Société entomologique de Belgique* 42: 390–439.

Fairmaire, L. 1898c. Matériaux pour la faune coléoptérique de la région malgache, 7e note. *Annales de la Société entomologique de Belgique* 42: 463–499.

Fairmaire, L. 1899. Matériaux pour la faune coléoptérique de la région malgache, 9e note. *Annales de la Société entomologique de France* 68: 466–507.

Fairmaire, L. & J.J.A. Laboulbène. 1855. Livraison 2, Pp. 181–370. In: *Faune entomologique française ou description des insectes qui se trouvent en France. Coléoptères. Tome premier.* [1854–1856]. Deyrolle, Paris, xxxv + 665 pp.

Fall, H.C. 1910. Miscellaneous notes and descriptions of North American Coleoptera. *Transactions of the American Entomological Society* 36: 89–197.

Fauvel, C.A.A. 1903. Faune analytique des coléoptères de la Nouvelle-Calédonie. *Revue d'Entomologie* 22: 203–378.

Fea, L. 1897. Viaggio di Leonardo Fea in Birmania e regioni vicini. LXXVI. Riassunto generale dei risulti zoologici. *Annali del Museo civico di Storia naturale di Genova* (2) 17: 385–658.

Fernández, J.M.D. 2014. Catálogo de los Coleoptera de la Sierra de Collserola (Barcelona, NE de España). *Arquivos Entomolóxicos* 10: 235–264.

Fierros-López, H.E. 1998. *Scaphidium mexicanum* Castelnau, 1840 (Coleoptera: Staphylinidae: Scaphidiinae). *Dugesiana* 5(2): 36–37.

Fierros-López, H.E. 2002. Descripción de dos especies nuevas de *Cyparium* Erichson, 1845 (Coleoptera: Staphylinidae) de México. *Dugesiana* 9(2): 7–14.

Fierros-López, H.E. 2005. Revisión del género *Scaphidium* Olivier, 1790 (Coleoptera, Staphylinidae) de México y Céntroamerica. *Dugesiana* 12: 1–152.

Fierros-López, H.E. 2006a. Four new species of *Scaphisoma* Leach with maculate elytra (Coleoptera: Staphylinidae: Scaphidiinae) from Mexico, with new records on *S. balteatum* Matthews. *Zootaxa* 1279: 53–68.

Fierros-López, H.E. 2006b. Datos nuevos de distributión de algunas especies de Scaphidiinae Neotropicales (Coleoptera: Staphylinidae). *Dugesiana* 13: 39–43.

Fierros-López, H.E. 2010. Description of two new species of *Baeocera* Erichson (Coleoptera: Staphylinidae: Scaphidiinae) from Mexico. *Journal of the Kansas Entomological Society* 83: 201–207.

Finkel, M., V. Chikatunov & E. Nevo 2002. *Coleoptera of "Evolution Canyon" II: Lower Nahal Keziv, Western Upper Galilee, Israel.* Pensoft, Sofia-Moscow, [4] +270 pp.

Fowler, C. 1889. *The Coleoptera of the British Islands. A descriptive account of the families, genera, and species indigenous to Great Britain and Ireland, with notes as to localities, habitats, etc. Vol. 3. Clavicornia (Leptinidae-Heteroceridae).*]., London, 399 pp., pls 71–98.

Franz, H. 1970. *Die Nordost-Alpen im Spiegel ihrer Landtierwelt. Eine Gebietsmonographie, umfassend: Fauna, Faunengeschichte, Lebensgemeinschaften und Beeinflussung der Tierwelt durch den Menschen. Band III. Coleoptera 1. Teil umfassend die Familien Cicindelidae bis Staphylinidae.* Universitätsverlag Wagner, Innsbruck-München, [4] + 1–501 pp.

Freude, H. 1971. 22. Familie: Scaphidiidae. Kahnkäfer. Pp. 343–347. In: H. Freude, K.W. Harde & G.A Lohse. *Die Käfer Mitteleuropas. Band 3 Adephaga 2 Palpicornia Histeroidea Staphylinoidea 1.* Goecke & Evers, Krefeld, 365 pp.

Frisch, J. 2017. On the distribution of *Scaphisoma obenbergeri* Löbl, 1963, with the first record from Hesse, Germany (Staphylinidae: Scaphidiinae). *Entomologische Zeitschrift* (Schwanfeld) 127(4): 245–249.

Fuente de la, M. 1924. Catalógo sistemático-geográfico de los coleópteros observados en la península ibérica, Pirineos propriamente dichos y Baleares (continuación). *Boletín de la Sociedad Entomológica de España* 7: 35–50.

Ganglbauer, L. 1899. *Die Käfer von Mitteleuropa. Die Käfer des österreichisch-ungarischen Monarchie, Deutschlands, der Schweiz, sowie des französischen und italienischen Alpengebietes. Dritter Band. Staphylinoidea, II Theil. Scydmaenidae, Silphidae, Clambidae, Leptinidae, Platypsyllidae, Corylophidae, Sphaeriidae, Trichopterygidae, Hydroscaphidae, Scaphidiidae, Histeridae. Familienreihe Clavicornia. Sphaeritidae, Ostomidae, Byturidae, Nitidulidae, Cucujidae, Erotylidae, Phalacridae, Thorictidae, Lathridiidae, Mycetophagidae, Colydiidae, Endomychidae, Coccinellidae.* Carl Gerold's Sohn, Wien, iii +1046 pp.

Garpebring, A. 2016. Smådjur i norrländska nationalparker. 8. Skalbaggar i Björnlandet. *Skörvnöpaarn* (Umeå) 8: 1–12.

Geiser, E. 1999. Neuentdeckte Käferarten im Bundesland Salzburg. *Mitteilungen der Gesellschaft für Salzburger Landeskunde* 139: 377–385.

Gemmelman, S.S. 1927. Spisok zhukov (Coleoptera) Pereslavskogo uezda Vlad. gub. *Trudy Pereslavl'-Zalesskogo istoriko-khudozhestvennogo i kraevedcheskogo muzeya. Pereslovl'* 4: 43–87.

Gemminger, M. & E. Harold 1868. *Catalogus Coleopterorum hucusque descriptorum synonymicus et systematicus. Tom. II. Dytiscidae, Gyrinidae, Hydrophilidae, Staphylinidae, Pselaphidae, Gnostidae, Paussidae, Scydmaenidae, Silphidae, Trichopterygidae, Scaphidiidae.* E.G. Gummi, Monachii, pp. 425–752 + [6].

Gestro, R. 1879a. Descrizioni di nuove specie di coleoptteri racolte nella regione Austro-Malese dal signor L.M. d'Albertis. *Annali del Museo civico di Storia naturale di Genova* 14: 552–565.

Gestro, R. 1879b. Note sopre alcuni coleotteri dell'Arcipelago Malese e specialmente delle isole della Sonda. *Annali del Museo civico di Storia naturale di Genova* [1879–1880] 15: 49–62.

Ghahari, H., S. Anlaş, H. Sakenin, H. Ostvan & M. Havaskary 2009. Biodivesity of rove beetles (Coleoptera: Staphylinoidea: Staphylinidae) from the Arasbaran biosphere reserve and vicinity, northwestern Iran. *Linzer biologische Beiträge* 41(2): 1949–1958.

Gmelin, J.F. 1790. *Caroli a Linné Systema Naturae per regna tria naturae, secundum classes, ordines, genera, species, cum characteribus, differentiis, synonymis, locis. Editio decima tertia, aucta, reformata. Tom. I. Pars IV.* Georg Emanuel Beer, Lipsiae, pp. 1517–2224.

Grebennikov, V.V., & A.F. Newton 2012. Detecting the basal dichotomies in the monophylum of carrion and rove beetles (Insecta: Coleoptera: Silphidae and Staphylinidae) with emphasis on the Oxyteline group of subfamilies. *Arthropod Systematics & Phylogeny* 70(3): 133–165.

Grosso-Silva, J.M. & P. Soares-Vieira 2011. The insects of the Gaia Biological Park (northern Portugal) (3rd note): Additions and new distributional data (Insecta: Coleoptera, Hemiptera, Hymenoptera, Mecoptera). *Arquivos Entomolóxicos* 5: 3–7.

Guéorguiev, B.V. & T. Ljubomirov 2009. Coleoptera and Hymenoptera (Insecta) from Bulgarian section of Maleshevska Planina Mountain: Study of an until recently unknown biodiversity. *Acta zoologica bulgarica* 61: 235–276.

Guérin-Méneville [Guèrin], F.E. 1834: Pl. 17, Fig. 14. In: *Iconographie du règne animal de G. Cuvier, ou représentation d'après nature de l'une des espèces les plus remarquables et souvent non encore figurées, de chaque genre d'animaux. Avec un texte descriptif mis au courant de la science. Ouvrage pouvant servir d'atlas à tous les traités de zoologie.* [1829–1837]. J.B. Baillière, Paris, 450 pls.

Guillebeau, F. 1893. Sur *Eledona agaricola* et *turcica*, et descriptions d'espèces algériennes. *Bulletin [bimensuell] de la Société entomologique de France* 1893: cccxxv–cccxxviii.

Gutfleisch, V. 1859. *Die Käfer Deutschlands. Nach des Verfassers Tode vervollständigt und herausgegeben von Dr. Fr. Chr. Bose.* Joh. Phil. Diehl, Darmstadt, xvi + 661 + [3] pp.

Gyllenhal, L. 1808. *Insecta Svecica Descripta. Classis I. Coleoptera sive Eleuterata. Tomus I.* F.J. Leverentz, Scaris, viii + [4] + 572 pp.

Handlirsch, A. 1906. *Die fossilen Insekten und die Phylogenie der rezenten Formen. Ein Handbuch für Paläontologen und Zoologen.* W. Engelmann, Leipzig. pp. i–vi + 1–640, pls. 1–36.

Hamet, A. & Z. Vancl 2016. Katalog brouků (Coleoptera) CHKO Broumovsko. Opravené a doplněné druhé vydání. Catalogue of beetles (Coleoptera) of the Broumovsko protected landscape area. Second completed and corrected edition. *Elateridarium* 10 (Supplementum): 1–137.

Hammond, P.M. & J.F. Lawrence. 1989. Appendix: Mycophagy in insects: a Summary. Pp. 275–324. In: N. Wilding, N.M. Collins, P.M. Hammond & J.F. Weber (eds). *Insect-fungus interactions, 14th Symposium of the Royal Entomological Society of London in collaboration with the British Mycological Society,* Academic Press, London, San Diego, xvi + 344 pp.

Hanley, R.S. 1996. Immature stages of *Scaphisoma castaneum* Motschulsky (Coleoptera: Staphylinidae: Scaphidiinae), with observation on natural history, fungal hosts and development. *Proceedings of the Entomological Society of Washington* 98: 36–43.

Hansen, M. 1996. Katalog over Danmarks biller. Catalogue of Coleoptera of Denmark. *Entomologiske Meddelelser* 64: 1–231 pp.

Hansen, M. 1997. Phylogeny and classification of the staphyliniform beetle families (Coleoptera). *Biologiske Skrifter, Kongelige Danske Videnskabernes Selskab* 48: 1–339.

Harrison, T. 2016. *Scaphisoma balcanicum* Tamanini, 1954 (Staphylinidae: Scaphidiinae) new to Britain. *The Coleopterist* 25(2): 63–65.

Harwood, P. 1918. *Scaphium immaculatum* Oliv. an additional genus and species to our list of British Coleoptera. *The Entomologist's Monthly Magazine* 54: 131–132.

Hatch, M.H. 1957. *The beetles of the Pacific Northwest. Part II: Staphyliniformia.* University of Washington Publications in Biology 16: ix + 384 pp, 37 pls.

Hawkeswood, T.J. 1989. New host records for adults of some fungus-feeding beetles (Coleoptera) from New South Wales and Queensland, Australia. *Naturalist Notes* 106(3): 93–95.

Hawkeswood, T.J. 1990. Neue Wirtsangaben für Imagines einiger pilzfressender Käfer (Coleoptera) von New South Wales und Queensland, Australien. *Entomologische Zeitschrift* 100(6): 93–112.

Hayashi, N. 1986. Larvae. Pp. 202–218, pls. 1–113. In: K. Morimota & N. Hayashi (eds). *The Coleoptera of Japan in color. Vol. 1.* Hoikusha Publishing Co., Osaka, vi + 323 pp., 113 pls. [in Japanese].

Hayashi, N., A. Fukuda & K. Kurosa 1959. Coleoptera. Pp. 392–545. In: T. Esaki, T. Yuasa, T. Ishii & T Motoki. (eds). *Illustrated insect larvae of Japan*. Hokuryakan Co., Tokyo, 712 + 50 pp.

He, Wen-Jia, Liang Tang & Li-Zhen Li. 2008a. Three new species of the genus *Scaphidium* Olivier (Coleoptera: Staphylinidae: Scaphidiinae) from China. *Entomological Review of Japan* 63: 103–108.

He, Wen-Jia, Liang Tang & Li-Zhen Li 2008b. Notes on the genus *Scaphidium* Olivier of China with description of a new species (Coleoptera: Staphylinidae: Scaphidiinae). *Entomological Review of Japan* 62: 177–182.

He, Wen-Jia, Liang Tang & Li-Zhen Li 2008c. A review of thew genus *Scaphidium* Oliver [sic] (Coleoptera, Staphylinidae, Scaphidiinae) from Tienmushan, East China. *Zootaxa* 1898: 55–62.

He, Wen-Jia, Liang Tang & Li-Zhen Li 2008d. New data on the genus *Ascaphium* Lewis (Coleoptera, Staphylinidae, Scaphidiinae) of China. *Zootaxa* 1923: 62–68.

He, Wen-Jia, Tang Liang & Li-Zhen Li 2009. A new species and a new record species of the genus *Scaphidium* Olivier (Coleoptera, Staphylinidae, Scaphidiinae) from China. *Acta Zootaxonomica Sinica* 34: 481–484.

Hebda, G. & K. Zając 2013. Nowe stanowisko *Caryoscapha limbata* (Erichson, 1845) (Coleoptera: Staphylinidae: Scaphidiinae) v Polsce. *Wiadomości entomologiczne* 32: 605

Heer, O. 1841. Fasciculus tertius et ultimus. Pp. 361–652. In: *Fauna coleopterorum Helvetica. Pars I.* [1838–1841]. Orellii, Fusslini et Sociorum, Turici, xii + 652 pp.

Heer, O. 1847. Die Insektenfauna der Tertiärgebilde von Oeningen und von Radoboj in Croatien. Erste Abtheilung: Käfer. *Neue Denkschriften der Allgemeinen schweizerischen Gesellschaft für die gesammte Naturwissenschaften* 8: 1–230, Tab. 8.

Heller, K.M. 1917. Scaphidiidae von den Philippinen. *Wiener entomologische Zeitung* 36: 41–50.

Herbst, J.F.W. 1793. *Natursystem aller bekannten in- und ausländischen Insekten, als eine Fortsetzung der von Büffonschen Naturgeschichte. Der Käfer. Fünfter Theil. Mit 16 illuminirten Kupfertafeln.* Joachim Pauli, Berlin, xvi + 392 pp., pls 44–59.

Heyden, L.F.J.D. von. 1880. Pp. 1–96. In: *Catalog der Coleopteren von Sibirien mit Einschluss derjenigen der Turanischen Länder, Turkestans und der chinesichen Grenzgebiete. Mit spezieller Angabe der einzelnen Fundorte in Sibirien und genauer Citierung der darauf bezüglichen einzelnen Arbeiten nach eigenem Vergleich, sowie mit besonderer Rücksicht auf die geographische Verbreitung der einzelnen Arten über die Grenzländer, namentlich Europa und Deutschland.* A.W. Schade, Berlin [, xxiv + 224 pp].

Heyden, L.F.J.D. von. 1893. *Catalog der Coleopteren von Sibirien mit Einschluss derjenigen des östlichen Caspi-Gebietes, von Turkmenien, Turkestan, Nord-Thibet und des Amur-Gebietes. Mit spezieller Angabe der einzelnen Fundorte und genauer Citierung der darauf bezüglichen Literatur. Nachtrag I.* A.W. Schade, Berlin, 217 pp.

Heyden, L.F.J.D. von. 1896. *Catalog der Coleopteren von Sibirien mit Einschluss derjenigen des östlichen Caspi-Gebietes, von Turkmenien, Turkestan, Nord-Thibet und des Amur-Gebietes. Mit spezieller Angabe der einzelnen Fundorte und genauer Citierung der darauf bezüglichen Literatur. Nachtrag II.* A.W. Schade, Berlin, 84 pp.

Heyden, C.H.G. & L.F.J.D. von. Heyden 1866. Käfer und Polypen aus der Braunkohle des Siebengebirges. *Palaeontographica* 15: 131–156, pls. 22–24.

Hisamatsu, S. 1977. Notes on Scaphidiidae in Hiroshima Prefecture. *Hiroshima Mushi-no-kai Kaiho* (16): 193–195 [in Japanese].

Holzer, E. 2016. Erstnachweise und Wiederfunde für die Käferfauna der Steiermark (XV) (Coleoptera). *Joannea Zoologie* 15: 59–75.

Holzschuh, C. 1977. Bemerkenswerte Käferfunde in Österreich II. *Koleopterologische Rundschau* 53: 27–69.

Horion, A. 1949. *Faunistik der mitteleuropäischen Käfer. Palpicornia-Staphylinoidea (ausser Staphylinidae). Band 2.* Vittorio Klostermann, Frankfurt am Main, xxiii + 388 pp.

Horn, G.H. 1894. The Coleoptera of Baja California. *Proceedings of the California Academy of Sciences. Second Series* 4: 302–449.

Horn, W., I. Kahle, G. Friese & R. Gaedike 1990. *Collectiones entomologicae. Ein Kompendium über den Verbleib entomologischer Sammlungen der Welt bis 1960. Teil II: L-Z.* Akademie der Landwirtschaften, Berlin, pp. 223–573.

Hoshina, H. 2001. Taxonomic notes on the genus *Scaphidium* (Coleoptera, Staphylinidae) from the Ryukyus, Japan. *Bulletin of the Institute of Environmental Science and Technology* 27: 99–106.

Hoshina, H. 2008a. New record of the genus *Scaphobaeocera* (Coleoptera, Staphylinidae, Scaphidiinae) from Yaeyama group, the Ryukyus, Japan, with description of a new species. *Japanese Journal of Systematic Entomology* 14: 141–144.

Hoshina, H. 2008b. New records of the genus *Scaphoxium* (Coleoptera: Staphylinidae: Scaphidiinae) from Yaeyama group, the Ryukyus, Japan, with description of a new species. *Entomological Review of Japan* 63: 57–61.

Hoshina, H. 2009. An additional new species of the genus *Scaphoxium* fauna (Coleoptera: Staphylinidae) of the Ryukyus, Japan. *Memoirs of the Faculty of Education and Regional Studies, University of Fukui. Series II: Natural Science* 59: 1–6.

Hoshina, H. 2010. Additional notes on Japanese Scaphidiinae. *Nejirebane* (Osaka) 127: 16–17 [in Japanese].

Hoshina, H. 2011. New record of *Scaphobaeocera dorsalis* (Coleoptera, Staphylinidae, Scaphidiinae) from Japan. *Elytra* N. S. 1(2): 196.

Hoshina, H. 2013. Scaphidiinae. Pp. 123–127. In: Y. Shibata, M. Maruyama, H. Hoshina, T. Kishimoto, S.-I. Naomi, S. Nomura, V. Puthz, T. Shimada, Y. Watanabe & S. Yamamoto (eds). Catalogue of Japanese Staphylinidae (Insecta: Coleoptera). *Bulletin of the Kyushu University Museum* 11: 69–218.

Hoshina, H. 2015. Additional note of the genus *Scaphidium* in Amami-Oshima Is., Ryukyus, Japan. *Sayabane*, New Series, (20): 46. [in Japanese].

Hoshina, H. & K.-J. Ahn 2005. New record of the genus and species *Pseudobironium ussuricum* Löbl (Coleoptera, Staphylinidae) from Korea. *Elytra* 33: 522.

Hoshina, H. & M. Maruyama 1999. An additional new species of the *Scaphidium* fauna (Coleoptera, Staphylinidae, Scaphidiinae) of the Ryukyus, Japan. *Elytra* 27: 479–484.

Hoshina, H. & K. Morimoto 1999. Description of three new species of the genus *Scaphidium* (Coleoptera, Staphylinidae, Scaphidiinae) from the Ryukyus, Japan. *Japanese Journal of Systematic Entomology* 5: 87–95.

Hoshina, H. & H. Sugaya 2003. New records of the genera *Scaphobaeocera* and *Scaphoxium* (Coleoptera: Staphylinidae) from the Ryukyus, Japan, with descriptions of two new species. *Entomological Review of Japan* 58: 35–41.

Hoshina, H., S.-J. Park & K.-J. Ahn 2009. Korean species of the genus *Scaphobaeocera* Csiki (Coleoptera: Staphylinidae: Scaphidiinae) in Korea. *Entomological Research* 39: 326–329.

Hoshina, H., S.-J. Park & K.-J. Ahn 2011. The Korean species of the genus *Baeocera* (Coleoptera: Staphylinidae: Scaphidiinae). *Journal of Entomological Sciences* 46: 46–51.

Hudson, G.V. 1934. *New Zealand beetles and their laevae*. Ferguson & Osborn, Wellington, 236 pp.

Hwang, W.-S. & K.-J. Ahn 2001. New records of *Cyparium* Erichson and *Scaphidium* Olivier species in Korea (Coleoptera, Staphylinidae, Scaphidiinae). *Insecta Koreana* 18: 369–372.

Iablokoff-Khnzorian, S.M. 1985. Zhuki-tchelnovidki (Coleoptera, Scaphidiidae) fauny SSSR. *Entomologicheskoe Obozrenie* 64: 132–143.

ICZN, 1999. *International Code of Zoological Nomenclature. Fourth Edition adopted by the International Union of Biological Sciences*. International Trust of Zoological Nomenclature, London, xxix + 306 pp.

Ihssen, G. 1935. Beiträge zur Kenntnis der Fauna von Südbayern. (3). *Entomologische Blätter* 31: 42–48.

Israelson, G. 1971. Notes on some North-European Coleoptera. *Entomologisk Tidskrift* 92: 66–73.

Jacquelin du Val, P.N.C. 1858. Famille des Scaphidiides, pp 121–123, pl. 34. In: *Manuel entomologique. Genera des coléoptères d'Europe, comprenant leur classification en familles naturelles, la description de tous les genres, des Tableaux synoptiques destinés à faciliter l'étude, le Cataloque de toutes les espèces, de nombreux dessins au trait de caractères. Tome deuxième*. [1857–1859]. A. Deyrolle, Paris, 286 + 53–124 pp., pls 67.

Jakobson, G. 1910. Pars 8, pp. 561–640, pls. 62–68, 70. In: *Zhuki Rossii i zapadnoy Evropy*. A.F. Devrjena, St. Petersburg, 1024 pp., 83 pls.

Jelínek, J. 1993. Check-list of Czechoslovak Insects IV (Coleoptera). Seznam československých brouků. *Folia Heyrovskyana. Supplementum* 1, Jaroslav Picka, Praha, 172 pp.

Jeremías, X. & J.J. Pérez De-Grerogio 2003. Coleópteros raros e interesantes de la fauna de Cataluña (Scaphidiidae, Lucanidae, Ochodaeidae, Malachiidae, Pyrochroidae, Buprestidae, Anthribidae). Nueva localidades y protección de sus microhábitats. *Sessió d'Entomològia ICHN-SCL* 12[2001]: 55–62.

Johnson, W.F. & J.N. Halbert 1902. A list of the beetles of Ireland. *Proceedings of the Royal Irish Academy* (3) 6(4): 535–827 + [2] pp.

Joy, N.H. *A practical gandbook of British beetles. Volume 1.* H.F. & G. Witherly, London, xxvii + 622 pp.

Kahlen, M. 1987. *Nachtrag zur Käferfauna Tirols. Ergänzung zu den bisher erschienenen faunistischen Arbeiten über die Käfer Nordtirols (1950, 1971 und 1976) und Südtirols (1977).* Tiroler Landesmuseum Ferdinandeum, Innsbruck, 288 pp.

Kapp, A. 2001. *Die Käfer des Hochschwalbgebietes und ihre Verbreitung in der Steirmark.* Erster Vorarlberger Coleopterologischer Verein, Bürs, [2] + 628 pp.

Kasule, F.K. 1966. The subfamilies of the larvae of Staphylinidae (Coleoptera) with keys to the larvae of British genera of Steninae and Proteininae. *Transactions of the Royal Entomological Society of London* 118: 261–283.

Kasule, F.K. 1968. The larval characters of some subfamilies of British Staphylinidae (Coleoptera) with keys to the known genera. *Transactions of the Royal Entomological Society of London* 120: 115–138.

Kimura, F. 1987. A new species of Scaphidiidae from Taiwan (Coleoptera). *The Entomological Review of Japan* 42 (Supplement): 9–11.

Kimura, F. 2008. Two new species of the genus *Scaphidium* (Coleoptera, Staphylinidae) from Japan. *Special Publication of the Japan Coleopterological Society* 2: 157–164.

Kimura, F. 2009. A new species of the genus *Ascaphium* (Coleoptera, Staphylinidae, Scaphidiinae) from Taiwan. *Japanese Journal of Systematic Entomology* 15: 227–229.

Kirby, W. 1837. *Fauna Boreali-Americana; or the zoology of the northern parts of British America: containing descriptions of the objects of natural history collected on the late northern land expeditions, under command of captain Sir John Franklin, R.N. by John Richardson, assisted by William Swainson, and the Reverend William Kirby. Illustrated by several coloured engravings. Published under the authority of the Right Honourable the Secretary of State for colonial affairs.* R.N.J. Fletcher, Norwich, xxxix + 325 pp, 8 pls.

Kirk, V.M. 1969. A list of beetles of South Carolina. *Technical Bulletin* (South Carolina Agricultural Experimental Station) 1033: 1–124.

Kirk, V.M. 1970. A list of beetles of South Carolina. Part 2 – Mountain, Piedmont, and southern Coastal Plain. *Technical Bulletin* (South Carolina Agricultural Experimental Station) 1038: 1–117.

Kirsch, T. 1873. Beiträge zur Kenntniss der peruanischen Käferfauna auf Dr. Abendroth's Sammlungen basirt. *Berliner entomologische Zeitschrift* 17: 121–152.

Klimaszewski, J. & S.B. Peck 1987. Succession and phenology of beetle faunas (Coleoptera) in the fungus *Polyporellus squamosus* (Huds.: Fr.) Karst. (Polyporaceae) in Silesia, Poland. *Canadian Journal of Zoology* 65: 542–550.

Koch, K. 1968. Käferfauna der Rheinprovinz. *Decheniana -Beihefte* 13: viii + 1–376 pp, 1 pl.

Koch, K. & W. Lucht 1962. Käferfauna des Siebengebirges und des Rodderbergs. *Decheniana -Beihefte* 10: 1–181, 4 pls.

Kocher, L. 1958. Catalogue commenté des coléoptères du Maroc. Fascicule II. Hydrocanthares Palpicornes Brachelytres. *Travaux de l'Institute Scientifique Chérifien, Série zoologique* 14: 1–244 + [2] pp.

Kofler, A. 1968. Die Arten der Gattung *Scaphisoma* Leach aus Nord- und Osttirol (Coleoptera, Scaphidiidae). *Zeitschrift der Arbeitsgemeinschafft österreichischer Entomologen* 20: 39–43.

Kofler, A. 1970. Die Arten der Gattung *Scaphisoma* Leach aus dem Landesmuseum Joanneum. *Mitteilungen der Abteilung für Zoologie und Botanik am Landesmuseum "Joanneum" in Graz* 35: 55–60.

Kofler, A. 1998. Xylobionte Porlinge aus Osttirol und ihre Insekten (Polyporaceae, Dermaptera, Heteroptera, Coleoptera, Hymenoptera, Lepidoptera, Diptera). *Stapfia* 55: 641–661.

Köhler, F. 2009. Die Totholzkäfer (Coleoptera) des Naturwaldreservates «Laangmuer». Untersuchungszeitraum 2007–2008. Pp. 48–115. In: D. Murat (ed.). *Naturwaldreservate in Luxemburg. Band 5. Zoologische und botanische Untersuchungen «Laangmuer» 2007–2008*. Naturverwaltung, Luxemburg, 227 pp.

Kolbe, H. 1897. *Coleopteren. Die Käfer Deutsch-Ost-Afrikas*. D. Reimer (E. Vohsen), Berlin. 368 pp, 4 pls.

Kompantsev, A.V. & V.A. Pototskaya 1987. Novye dannye po lichikam zhukov-chelnovidok (Coleoptera, Scaphidiidae). Pp. 87–100. In: *Ekologiya i morfologiya nasekomychobyvateley gribnych substratov*. Nauka, Moskva, 120 pp.

Konvička, O., P. Boža & M. Mantič 2009. Coleoptera. Faunistic records from the Czech Republic – 271. *Klapalekiana* 45(8): 8.

Kraatz, G. 1895. Scaphidiidae aus Togo. *Deutsche entomologische Zeitschrift* 1895: 154.

Krasutskij, B.V. 1996a. *Mycetofilnye zhestkokrylie Urala i Zauralya. Tom 1.* Izdatelstvo Ekaterinburg, Ekaterinburg, 148 pp.

Krasutskij, B.V. 1996b. Zhestkokrylye-micetobionty (Coleoptera) osnovnykh derevorazrushayushchikh gribov lesostepnogo zauralya. *Entomologicheskoe Obozrenie* 75(2): 274–277.

Krasutskij, B.V. 1997a. Zhestkokrylye (Coleoptera) mycetobionty osnovnykh derevorazrushayushchikh gribov yuzhnoy podzony zapadnosibirskoy tajgi. *Entomologicheskoe Obozrenie* 76(2): 302–308.

Krasutskij, B.V. 1997b. Zhestkokrylye-micetobionty (Coleoptera) osnovnykh derevorazrushayushchikh gribov podzony sredney taygi zapadnoy Sibiri. *Entomologicheskoe Obozrenie* 76(4): 770–775.

Krasutskij, B.V. 2000. Soobshchestva zhestkokrylikh, svayzannye s osnovnymi derevorazrushayushchikh gribami chelyabinskoy oblasti. *Trudy Instituta bioresursov i prikladnoy ekologii* 1: 76–89.

Krasutskij, B.V. 2005. *Mycetofilnye zhestkokrylie Urala i Zauralya. Tom 2. Sistema «Gryby-nasekomye»*. Russkoe Entomologicheskoe Obshchetvo, Chelyabinsk, 212 pp.

Krasutskij, B.V. 2010. Zhestkokrylye (Coleoptera), svyazanie s trutovikom *Trichaptum biforme* (Fr. in Klotzsch) (Basidiomycetes, Aphyllophorales) v lesakh Urala i Zaurala. *Entomologicheskoe Obozrenie* 89(2): 367–379.

Krivosheyev, R.E. 2009. New records of xylomycetobiotic beetles (Insecta, Coleoptera) from Kyiv region (Ukraine). *Vestnik Zoologii* 43(6): e-13–e17.

Kuhnt, P. 1912. *Illustrierte Bestimmungs-Tabellen der Käfer Deutschlands. Handbuch zum genauen und leichten Bestimmen aller in Deutschland vorkommende Käfer. Mit über 10000, alle wichtigen Bestimmungsmerkmale illustrierten Text-Abbildungen.* E. Schweizerbart'sche Verlagsbuchhandlung, Stuttgart, vii + [2] + 1138 pp.

Kuschel, G. 1990. Beetles in a suburban environment: a New Zealand case study. The identity and status of Coleoptera in the natural and modified habitats of Lynfied, Auckland (1974–1989). *DSIR Plant Protenction Report* No. 3, Auckland, 118 pp.

Kuthy, D. 1897. Ordo. Coleoptera, 214 pp, 1 map. In: J. Paszlavszky (ed.). *A Magyar Birodalom állatvilága. A Magyar Birodalomból eddig ismert állatok rendszeres lajstroma. III. Arthropoda. (Insecta. Coleoptera)*. Királyi Magyar Természettudományi Társulat, Budapest.

Lacordaire, T. 1854. *Histoire Naturelle des Insectes. Genera des Coléoptères ou exposé méthodique et critique de tous les genres proposés jusqu'ici dans cet ordre d'insectes. Tome deuxième contenant les familles des Paussides, Staphyliniens, Psélaphiens, Scydménides, Silphales, Sphériens, Trichoptérygiens, Scaphidiles, Histériens, Phalacrides, Nitidulaires, Trogisitaires, Colydiens, Rhysodides, Cucujipes, Cryptophagides, Lathridiens, Mycétophagides, Thorictides, Dermestins, Byrrhiens, Géoryssins, Parnides, Hétérocérides*. Librairie Encyclopédique de Roret, Paris, 548 pp.

Lafer, G.Sh. 1989. 25.Sem. Scaphidiidae – Tchelnovidki. In: P.A. Ler (ed.). *Opredelitel nasekomych dalnego vostoka SSSR v shesti tomakh. Tom III. Zhestkokrylye ili Zhuki, Tchast I*. Nauka, Leningrad, 572 + [1] pp.

Laporte, F.L.N. de Caumont (Comte de Castelnau) 1840. *Histoire naturelle des insectes coléoptères; avec une introduction renfermant l'anatomie et la physiologie des animaux articulés, par M. Brullé; ouvrage accompagné de* 155 planches gravées sur acier représentant plus de 800 sujets. *Tome deuxième*. P. Duménil, Paris, 563 + [1] pp., 38 pls.

Latreille, P.A. 1804. *Histoire naturelle, générale et particulière, des crustacés et des insectes. Ouvrage faisant suite aux oeuvres de Leclerc de Buffon, et partie du cours complet d'histoire naturelle rédigé par C.S. Sonnini, membre de plusieurs sociétés savantes. Tome neuvième.* F. Dufart, Paris, 416 pp, pls 67–73.

Latreille, P.A. 1806. *Genera Crustaceorum et Insectorum, secundem ordinem naturalem in familias dispositas, iconibus exemplisque plurimus explicata. Tomus secundus.* [1807]. Amand Koenig, Paris, 280 pp.

Latreille, P.A. 1810. *Considérations générales sur l'ordre naturel des animaux composant les classes des crustacés, des arachnides, et des insectes; avec un tableau méthodique de leurs genres, disposés en familles.* Schoell, Paris, 444 pp.

Latreille, P.A. 1829. *Les crustacés, arachnides et les insectes distribués en familles naturelles, ouvrage formant les tomes 4 et 5 de celui de M. le Baron Cuvier sur le règne animal (deuxième édition). Tome premier.* Déterville, Paris, xxvii + 584 pp.

Lawrence, J.F. 1989. Mycophagy in the Coleoptera: Feeding strategies and morphological adaptations. Pp. 1–23. In: N. Wilding, N.M. Collins, P.M. Hammond & J.F. Webber (eds). *Insect-fungus interactions. 14th Symposium of the Royal Entomological Society of London in collaboration with the British Mycological Society.* Academy Press, London, etc., xvi + 344 pp.

Lawrence, J.F. & A.F. Newton Jr. 1980. Coleoptera associated with the fruiting bodies of slime molds (Myxomycetes). *The Coleopterists Bulletin* 34: 129–143.

Lawrence, J.F. & A.F. Newton Jr. 1982. Evolution and classification of beetles. *Annual Review of Ecology and Systematics* 13: 261–290.

Lawrence, J.F. & A.F. Newton 1995. Families and subfamilies of Coleoptera (with selected genera, notes, references and data on family-group names). Pp. 779–1007 + [1]-[48]. In: J. Pakaluk & S.A. Słipiński (eds). *Biology, phylogeny, and classification of Coleoptera. Papers celebrating the 80th birthday of Roy A. Crowson. Volume two.* Muzeum i Instytut Zoologii PAN, Warszawa, vi + 559–1092 pp.

Lawrence, J.F. & S.A. Słipiński 2013. *Australian Beetles. Volume 1. Morphology, classification and keys.* CSIRO, Collingwood, 576 pp.

Lea, A.M. 1926. On some Australian Coleoptera collected by Charles Darwin during the voyage of the "Beagle". *Transactions of the Royal entomological Society of London* 74: 279–288.

Leach, W.E. 1815. Entomology. In Brewster, D. (ed.). *Edingburg Encyclopaedia* 9: 57–172.

LeConte, J.L. 1860. Synopsis of the Scaphidiidae of the United States. *Proceedings of the Academy of Natural Sciences of Philadelphia* (1860): 321–324.

LeConte, J.L. 1869. Synonymical notes on Coleoptera of the United States, with description of new species, from the MSS. of late Dr. C. Zimmermann. *Transactions of the American entomological Society* 2: 243–259.

LeConte, J.L. & G.H. Horn 1883. Classification of the Coleoptera of North America. 2. edit. *Smithonian Miscellaneous Collections* Ser. No. 507: xxxviii + 567 pp.

Lecoq, J.-C. 2015. Staphylinidae. Pp. 169–232. In: Y. Gomy, R. Lemagnen & J. Poussereau (eds). *Les coléoptères de l'île de La Réunion.* Editions Orphie, Saint-Denis, [1–5] + 6–759 pp.

Leng, C.W. 1920. *Catalogue of the Coleoptera of America, North of Mexico*. John D. Sherman, Mont Vernon, x + 470 pp.

Leng, W.C. & A.J. Mutchler 1927. *Supplement 1919 to 1924 (inclusive) to Catalogue of the Coleoptera of America, North of Mexico*. John D. Sherman, Jr., Mount Vernon, 78 pp.

Leonard, M.D. 1928. A list of the insects of New York with a list of the spiders and certain other allied groups. *Cornell University, Agricultural Experimental Station, Memoir* 101: 5–1001.

Leschen, R.A.B. 1988a. The natural history and immatures of *Scaphisoma impunctatum* (Coleoptera: Scaphidiidae). *Entomological News* 99: 225–231.

Leschen, R.A.B. 1988b. *Coleoptera/Basidiomycete relationships: Arkansas fauna*. Master Thesis. University of Fayetteville, Fayetteville.

Leschen, R.A.B. 1993. Evolutionary patterns of feeding in selected Staphylinoidea (Coleoptera): Shifts among food texture. In: C.W. Schaefer & R.A.B. Leschen (eds). *Functional morphology of insect feeding*. Thomas Say Publications in Entomology: Proceedings. Entomological Society of America, Lanham.

Leschen, R.A.B. 1994. Retreat-building by larval Scaphidiinae (Staphylinidae). *Mola* 4: 3–5.

Leschen, R.A.B. & I. Löbl 1995. Phylogeny of Scaphidiinae with redefinition of tribal and generic limits (Coleoptera: Staphylinidae). *Revue suisse de zoologie* 102: 425–474.

Leschen, R.A.B. & I. Löbl 2005. Phylogeny and classification of Scaphisomatini Staphylinidae: Scaphidiinae with notes on mycophagy, termitophily, and functional morphology. *Coleopterists Society Monographs* 3: 1–63.

Leschen, R.A.B., T.R. Buckley, H.M. Harman & J. Shulmeister 2008. Determining the origin and age of the Westland beech (*Notophagus*) gap, New Zealand, using fungus beetle genetics. *Molecular Ecology* Doi: 10:111/j.1365-294X2007.03630.x.

Leschen, R.A.B., I. Löbl & K. Stephen 1990. Review of the Ozark Highland *Scaphisoma* (Coleoptera: Scaphidiidae). *The Coleopterists Bulletin* 44: 274–294.

Levey, B. 2010. Some unpublished records of *Scaphisoma assimile* Erichson (Staphylinidae). *The Coleopterist* 19(3): 117–118.

Lewis, G. 1879. On certain new species of Coleoptera from Japan. *Annals and Magazine of Natural History* (5) 4: 459–467.

Lewis, G. 1893. On some Japanese Scaphidiidae. *Annals and Magazine of Natural History* (6) 11: 288–294.

Li, J. 1992. *The Coleoptera Fauna of Northeast China*. Jilin Education Publishing House, 205 pp.

Li, J. 2015. Family Scaphidiidae distribution from Heilongjiang province, China (Coleoptera). Pp. 37–40. In: *Coleopterorum study of China* (1). Author, 48 pp. [in Chinese and English].

Li, J., L. Zhang & X. Zhang 2015. *Color illustrations of soil insects from northeast China (Staphylinoidea, Tenebrionoidea)*. Harbin Map Publishing House, Harbin, 199 pp [in Chinese].

Linnaeus, C. 1758. *Systema Naturae per regna tria naturae secundum classes, ordines, genera, species, cum characteribus, differentiis, synonymis, locis. tomus I. Editio decima, reformata.* Laurentii Salvii, Holmiae, [5] + 6–824 pp.

Linnaeus, C. 1760. *Fauna Svecica sistens animalia sveciae regni: Mammalia, Aves, Amphibia, Pisces, Insecta, Vermes. Distributa per classes & ordines, genera & species, cum differentiis specierum, synonymis auctorum, nominibus incolarum, locis natalium, descriptionibus insectorum. Editio altera, auctior.* [1761]. Laurentii Salvii, Stockholmiae, [48] + 578.

Löbl, I. 1963a. *Scaphosoma assimile* Erichson, selection of a lectotype, and its relation to *Sc. curvistria* Reitter (Col., Scaphidiidae). *Biologia. Bratislava* 18(9): 704–705.

Löbl, I. 1963b. Eine neue Art der Gattung *Scaphosoma* Leach (Coleoptera, Scaphidiidae). *Reichenbachia* 1: 273–275.

Löbl, I. 1964a. *Scaphosoma balcanicum* Tam., eine neue Art der Gattung für die Fauna Mitteleuropas. *Acta entomologica bohemoslovaca* 61: 71.

Löbl, I. 1964b. *Scaphosoma corcyricum* sp.n., ein neuer Scaphidiide (Col.) aus Süd-Europa. *Annotationes zoologicae et botanicae. Bratislava* 1: 1–4.

Löbl, I. 1964c. Eine neue mittelasiatische Art der Gattung *Scaphosoma* Leach (Col.). *Annalen des Naturhistorischen Museums in Wien* 67: 487–488.

Löbl, I. 1964d. Nachträge zut geographischen Verbreitung einiger *Scaphosoma*-Arten (Col., Scaphidiidae). 4. Beitrag zur Kenntnis der Gattung *Scaphosoma* Leach. *Acta rerum naturalium Musei nationalis slovenici* 10: 49–50.

Löbl, I. 1965a. Beitrag zur Kenntnis der japanischen Arten der Gattung *Scaphosoma* Leach (Scaphidiidae). *Entomologische Blätter für Biologie und Systematik der Käfer* 61: 44–58.

Löbl, I. 1965b. Eine neue Art der Gattung *Caryoscapha* Ganglbauer aus Japan (Col.). *Annotationes zoologicae et botanicae. Bratislava* 17: 1–3.

Löbl, I. 1965c. *Scaphidium bodemeyeri* Reitter, species propria (Col., Scaphidiidae). *Annotationes zoologicae et botanicae. Bratislava* 19: 1–3.

Löbl, I. 1965d. Bemerkungen zu einigen westpalaearktischen Scaphidiiden (Coleoptera). *Acta entomologica bohemoslovaca* 62: 334–339.

Löbl, I. 1965e. Beitrag zur Kenntnis des *Scaphosoma*-Arten Chinas (Coleoptera, Scaphidiidae). *Reichenbachia* 6: 25–31.

Löbl, I. 1965f. A new species of *Scaphosoma* Leach (Coleoptera, Scaphidiidae) from Afghanistan. *Annales historico-naturales Musei nationalis Hungarici* 57: 267–268.

Löbl, I. 1965g. Ergebnisse der Albanien-Expedition 1961 des Deutschen Entomologischen Institutes. 40. Beitrag Coleoptera: Scaphidiidae. *Beiträge zur Entomologie* 15: 731–734.

Löbl, I. 1965h. Zwei neue Arten der Gattung *Scaphosoma* Leach (Col., Scaphidiidae). *Annotationes zoologicae et botanicae. Bratislava* 23: 1–4.

Löbl, I. 1966a. *Baeocera myrmidon* (Achard, 1923) comb.n. (Col., Scaphidiidae). *Annotationes zoologicae et botanicae. Bratislava* 31: 1–3.

Löbl, I. 1966b. Neue und interessante paläarktische Scaphidiidae aus dem Museum G. Frey (Col.). *Entomologische Arbeiten aus dem Museum Georg Frey* 17: 129–134.

Löbl, I. 1967a. Beitrag zur Kenntnis der neotropischen Arten der Gattung *Baeocera* Er. *Opuscula zoologica* 97: 1–3.

Löbl, I. 1967b. Neue und wenig bekannte paläarktische Arten der Gattung *Scaphosoma* Leach (Col., Scaphidiidae). *Acta entomologica bohemoslovovaca* 64: 105–111.

Löbl, I. 1967c. Neue und wenig bekannte Scaphidiiden aus Japan (Coleoptera). *Reichenbachia* 8: 129–132.

Löbl, I. 1967d. Über die europäischen Arten der *Scaphosoma agaricinum*-Gruppe (Col., Scaphidiidae). *Norsk entomologisk Tidsskrift* 14: 33–36.

Löbl, I. 1967e. *Scaphosoma balcanicum* Tamanini, 1954 (Col. Scaphidiidae) in Sweden. *Opuscula entomologica* 32: 8.

Löbl, I. 1968a. Beitrag zur Kenntnis der japanischen Arten der Gattung *Eubaeocera* Cornell (Col., Scaphidiidae). *Annotationes zoologicae et botanicae. Bratislava* 46: 12–2.

Löbl, I. 1968b. Beitrag zur Kenntnis der Scaphidiidae und Pselaphidae von Korea (Coleoptera). *Annales Zoologici* 25: 419–423.

Löbl, I. 1968c. Description of *Scaphidium comes* sp.n. and notes on some other Palaearctic species of the genus *Scaphidium* (Coleoptera, Scaphidiidae). *Acta entomologica bohemoslovaca* 65: 386–390.

Löbl, I. 1969a. Contribution à la connaissance des Scaphisomini de la Nouvelle-Calédonie. *Bulletin de l'Institut royal des Sciences naturalles de Belgique* 45: 1–4.

Löbl, I. 1969b. Revision der paläarktischen Arten der Gattungen *Pseudobironium* Pic, *Scaphischema* Reitter und *Eubaeoceera* Cornell der Tribus Scaphisomini (Col. Scaphidiidae). *Mitteilungen der Schweizerischen entomologischen Gesellschaft* 42: 321–343.

Löbl, I. 1969c. Revision der paläarktischen Arten der Tribus Toxidiini (Col. Scaphidiidae). *Mitteilungen der Schweizerischen entomologischen Gesellschaft* 42: 344–350.

Löbl, I. 1970a. Scaphidiidae. In: *Klucze do oznaczania owadow polski, XIX. Chrzaszcze – Coleoptera*. Zeszyt 23: 1–16. Panstwowe Wydawnictwo Naukowe, Warszawa.

Löbl, I. 1970b. Über einige Scaphidiidae (Coleoptera) aus der Sammlung des Muséum National d'Histoire Naturelle de Paris. *Mitteilungen der schweizerischen entomologischen Gesellschaft* 43: 125–132.

Löbl, I. 1970c. Revision der paläarktischen Arten der Gattungen *Scaphisoma* Leach und *Caryoscapha* Ganglbauer der Tribus Scaphisomini (Col. Scaphidiidae). *Revue suisse de zoologie* 77: 727–799.

Löbl, I. 1971a. Scaphidiidae der Noona Dan Expedition nach den Philippinen und Bismark Inseln (Insecta, Coleoptera). *Steenstrupia* 1: 247–253.

Löbl, I. 1971b. Zwei neue Arten der Gattung *Scaphidium* Ol. von Süd-Indien (Coleoptera, Scaphidiidae). *Opuscula zoologica* 118: 1–3.

Löbl, I. 1971c. Scaphidiidae von Ceylon (Coleoptera). *Revue suisse de zoologie* 78: 937–1006.

Löbl, I. 1972a. Beitrag zur Kenntnis der Scaphidiidae (Coleoptera) von China und Japan. *Notulae entomologicae* 52: 115–118.

Löbl, I. 1972b. Beitrag zur Kenntnis der Scaphidiidae (Coleoptera) von den Philippinen. *Mitteilungen der Schweizerischen entomologischen Gesellschaft* 45: 79–109.

Löbl, I. 1973a. Über einige orientalische Scaphidiidae (Coleoptera) aus dem Museo Civico di Storia Naturale di Genova und Muséum National d'Histoire Naturelle de Paris. *Nouvelle Revue d'Entomologie* 3: 149–160.

Löbl, I. 1973b. Scaphidiidae (Coleoptera) von Neu Kaledonien. *Archives des sciences* 25 [1972]: 309–334.

Löbl, I. 1973c. Neue orientalische Arten der Gattung *Eubaeocera* Cornell (Coleoptera, Scaphidiidae). *Mitteilungen der Schweizerischen entomologischen Gesellschaft* 46: 157–174.

Löbl, I. 1973d. Über die Identität von *Scaphisoma madeccasa* Brancsik (Coleoptera Scaphidiidae). *Archives des sciences* 26: 19–22.

Löbl, I. 1974a. Bemerkenswerte Funde einiger Scaphidiiden-Arten. Kurze Mitteilungen 1914. *Entomologische Blätter für Biologie und Systematik der Käfer* 70: 61.

Löbl, I. 1974b. Une nouvelle espèce du genre *Scaphisoma* Leach de la Nouvelle Calédonie (Coleoptera, Scaphidiidae). *Revue suisse de zoologie* 81: 405–407.

Löbl, I. 1974c. Contribution à la connaissance du genre *Pseudobironiella* Löbl (Coleoptera, Scaphidiidae). *Mitteilungen der Schweizerischen entomologischen Gesellschaft* 47: 315–317.

Löbl, I. 1974d. New species of the genus *Amalocera* Erichson from Brazil (Coleoptera, Scaphidiidae). *Studies on the Neotropical Fauna* 9: 39–45.

Löbl, I. 1975a. Beitrag zur Kenntnis der Scaphidiidae (Coleoptera) von Neuguinea. *Revue suisse de Zoologie* 82: 369–420.

Löbl, I. 1975b. Beitrag zur Kenntnis der orientalischen Scaphisomini (Coleoptera, Scaphidiidae). *Mitteilungen der Schweizerischen entomologischen Gesellschaft* 48: 269–290.

Löbl, I. 1976a. Drei neue Arten der Gattung *Scaphisoma* Leach (Coleoptera, Scaphidiidae) aus Indonesien. *Entomologische Berichten* 36: 8–11.

Löbl, I. 1976b. Eine neue bemerkenswerte Art der Gattung *Scaphidium* Olivier (Coleoptera, Scaphidiidae) von Neuguinea. *Zoologische mededeelingen* 49: 317–320.

Löbl, I. 1976c. New species of the genus *Sciatrophes* Blackburn from Arizona (Coleoptera: Scaphidiidae). *The Coleopterists Bulletin* 30: 207–211.

Löbl, I. 1976d. Sur l'identité du *Baeocera falsata* Achard (Coleoptera, Scaphidiidae). *Revue suisse de zoologie* 83: 777–778.

Löbl, I. 1976e. The Australian species of *Scaphidium* Olivier (Coleptera: Scaphidiidae). *Journal of the Australian Entomological Society* 15: 285–295.

Löbl, I. 1977a. Ergebnisse der Bhutan-Expedition 1972 des Naturhistorischen Museums in Basel. Coleoptera: Fam. Scaphidiidae Genus *Baeocera* Er. unter Berücksichtigung einiger Arten aus benachbarten Gebieten. *Entomologica basiliensia* 2: 251–258.

Löbl, I. 1977b. *Baeocera galapagoensis* nov. spec. a new scaphidiid beetle from the Galapagos Islands (Coleoptera, Scaphidiidae). *Studies on Neotropical Fauna and Environment* 12: 249–252.

Löbl, I. 1977c. Beitrag zur Kenntnis der vietnamesischen Arten der Gattung *Scaphisoma* Leach (Coleoptera, Scaphidiidae). *Archives des sciences* 29 [1976]: 221–226.

Löbl, I. 1977d. Beitrag zur Kenntnis der Gattung *Bironium* Csiki (Coleoptera, Scaphidiidae). *Mitteilungen der Schweizerischen entomologischen Gesellschaft* 50: 59–61.

Löbl, I. 1977e. Beitrag zur Kenntnis der Scaphidiidae (Coleoptera) Australiens. *Revue suisse de zoologie* 84: 3–69.

Löbl, I. 1977f. Les Scaphidiidae (Coleoptera) de l'île de la Réunion. *Nouvelle Revue d'entomologie* 7: 39–52.

Löbl, I. 1977g. Contribution to the knowledge of *Scaphisoma* Leach (Coleoptera, Scaphidiidae) from the Ryukyu Islands. *Bulletin of the National Sciences Museum, Tokyo, Ser. A (Zoology)* 3: 163–165.

Löbl, I. 1977h. *Baeocera* Erichson, 1845 (Coleoptera, Scaphidiidae) Request for the designation of type species in harmony with the intention of its author. Z. N. (S) 2194. *Bulletin of Zoological Nomenclature* 34: 101–103.

Löbl, I. 1977i. Wenig bekannte und neue Scaphidiidae (Coleoptera) von Neukaledonien, Samoa und von den Fidschiinseln. *Revue suisse de zoologie* 84: 817–829.

Löbl, I. 1978a. Beitrag zur Kenntnis der Gattung *Sapitia* Achard (Coleoptera, Scaphidiidae). *Mitteilungen der Schweizerischen entomologischen Gesallschaft* 51: 53–57.

Löbl, I. 1978b. Two new Scaphidiidae (Coleoptera) from the New Hebrides. *Pacific Insects* 19: 109–111.

Löbl, I. 1978c. Contribution to the knowledge of the New Guinean species of *Scaphidium* (Coleoptera, Scaphidiidae). *Pacific Insects* 19: 113–119.

Löbl, I. 1979a. Die Scaphidiidae (Coleoptera) Südindiens. *Revue suisse de zoologie* 86: 77–129.

Löbl, I. 1979b. Two new Sumatran Scaphidiidae associated with termites and one new species of the genus *Scaphisoma* Leach from Java. *Sociobiology* 4: 321–328.

Löbl, I. 1980a. Beitrag zur Kenntnis der Scaphidiidae (Coleoptera) Taiwans. *Revue suisse de zoologie* 87: 91–123.

Löbl, I. 1980b. Beitrag zur Kenntnis der Scaphidiidae (Coleoptera) Neuirlands. *Mitteilungen der Schweizerischen entomologischen Gesellschaft* 53: 221–224.

Löbl, I. 1980c. Scaphidiidae (Coleoptera) of Fiji. *New Zealand Journal of Zoology* 7: 379–398.

Löbl, I. 1981a. Les Scaphidiidae (Coleoptera) de la Nouvelle-Calédonie. *Revue suisse de zoologie* 88: 347–397.

Löbl, I. 1981b. Insects of Micronesia Coleoptera Scaphidiidae. *Insects of Micronesia* 15: 69–80.

Löbl, I. 1981c. Über die japanische Arten der Gattungen *Scaphobaeocera* Csiki und *Scaphoxium* Löbl (Col., Scaphidiidae). *Mitteilunger der Schweizerischen entomologischen Gesellschaft* 54: 229–244.

Löbl, I. 1981d. Über die Arten-Groupe *Rouyeri* der Gattung *Scaphisoma* Leach (Coleoptera Scaphidiidae). *Archives de sciences* 34: 153–168.

Löbl, I. 1981e. Zwei neue Arten der Gattung *Scaphoxium* Löbl (Coleoptera, Scaphidiidae). *Annales historico-naturales Musei nationalis Hungarici* 73: 101–104.

Löbl, I. 1981f. Über einige Arten der Gattung *Scaphisoma* Leach (Coleoptera, Scaphidiidae) aus Vietnam und Laos. *Annales historico-naturales Musei nationalis Hungarici* 73: 105–112.

Löbl, I. 1982a. Weitere Arten der Gattung *Scaphisoma* Leach aus Japan (Coleoptera, Scaphidiidae). *Archives des sciences* 34 [1981]: 327–334.

Löbl, I. 1982b. Über die Scaphidiidae (Coleoptera) der japanischen Ryukyu-Inseln. *Mitteilungen der Schweizerischen entomologischen Gesellschaft* 55: 101–105.

Löbl, I. 1982c. Two new termitophilous Scaphidiidae (Coleoptera) from Sulawesi. *Sociobiology* 7: 29–39.

Löbl, I. 1982d. Sur l'identité de trois "*Amalocera*" de Bornéo (Coleoptera, Scaphidiidae). *Revue suisse de zoologie* 89: 789–795.

Löbl, I. 1982e. Contribution à la connaissance des *Pseudobironium* Pic de l'Inde (Coleoptera, Scaphidiidae). *Archives de sciences* 35: 157–160.

Löbl, I. 1982f. Little known and new Oriental species of the genus *Scaphisoma* Leach (Coleoptera, Scaphidiidae). Pp. 5–16. *Special Issue to the Memory of retirement of Emeritus Professor Michio Chûjô*. Association of the Memorial Issue of Emeritus Professor Michio Chûjô, Nagoya, 145 pp.

Löbl, I. 1982g. On the Scaphidiidae (Coleoptera) of Israel. *Israel Journal of Entomology* 16: 47–48.

Löbl, I. 1982h. Further remarkable new species of the genus *Scaphisoma* Leach from Japan (Coleoptera: Scaphidiidae). *Transactions of the Shikoku Entomological Society* 16: 19–22.

Löbl, I. 1983a. Sechs neue Scaphidiidae (Coleoptera) von Sulawesi, Indonesien. *Mitteilungen der Schweizerischen entomologischen Gesellschaft* 56: 285–293.

Löbl, I. 1983b. On the Scaphidiidae (Coleoptera) of Chile. *Entomologische Arbeiten aus dem Museum Georg Frey* 31/32: 161–168.

Löbl, I. 1984a. Les Scaphidiidae (Coleoptera) du nord-est de l'Inde et du Bhoutan I. *Revue suisse de zoologie* 91: 57–107.

Löbl, I. 1984b. Contribution à la connaissance des *Baeocera* du Japon (Coleoptera, Scaphidiidae). *Archives des sciences* 37: 181–192.

Löbl, I. 1984c. Scaphidiidae (Coleoptera) de Birmanie et de Chine nouveaux ou peu connus. *Revue suisse de zoologie* 91: 993–1005.

Löbl, I. 1986a. Les Scaphidiidae (Coleoptera) du nord-est de l'Inde et du Bhoutan II. *Revue suisse de zoologie* 93: 133–212.

Löbl, I. 1986b. Scaphidiidae (Coleoptera) nouveaux ou peu connus de l'Asie du sud-est. *Archives des sciences* 39: 87–102.

Löbl, I. 1986c. Contribution à la connaissance des Scaphidiidae (Coleoptera) du nord-ouest de l'Inde et du Pakistan. *Revue suisse de zoologie* 93: 341–367.

Löbl, I. 1986d. Three new Scaphidiidae (Coleoptera) from Queensland. *Mitteilungen der Schweizerischen entomologischen Gesellschaft* 59: 465–469.

Löbl, I. 1986e. *Scaphisoma viti* sp.n., un Scaphidiidae (Coleoptera) nouveau du Pakistan. *Archives des sciences* 39: 387–390.

Löbl, I. 1987a. Scaphidiidae (Coleoptera) nouveaux de Bornéo. *Revue suisse de zoologie* 94: 85–107.

Löbl, I. 1987b. Notes synonymiques sur trois Scaphidiidae (Coleoptera) néarctiques. *Mitteilungen der Schweizerischen entomologischen Gesellschaft* 60: 315–317.

Löbl, I. 1987c. Contribution to the knowledge of the genus *Caryoscapha* Ganglbauer (Coleoptera: Scaphidiidae). *The Coleopterists Bulletin* 41: 385–391.

Löbl, I. 1987d. Contribution à la connaissance des *Baeocera* d'Afrique et de Madagascar (Coleoptera, Scaphidiidae). *Revue suisse de zoologie* 94: 841–860.

Löbl, I. 1988a. Description de deux Scaphidiidae (Coleoptera) nouveaux du Nord de l'Inde. *Mitteilungen der Schweizerischen entomologischen Gesellschaft* 61: 373–376.

Löbl, I. 1988b. *Scaphisoma sakai* n. sp., un Scaphidiidae (Coleoptera) nouveau de l'archipel de Ryukyu. *Revue suisse de zoologie* 95: 1133–1136.

Löbl, I. 1989a. Über die Scaphidiidae Algeriens (Coleoptera). *Mitteilunges des internationalen entomologischen Vereins* 14: 9–12.

Löbl, I. 1989b. Scaphidiidae (Coleoptera) nouveaux ou méconnus de l'Afrique intertropicale. *Revue de zoologie africaine* 103: 277–283.

Löbl, I. 1989c. Sur les *Bironium* (Coleoptera, Scaphidiidae) de la Nouvelle-Guinée. *Mitteilungen der Schweizerischen entomologischen Gesellschaft* 62: 367–374.

Löbl, I. 1990a. Contribution à la connaissance des *Scaphisoma* (Coleoptera, Scaphidiidae) de l'Himachal Pradesh, Inde. *Archives des sciences* 43: 117–123.

Löbl, I. 1990b. Review of the Scaphidiidae (Coleoptera) of Thailand. *Revue suisse de zoologie* 97: 505–621.

Löbl, I. 1990c. *Cyparium javanum* sp.n., a new Scaphidiidae (Coleoptera) from Indonesia. *Elytron* 4: 125–129.

Löbl, I. 1992a. The Scaphidiidae (Coleoptera) of the Nepal Himalaya. *Revue suisse de zoologie* 99: 471–627.

Löbl, I. 1992b. On some Scaphidiinae (Coleoptera, Staphylinidae) from Mexico and continental Central America. *Mitteilungen der Schweizerischen entomologischen Gesellschaft* 65: 379–384.

Löbl, I. 1993. Contribution to the knowledge of the Scaphidiinae (Coleoptera, Staphylinidae) of the Far Eastern Region of Russia. *Russian entomological Journal* 2: 35–40.

Löbl, I. 1994. Note sur les *Scaphisoma* suisses (Coleoptera, Staphylinidae, Scaphidiinae). *Bulletin romand d'Entomologie* 12: 50.

Löbl, I. 1997. Catalogue of the Scaphidiinae (Coleoptera: Staphylinidae). *Instrumenta biodiversitatis* 1: xx + 190 pp.

Löbl, I. 1999. A review of the Scaphidiinae (Coleoptera: Staphylinidae) of the People's Republic of China, I. *Revue suisse de zoologie* 106: 691–744.

Löbl, I. 2000. A review of the Scaphidiinae (Coleoptera: Staphylinidae) of the People's Republic of China, II. *Revue suisse de zoologie* 107: 601–656.

Löbl, I. 2001. Four new Asian species of Scaphidiinae (Coleoptera, Staphylinidae). *Veröffentlichungen des Naturkundemuseum Erfurt* 20: 181–187.

Löbl, I. 2002a. On *Baeocera* (Coleoptera: Staphylinidae: Scaphidiinae) of New Guinea. *Mitteilungen der Schweizerischen entomologischen Gesellschaft* 75: 1–20.

Löbl, I. 2002b. Two new species of *Episcaphium* Achard (Coleoptera, Staphylinidae, Scaphidiinae). *Special Bulletin of the Japanese Society of Coleopterology* 5: 289–295.

Löbl, I. 2002c. Three new species of *Scaphoxium* (Coleoptera, Staphylinidae, Scaphidiinae) from New Guinea. *Revue suisse de zoologie* 109: 469–474.

Löbl, I. 2003a. A supplement to the knowledge of the Scaphidiines of China (Coleoptera: Staphylininae). *Mitteilungen der münchener entomologischen Gesellschatf* 93: 61–76.

Löbl, I. 2003b. Two new species of *Scaphisoma* Leach from Nepal and North India (Coleoptera: Staphylinidae: Scaphidiinae). *Mitteilungen der schweizerischen entomologischen Gesellschaft* 76: 155–160.

Löbl, I. 2003c. New species of *Scaphisoma* Leach (Coleoptera: Staphylinidae: Scaphidiinae) from Mt. Wilhelm, Papua New Guinea. *Acta Zoologica Academiae scienttiarum hungaricae* 48: 181–189.

Löbl, I. 2003d. Descriptions of two new Scaphiinae from South-India (Coleoptera, Staphylinidae). *Mitteilungen des internationalen entomologischen Vereins* 28: 93–98.

Löbl, I. 2004a. On the Scaphidiinae (Coleoptera, Staphylinidae) of Central India. *Mitteilungen der Schweizerischen entomologischen Gesellschaft* 77: 345–349.

Löbl, I. 2004b. Scaphidiinae. Pp. 495–505. In: I. Löbl & A. Smetana (eds). *Catalogue of Palaearctic Coleoptera. Volume 2 Hydrophiloidea – Histeroidea – Staphylinoidea.* Apollo Books, Stenstrup, 942 pp.

Löbl, I. 2005. On a collection of Scaphidiinae (Coleoptera, Staphylinidae) from Nepal. *Veröffentlichungen des Naturkundemuseum Erfurt* 24: 177–181.

Löbl, I. 2006. On the Philippine species of Cypariini and Scaphidiini (Coleoptera: Staphylinidae: Scaphidiinae). *Revue suisse de zoologie* 113: 23–49.

Löbl, I. 2010a. *Sphaeroscapha punctata*, a new species from New Caledonia (Coleoptera: Staphylinidae: Scaphidiinae). *Mitteilungen der Schweizerischen entomologischen Gesellschaft* 83: 37–39.

Löbl, I. 2010b. On *Scaphoxium* (Coleoptera: Staphylinidae: Scaphidiinae) from Africa and Madagascar. *Mitteilungen der Schweizerischen entomologischen Gesellschaft* 83: 119–129.

Löbl, I. 2011a. On the Scaphisomatini (Coleoptera: Staphylinidae: Scaphidiinae) of the Philippines. *Studies and Reports Taxonomic Series* 7: 301–314.

Löbl, I. 2011b. Notes on some Taiwanese Scaphidiinae (Coleoptera, Staphylinidae) described by Miwa and Mitono, with description of a new species and new records. *Japanese Journal of systematic Entomology* 17: 199–207.

Löbl, I. 2011c. A new species of *Scaphisoma* Leach (Coleoptera: Staphylinidae: Scaphidiinae) from Taiwan. *Mitteilungen der Schweizerischen entomologischen Gesellschaft* 84: 109–112.

Löbl, I. 2011d. A new species and a new record of Scaphidiinae (Coleoptera: Staphylinidae) from Bhutan. *Vernate* 30: 183–184.

Löbl, I. 2011e. On the Scaphisomatini (Coleoptera: Staphylinidae: Scaphidiinae) of the Philippines, II. *Revue suisse de zoologie* 118: 695–721.

Löbl, I. 2012a. On Taiwanese species of *Baeocera* Erichson (Coleoptera: Staphylinidae: Scaphidiinae). *Zoological Studies* 51: 118–130.

Löbl, I. 2012b. On a collection of Scaphisomatini (Coleoptera: Staphylinidae: Scaphidiinae) from West Malaysia. *Acta entomologica Musei nationalis Pragae* 52: 173–184.

Löbl, I. 2012c. Two new species of *Scaphisoma* Leach (Coleoptera: Staphylinidae: Scaphidiinae) from the Andaman Island. *Mitteilungen der Schweizerischen entomologischen Gesellschaft* 85: 85–90.

Löbl, I. 2012d. On the Scaphisomatini (Coleoptera: Staphylinidae: Scaphidiinae) of the Philippines, III: the genus *Baeocera* Erichson. *Revue suisse de zoologie* 119: 351–383.

Löbl, I. 2012e. *Baeocera socotrana*, the first species of Scaphidiinae (Coleoptera: Staphylinidae) reported from Socotra. *Acta entomologica Musei nationalis Pragae* 52 (Suppl. 2): 141–145.

Löbl, I. 2012f. *Irianscapha dimorpha*, an unusual new Scaphisomatini from Western New Guinea, and two new Indonesian species of *Scaphisoma* Leach (Coleoptera: Staphylinidae: Scaphidiinae). *Vernate* 31: 309–317.

Löbl, I. 2012g. Unterfamilie Scaphidiinae, pp. 201–205. In: V. Assing & M. Schülke (eds). Freude-Harde-Lohse-Klausnitzer: *Die Käfer Mitteleuropas, Band 4: Staphylinidae I. Zweite neubearbeitete Auflage*. Spektrum, Akademischer Verlag, Heidelberg, xii + 560 pp.

Löbl, I. 2014a. On the Scaphidiinae (Coleoptera, Staphylinidae) of the Moluccas. *Mitteilungen der Schweizerischen entomologischen Gesellschaft* 87: 49–60.

Löbl, I. 2014b. A new apterous species of *Scaphisoma* Leach (Coleoptera: Staphylinidae: Scaphidiinae) from Western New Guinea. *Klapalekiana* 50: 61–64.

Löbl, I. 2015a. On the Scaphidiinae (Coleoptera: Staphylinidae) of the Lesser Sunda Islands. *Revue suisse de zoologie* 122: 65–120.

Löbl, I. 2015b. Contribution to the knowledge of the Scaphidiinae (Coleoptera: Staphylinidae) of the Moluccas. *Stuttgarter Beiträge zur Naturkunde A, Neue Serie* 8: 165–187.

Löbl, I. 2015c. Notes on *Scaphisoma* (Coleoptera: Staphylinidae: Scaphidiinae) of Kalimantan. *Acta entomologica Musei nationalis Pragae* 55: 129–144.

Löbl, I. 2015d. *Scaphisoma poussereaui* sp. nov. from La Réunion (Coleoptera: Staphylinidae: Scaphidiinae), a range extension of the *S. tricolor* group. *Mitteilungen der Schweizerischen entomologischen Gesellschaft* 88: 367–370.

Löbl, I. 2015e. Staphylinidae: Scaphidiinae, p. 21. In: I. Löbl & D. Löbl (eds). *Catalogue of Palaearctic Coleoptera. Volume 2. Revised and Updated Edition. Hydrophiloidea – Staphylinoidea*. Brill, Leiden / Boston, xxv + 1702 pp.

Löbl, I. 2017. New species and records of *Scaphobaeocera* Csiki from New Guinea (Coleoptera: Staphylinidae: Scaphidiinae). *Mitteilungen der Münchner entomologischen Gesellschaft* 107: 33–41.

Löbl, I. & A. Faille 2017. *Toxidium cavicola* sp. nov., a new cave dwelling Malagasy Scaphidiinae (Coleoptera: Staphylinidae). *Annales Zoologici* 67(2): 345–348.

Löbl, I. & R.A.B. Leschen 2003a. Redescription and new species of *Alexidia* (Coleoptera: Staphylinidae: Scaphidiinae). *Revue suisse de zoologie* 110: 315–324.

Löbl, I. & R.A.B. Leschen 2003b. Scaphidiinae (Insecta: Coleoptera: Staphylinidae). *Fauna of New Zealand* 48, 94 pp.

Löbl, I. & R.A.B. Leschen 2010. Notes on the *Toxidium* group (Coleoptera: Staphylinidae: Scaphidiinae). *Folia Heyrovskyana*, series A 18: 71–93.

Löbl, I. & R. Ogawa 2016a. Contribution to the knowledge of the Himalayan and North Indian species of *Scaphidium* (Coleoptera, Staphylinidae). *Revue suisse de zoologie* 123: 159–163.

Löbl, I. & R. Ogawa 2016b. On the Scaphisomatini (Coleoptera, Staphylinidae, Scaphidiinae) of the Philippines, IV: the genera *Sapitia* Achard and *Scaphisoma* Leach. *Linzer biologische Beiträge* 48(2): 1339–1492.

Löbl, I. & R. Ogawa 2016c. *Persescaphium pari* new genus and species, with an ov erview of Iranian Scaphidiinae (Coleoptera, Staphylinidae). *Entomologische Blätter und Coleoptera* 112: 35–40.

Löbl, I. & R. Ogawa 2017. A new species of *Scaphisoma* Leach, 1815 from New Guinea and a new replacement name. Pp. 415–417, Fig. 1, Pl. 82. In: D. Telnov, M.V.L. Barclay & O.S.G. Pauwels (eds). *Biodiversity, biogegraphy and nature conservation in Wallacea. Volume III*. Entomological Society of Latvia, Riga.

Löbl, I. & J. Růžička 2000. Faunistic records from the Czech Republik – 117. Coleoptera: Staphylinidae: Scaphidiinae. *Klapalekiana* 36: 289.

Löbl, I. & K. Stephan 1993. A review of the species of *Baeocera* Erichson (Coleoptera, Staphylinidae, Scaphidiinae) of America north of Mexico. *Revue suisse de zoologie* 100: 675–733.

Löbl, I. & L. Tang. 2013. A review of the genus *Pseudobironium* Pic. (Coleoptera: Staphylinidae: Scaphidiinae). *Revue suisse de zoologie* 120: 665–734.

Lott, D.A. 2009. *The Staphylinidae (rove beetles) of Britain and Ireland Part 5: Scaphidiinae, Piestinae, Oxytelinae Pt. (Handbooks for the Indentification of British Insects)*. Royal Entomological Society, 105 pp.

Lundblad, O. 1952. Die schwedischen *Scaphosoma*-Arten. *Entomologisk Tidskrift* 73: 27–32.

Macleay, W.J. 1871. Notes on a collection of insects from Gayndah. *Transactions of the Entomological Society of New South Wales* 2: 79–205.

Majka, C.G., D.S. Chandler & C.P. Dohahue 2011. *Checklist of the beetles of Maine, USA*. Empty Mirrors Press, Halifax, 328 pp.

Majzlan, O. 2016. Chrobáky (Coleoptera) vybraných lokalít z oblasti Východné Karpaty. *Naturae tutela* 20(2): 101–127.

Márquez, J. 2006. Primeros registros estatales y datos de distribución geográfica de especies mexicanas de Staphylinidae (Coleoptera). *Boletín de la Sociedad Entomológica Aragonesa* 38: 181–198.

Márquez, J. 2007. Preliminary analysis of the color variation in *Cyparium terminale* from Mexico, with comments on *C. palliatum*, and a new record for *C. yapalli* (Coleoptera: Staphylinidae, Scaphidiinae). *Entomological News* 118(1): 1–10.

Márquez Luna, J. & J.L. Navarrete Heredia 1995. Especie de Staphylinidae (Insecta: Coleoptera) asociadas a detritos de *Atta mexicana* (F. Smith) (Hymenoptera: Formicidae) en dos localidades de Morelos, Mexico. *Folia Entomológica Mexicana* 91: 31–46.

Marsham, T. 1802. *Entomologia Britannica sistens insecta Britanniae indigena secundum methodum linnaeanam disposita. Tomus I. Coleoptera*. Wilks et Taylor, London, xxxi + 547 + [1] pp.

Matějíček, J. & J. Boháč 2003. Faunistic records from the Czech Republic – 166. *Klapalekiana* 39: 131–135.

Mateleshko, O. 2005. Tverdokrili (Insecta, Coleoptera) – micetobionti gryb z rody *Pleurotus* (Fr.) Kumm. Ukrainskikh Karpat. *Naukoviy Visnik Uzgorodskogo Universitetu. Seria Biologiya* 17: 127–130.

Matthews, A. 1888. Fam. Scaphidiidae. Pp. 158–181, pls 3, 4. In: *Biologia Centrali-Americana. Insecta, Coleoptera. Vol. 2, Part 1.* [1887–1888]. Taylor and Francis, London, xii + 717 pp., 10 pls.

Mazumder, N., S.K. Dutta & R. Gogoi 2001. A new record of *Scaphisoma* (Coleoptera: Scaphidiidae) as a pest of oyster mushroom. *Mushroom Research* 10(1): 59.

Mazumder, N., S.K. Dutta R. Gogoi & Y. Rathaiah 2008. Seasonal abundance of *Scaphisoma tetrastictum* Champ. on oyster mushroom and its relation to meteorological factors. *Acta Phytopathologica et Entomologica Hungarica* 43(1): 63.

McKenna, D.D., B.D. Farrell, M.S. Caterino, C.W. Farnum, D.C. Hawks, D.R. Maddison, A.E. Seago, A.E.Z. Short, A.F. Newton, M.K. Thayer 2015. Phylogeny and evolution of Staphyliniformia and Scarabaeiformia: forest litter as a stepping stone for diversification of nonphytophagous beetles. *Systematic Entomology* 40(1): 35–60.

Melsheimer, F.E. 1846. Descriptions of new species of Coleoptera of the United States. *Proceedings of the Academy of Natural Sciences of Philadelphia* 2: 98–118.

Merkl, O. 1987. Scydmaenidae, Corylophidae, Sphaeriidae, Ptiliidae, Scaphidiidae, Pselaphidae and Histeridae of the Kiskunság National Park (Coleoptera). Pp. 111:119. In: S. Mahunka (ed.). *The Fauna of the Kiskunság National Park, 1987. Voluzme 2.* Akadémiai Kladó, Budapest.

Merkl, O. 1996. Histeridae and Scaphidiidae (Coleoptera) from the Bükk National Park. Pp. 259–262. In: S. Mahunka (ed.). *The Fauna of the Bükk National Park. Volume II.* [1993–1996]. Hungarian Natural History Museum, Budapest, 655 pp.

Miwa, Y. & T. Mitono 1943. Scaphidiidae of my country [= Scaphidiidae of Japan and Formosa]. *Transactions of the Natural History Society of Formosa* 33: 512–555 [in Japanese].

Monsevičius, V. 2013: 27. New and littler known for the Lithuanian fauna species of beetles (Coleoptera), found in 2002, 2011–2012. *New and rare for Lithuania insect species* 25: 24–30.

Morimoto, K. 1985. Scaphidiidae, pp. 252–258. pl. 45. In: S. Uéno, Y. Kurosawa & M. Satô (eds.): *The Coleoptera of Japan in color. Vol. 2.* Hoikusha Publ. Co., Osaka, viii + 479 pp., 80 pls [in Japanese].

Motschulsky, V. de 1845. Observations sur le Musée Entomologique de l'Université Impériale de Moscou. *Bulletin de la Société impériale des Naturalistes de Moscou* 18: 332–387.

Motschulsky, V. de. 1860. Insectes des Indes orientales, et de contrées analogues. *Etudes entomologiques* 8[1859]: 24–118.

Motschulsky, V. de. 1863. Essai d'un catalogue de l'île de Ceylan. *Bulletin de la Société impériale des Naturalistes de Moscou* 36: 421–532.

Nakane, T. 1955a. Nihon no kôchu (26). *Shin-konchū* 8(8): 53–56 [in Japanese].

Nakane, T. 1955b. Nihon no kôchu (27). *Shin-konchū* 8(9): 49–53 [in Japanese].

Nakane, T. 1955c. Nihon no kôchu (28). *Shin-konchū* 8(10): 54–57 [in Japanese].

Nakane, T. 1956. New or little-known Coleoptera from Japan and its adjancent Regions, XIII. *Scientific Reports of the Saikyo University* 2 (A): 159–174.

Nakane, T. 1963a. New or little known Coleoptera from Japan and its adjacent regions. XVII. *Fragmenta coleopterologica* 5: 21–22.

Nakane, T. 1963b. Scaphidiidae. Pp. 78–80, pls. 39–40. In: T. Nakane, K. Ohbayashi S. Nomura & Y. Kurosawa (eds). *Iconographia Insectorum Japonicorum colore naturali edita Volumen II* (*Coleoptera*) (1st edition). Hokuryu-kan, Tokyo, 443 pp., pls. 1–192 [in Japanese].

Naomi, S-I. 1985. The phylogeny and higher classification of the Staphylinidae and their allied groups (Coleoptera, Staphylinoidea). *Esakia* 23: 1–27.

Navarrete-Heredia, J.L. 1991. Analisis preliminar de los coleopteros micetocolos de Basidiomycetes de San Jose de Los Laureles, Mor. Mexico. Pp. 115–149. In: J.L. Navarrete-Heredia & G.A. Quitroz-Rocha (eds). *I simposio national sobre la interaccion insecto-hongo. Memorias*. Veracruz, Mexico, 192 pp.

Navarrete-Heredia, J.L., A.F. Newton & M.K. Thayer, 2002. Scaphidiinae. Pp. 201–206. In: J.L. Navarrete-Heredia, A.F. Newton, M.K. Thayer, J.S. Ashe & D.S. Chandler: *Guía ilustrada para los géneros de Staphylinidae* (*Coleoptera*) *de Méchico. Illustrated guide to the genera of Staphylinidae* (*Coleoptera*) *of Mexico*. CONABIO, Mexico, viii +401 pp.

Newton, A.F. Jr. 1984. Mycophagy in Staphylinoidea (Coleopetra). Pp. 302–353. In: Q. Wheeler & M. Blackwell (eds). *Fungus/insect relationships. Perspectives in ecology and evolution*. Columbia University Press, New York, 514 pp.

Newton, A.F. Jr. 1991. Scaphidiidae (Staphylinoidea). Pp. 337–339. In: F.W. Stehr (ed.). *Immature Insects. Volume 2*. Kendall/Hunt Publishing Company, Dubuque, xvi + 975 pp.

Newton, A.F. 1996. In: J. Klimaszewski, A.F. Newton & M.K. Thayer A review of the New Zealand rove beetles (Coleoptera: Staphylinidae). *New Zealand Journal of Zoology* 23: 143–160.

Newton, A.F. Jr. 2017. Nomenclatural and taxonomic changes in Staphyliniformia (Coleoptera). *Insecta Mundi* 595: 1–52.

Newton, A.F. Jr. & S.L. Stephenson 1990. A beetle/slime mold assemblage from northern India (Coleoptera; Myxomycetes). *Oriental Insects* 24: 197–218.

Newton, A.F. Jr. & M.K. Thayer 1992. Current classification and family-group names in Staphyliniformia (Coleoptera). *Fieldiana: Zoology* N.S. 67: 1–92.

Newton, A.F., M.K. Thayer, J.S. Ashe & D.S. Chandler 2000. 22. Superfamily Staphylinoidea Latreille, 1802 Staphyliniformia Lameere, 1900; Brachelytra auctorum. 22. Staphylinidae Latreille, 1802. Pp. 272–418. In: R.H. Arnett & M.C. Thomas (eds). *American beetles. Volume 1. Archostemata, Myxophaga, Adephaga, Polyphaga: Staphyliniformia*. CRC Press; Boca Raton, London, New York, Washington, D.C., xv + 443 pp.

Nikitsky, N.B. & D.S. Schigel 2004. Beetles on polypores of the Moscow region: checklist and ecological notes. *Entomologia Fennica* 15: 6–22.

Nikitsky, N.B., A.R. Bibin & M.M. Dolgin 2008. *Ksilofilnye zhesktokrylye (Coleoptera) Kavkazskogo Gosudarstvennogo Prirodnogo Biosfernogo Zapovednika i sopredelnykh territorii.* Syktyvkar, 452 pp.

Nikitsky, N.B., S.N. Mamontov & A.S. Zamotajlov 2016. Novye dannye o zhestkokrylykh (Coleoptera) zasechnykh lesov Tulskoy Oblasti, sobrannykh okonnymi lovushkami. Chast 1. Carabidae-Sphindidae. New data on beetles (Coleoptera) of the abatis forests of Tula Province collected by the aid of window traps. Part 1. Carabidae-Sphindidae. *Trudy Kubanskogo gosudarstvennogo agrarnogo universiteta* 1 (58): 134–144.

Nikitsky, N.B., I.N. Osipov, M.V. Chemeris, V.B. Semenov & A.A. Gusakov 1996. Zhestkokrylye-ksilobionnty, micetobionty i plastichatousye Prioksko-Terrasnogo Biosfernogo Zapovednika (s obzorom fauny etikh grupp Moskovskoy oblasti). *Sbornik Trudov zoologicheskogo muzeya MGU* 36, 196 + [2] pp.

Nikitsky, N.B., V.B. Semenov & M.M. Dolgin 1998. Zhestkokrylye-ksilobionnty, micetobionty i plastichatousye Prioksko-Terrasnogo Biosfernogo Zapovednika (s obzorom fauny etikh grupp Moskovskoy oblasti. Dopolnenie 1). *Sbornik Trudov Zoologicheskogo Muzeya MGU* 36, Suppl. 1, 55 + 4 pp.

Normand, H. 1934. Contribution au catalogue des coléoptères de la Tunésie. Fascicule 3(1). *Bulletin de la Société d'Histoire naturelle de l'Afrique du Nord* 25: 35–45.

Novak, P. 1952. *Kornjaši jadranskov primorja (Coleoptera).* Jugoslavska Akademia Znanost i Umjetnosti, 621 + [2] pp.

Nuss, I. 1975. *Zur Ökologie der Porlinge. Untersuchungen über die Sporulation einiger Porlinge und die an ihnen gefundenen Käferarten.* Bibliotheca Mycologica. J. Cramner, Vaduz, 258 pp.

Oberthür, R. 1883. Scaphidiides nouveaux. Pp. 5–16. In: *Coleopterorum Novitates. Recueil spécialement consacré à l'étude des coléoptères, Rennes* 1: 1–80, 2 pls.

Ogawa, R. 2015. *Phylogeography of the subfamily Scaphidiinae (Coleoptera, Staphylinidae) in Sulawesi, with its systematic revision.* Graduate School of Agricultural Science, Kobe University, 180 pp. [PhD thesis, unpublished].

Ogawa, R. & M. Sakai 2011. A review of the genus *Cyparium* Erichson (Coleoptera, Staphylinidae, Scaphidiinae) of Japan. *Japanese Journal of Systematic Entomology* 17: 129–136.

Ogawa, R. & H. Hoshina 2012. Notes an the tribe Scaphisomatini (Coleoptera, Staphylinidae, Scaphidiinae) of Japan. *Elytra* (N.S.) 2: 263–266.

Ogawa, R. & I. Löbl 2013. A revision of the genus *Baeocera* in Japan, with a new genus of the tribe Scaphisomatini (Coleoptera, Staphylinidae, Scaphidiinae). *Zootaxa* 3652: 301–326.

Ogawa, R. & K. Maeto 2015. The termitophilous Scaphidiinae (Coleoptera: Staphylinidae) from Sulawesi, Indonesia. *The Coleopterists Bulletin* 69: 301–304.

Ogawa, R. & I. Löbl 2016a. A review of the genus *Xotidium* Löbl, 1992 (Coleoptera, Staphylinidae, Scaphidiinae), with description of five new species. *Deutsche entomologische Zeitschrift* 63: 155–169.

Ogawa, R. & I. Löbl 2016b. [new taxon] *In*: Ogawa, R., Löbl, I. & Maeto, K.: A new species of the genus *Cyparium* from northern Sulawesi, Indonesia (Coleoptera: Staphylinidae: Scaphidiinae). *Acta entomologica Musei nationalis Pragae* 56(1): 195–201.

Ogawa, R., I. Löbl & K. Maeto 2014. Three new species of the genus *Scaphicoma* Motschulsky, 1863 (Coleoptera, Staphylinidae, Scaphidiinae) from Northern Sulawesi, Indonesia. *ZooKeys* 403: 1–13 [DOI: 10.3897/zookeys.403.7200].

Ogawa, R., S. Matsuo, H. Hoshina & K. Maeto 2016. Allopatric color forms of *Scaphidium morimotoi* Löbl, 1982 (Coleoptera, Staphylinidae, Scaphidiinae) emdemic to the Amami Islands, the Ryukyus, Japan. *Zootaxa* 4175(1): 64–74.

Olivier, G.-A. 1790. *Entomologie, ou Histoire Naturelle des Insectes, Avec leurs caractères, génériques et spécifiques, leur description, leur synonymie, et leur figure enluminée. Coléoptéres. Tom 2*. Baudouin, Paris [Nos. 9-34, [4] + 458 pp, 28 genera, each separately paginated].

Ostrovsky, A.M. 2016. A preliminary list of beetle species (Insecta, Coleoptera) of the South-Eastern part of the Republic of Belarus. *Euroasian Entomological Journal* 15: 379–386.

Palm, T. 1951. Die Holz- und Rinden-Käfer der nordschwedischen Laubbäume. De nordsvenska lövträdens ved- och barkskalbaggar. *Meddelanden från Statens Skogsforskninginstitut* 40(2): 1–242, 1 pl.

Palm, T. 1953. Anteckningar om svenska skalbaggar. VIII. *Entomologisk Tidskrift* 74: 171–186.

Palm, T. 1959. Die Holz- und Rinden-Käfer der Süd- und Mittelschwedischen Laubbäume. *Opuscula entomologica, Supplement* 16: 1–374, 47 pls.

Palm, T. 1966. Anteckningar om svernska skalbaggar. XVII. *Entomologisk Tidskrift* 87: 43–46.

Palm, T. 1971. Notes on some North-European Coleoptera. *Entomologisk Tidskrift* 92: 66–73.

Panzer, G.W.F. 1792. *Fauna insectorum Germanicae initia oder Deutschlands Insecten. Heft 2*. Felsecker, Nürnberg, 24 pp., 24 pls.

Panzer, G.W.F. 1793. *Fauna insectorum Germanicae initia oder Deutschlands Insecten. Heft 12*. Felsecker, Nürnberg, 24 pp., 24 pls.

Pascoe, F.P. 1863. Notices of new or little-known genera and species of Coleoptera. Part IV. *The Journal of Entomology* 2: 26–56.

Pascoe, F.P. 1876. Descriptions of new genera and species of New-Zealand Coleoptera. Part II. *The Annals and Magazine of Natural History* (4) 17: 48–60, pl. 2.

Paulian, R. 1941. Les premiers états des Staphylinoidea (Coleoptera). Etude de la morphologie comparée. *Mémoires du Muséum national d'Histoire naturelle* (Nouvelle Série) 15: 1–361, 3 pls.

Paulian, R. 1943. Observation sur la larve de *Scaphidium quadrimaculatum* Ol. (Col. Scaphidiidae). *Bulletin de la Société entomologique de France* 48: 147–148.

Paulian, R. 1951. Les "*Scaphidiolum*" Achard d'Afrique Noire (Col. "Scaphidiidae"). *Actas da Conferência Internacional dos Africanistas Ocidentais em Bissau 1947* (Lisboa), 3(2): 195–200.

Paulian, R. & A. Villiers 1940. Les coléoptères des lobelias des montagnes du Caméroun. *Revue française d'entomologie* 7: 72–83.

Paulino de Oliveira, M. 1895. *Catalogue des insectes du Portugal Coleopteres.* Impressa da Universidade, Coimbra, 393 pp.

Paykull, G. von. 1800. *Fauna Svecica. Insecta. Tomus III.* Joh. F. Edman, Upsaliae, 459 pp.

Peck, S.B. & M.C. Thomas 1998. A distributional checklist of the beetles (Coleoptera) of Florida. *Arthropods of Florida and neighboring land areas*, 16: viii + 180 pp.

Pelletier, A.L.M. & J.G. Audinet-Serville 1825. In: P.A. Latreille, A.L.M. Lepeletier, J.G. Audinet-Serville & F.E. Guérin-Méneville *Encyclopédie méthodique, ou par ordre de matières; par une société de gens de lettres, de savants et d'artistes; précédée d'un vocabulaire universel, servant de table pour tout l'ouvrage, ornée des portraits de Mm. Diderot & d'Alembert, premiers éditeurs de l'Encyclopédie. Histoire naturelle. Entomologie, ou histoire naturelle des crustacés, des arachnides et des insectes. Tome dixième. Pars 1.* Mme Veuve Agasse, Paris, [6] + 344 pp.

Perris, É. 1876. Larves de coléoptères. *Annales de la Société linnéenne de Lyon* 22: 259–418.

Perty, J.A.M. 1830. Fasc. 1, pp. 1–60, pls 1–12. In: *Delectus animalium articulatorum, quae in itinere per Brasiliam annis MDCCCXVII-MDCCCXX jussu et auspiciis Maximiliani Josephi I. Bavariae Regis augustissimi peracto collegerunt Dr. J.B. de Spix et Dr. C.F.Ph. de Martius. Digessit, descripsit, pingenda curavit Dr. Maximilianus Perty, praefatus est et edidit Dr. C.F.Ph. de Martius.* Monachii [1830–1833], 224 pp, 39 pls.

Petz, J. 1905. Coleopterologische Notizen. *Wiener entomologische Zeitung* 24: 100.

Pic, M. 1905a. Descriptions abrégées et notes diverses. *L'Echange, Revue linnéenne* 21: 128–131.

Pic, M. 1905b. Captures diverses, noms nouveaux et diagnoses (Coléoptères). *L'Echange, Revue linnéenne* 21: 169–171.

Pic, M. 1915a. Nouvelles espèces de diverses familles. *Mélanges exotico-entomologiques* 15: 1–24.

Pic, M. 1915b. Diagnoses de nouveaux genres et nouvelles espèces de Scaphidiides. *L'Echange, Revue linnéenne* 31: 30–32.

Pic, M. 1915c. Diagnoses de nouveaux genres et nouvelles espèces de scaphidiides. *L'Echange, Revue linnéenne* 31: 35–36.

Pic, M. 1915d. Diagnoses de nouveaux genres et nouvelles espèces de scaphidiides. *L'Echange, Revue linnéenne* 31: 40.

Pic, M. 1915e. Genres nouveaux, espèces et variétés nouvelles. *Mélanges exotico-entomologiques* 16: 2–13.

Pic, M. 1915f. Diagnoses de nouveaux genres et nouvelles espèces de Scaphidiides (fin). *L'Echange, Revue linnéenne* 31: 43–44.

Pic, M. 1916a. Coléoptères exotiques en partie nouveaux. *L'Echange, Revue linnéenne* 32: 3–4.

Pic, M. 1916b. Notes et descriptions abrégées diverses. *Mélanges exotico-entomologiques* 17: 2–8.

Pic, M. 1916c. Diagnoses spécifiques. *Mélanges exotico-entomologiques* 17: 8–20.

Pic, M. 1916d. Diagnoses génériques et spécifiques. *Mélanges exotico-entomologiques* 18: 2–20.

Pic, M. 1916e. Notes relatives à divers Scaphidiidae (Col.). *Bulletin de la Société entomologique de France* 1916: 49.

Pic, M. 1917a. Descriptions abregées diverses. *Mélanges exotico-entomologiques* 22: 2–20.

Pic, M. 1917b. Descriptions abregées diverses. *Mélanges exotico-entomologiques* 24: 2–24.

Pic, M. 1917c. Descriptions abrégées diverses. *Mélanges exotico-entomologiques* 26: 2–24.

Pic, M. 1918. Courtes descriptions diverses. *Mélanges exotico-entomologiques* 27: 1–24.

Pic, M. 1920a. Note diverses, descriptions et diagnoses. *L'Echange, Revue linnéenne* 36: 5–8.

Pic, M. 1920b. Nouveautés diverses. *Mélanges exotico-entomologiques* 32: 1–28.

Pic, M. 1920c. Notes diverses, descriptions et diagnoses. *L'Echange, Revue linnéenne* 36: 13.

Pic, M. 1920d. Diagnoses de coléoptères exotiques. *L'Echange, Revue linnéenne* 36: 15–16.

Pic, M. 1920e. Scaphidiides nouveaux de diverses origines. *Annali del Museo civico di Storia naturale di Genova* (3) 9: 93–97.

Pic, M. 1920f. Coléoptères exotiques en partie nouveaux. *L'Echange, Revue linnéenne* 36: 22–24.

Pic, M. 1920g. Note sur divers scaphidiides. *Annales de la Société entomologique de Belgique* 60: 188–189.

Pic, M. 1920h. Nouveaux scaphidiides de Sumatra (Col.). *Bulletin de la Société entomologique de France* 1920: 242.

Pic, M. 1920i. Sur *Baeocera argentina* Pic (Col. Scaphidiidae). *Bulletin de la Société entomologique de France* 1920: 50.

Pic, M. 1921a. Scaphidiides recueillis par feu L. Fea. *Annali del Museo civico di Storia naturali di Genova* (3) 9: 158–167.

Pic, M. 1921b. Notes diverses, descriptions et diagnoses. *L'Echange, Revue linnéenne* 37: 1–4.

Pic, M. 1921c. Nouveautés diverses. *Mélanges exotico-entomologiques* 33: 1–32.

Pic, M. 1922a. Nouveautés diverses. *Mélanges exotico-entomologiques* 36: 1–32.

Pic, M. 1922b. Nouveautés diverses. *Mélanges exotico-entomologiques* 37: 1–32.

Pic, M. 1923a. Coléoptères exotiques en partie nouveaux. *L'Echange, Revue linnéenne* 39: 4.

Pic, M. 1923b. Nouveautés diverses. *Mélanges exotico-entomologiques* 38: 1–32.

Pic, M. 1923c. Nouveaux coléoptères du Tonkin. *Bulletin de la Société zoologique de France* 48: 269–271.

Pic, M. 1923d. Scaphidiides exotiques nouveaux (Col.). *Bulletin de la Société entomologique de France* 1923: 194–196.

Pic, M. 1925a. Notes sur les coléoptères scaphidiides. *Annales de la Société entomologique de Belgique* 64 [1924]: 193–196.

Pic, M. 1925b. Coléoptères exotiques en partie nouveaux. *L'Echange, Revue linnéenne* 41: 8.

Pic, M. 1926a. Nouveaux coléoptères exotiques. *Bulletin de la Société entomologique de France* 1925: 322–324.

Pic, M. 1926b. Nouveautés diverses. *Mélanges exotico-entomologiques* 45: 1–32.

Pic, M. 1926c. Note diverses, descriptions et diagnoses. *L'Echange, Revue linnéenne* 42: 5–6.

Pic, M. 1926d. Nouveaux coléoptères du Tonkin (2e article). *Bulletin de la Société zoologique de France* 51: 45–48.

Pic, M. 1926e. Nouveaux coléoptères du Tonkin. III. *Bulletin de la Société zoologique de France* 51: 143–145.

Pic, M. 1927. Coléoptères de l'Indochine. *Mélanges exotico-entomologiques* 49: 1–36.

Pic, M. 1928a. Notes et descriptions. *Mélanges exotico-entomologiques* 51: 1–36.

Pic, M. 1928b. Scaphidiidae du Congo Belge. *Revue de Zoologie et de Botanique africaines* 16: 33–44.

Pic, M. 1928c. Nouveautés diverses. *Mélanges exotico-entomologiques* 52: 1–32.

Pic, M. 1928d. Nouveaux coléoptères de la République Argentine. *Revista de la Sociedad entomologica argentina* 2: 49–52.

Pic, M. 1928e. Nové druhy koleopter z Brasilie. Nouveaux coléoptères du Brésil. *Sborník entomologického oddělení Národního musea v Praze* 6: 74–76.

Pic, M. 1930a. Scaphidiidae recueillis au Congo Belge par A. Collart. *Revue de Zoologie et de Botanique africaines* 20: 87–89.

Pic, M. 1930b. Coléoptères asiatiques nouveaux. *Sborník entomologického oddělení Národního musea v Praze* 8: 58–59.

Pic, M. 1930c. Coléoptères nouveaux de la République Argentine. *Bulletin de la Société zoologique de France* 55: 175–179.

Pic, M. 1931a. Nouveautés diverses. *Mélanges exotico-entomologiques* 57: 1–36.

Pic, M. 1931b. Résultats de la mission scientifique suisse en Angola. Coléoptères (Clavicornes, Clérides, Malacodermes, Hétéromères, Bruchides, Phytophages) d'Angola. *Revue suisse de zoologie* 38: 419–427.

Pic, M. 1933a. Neue Coleopteren-Clavicornia. *Entomologisches Nachrichtenblatt* 7: 71–72.

Pic, M. 1933b. Materiali per lo studio della fauna Erithea raccolti nel 1901–1903 dal Dr. A. Andreini. *Bollettino della Società entomologica Italiana* 65: 119–130.

Pic, M. 1935. Nouveaux coléoptères exotiques. *Annals and Magazine of Natural History* (10)16: 470–473.

Pic, M. 1937. Neue Scaphidiidae aus dem Hamburger Zoologischen Museum. *Entomologische Rundschau* 54: 206–207.

Pic, M. 1940a. Diagnoses de coléoptères exotiques. *L'Echange, Revue linnéenne* 56: 2–4.

Pic, M. 1940b. Nouvelle série de coléoptères d'Angola. *Revue suisse de zoologie* 47: 359–365.

Pic, M. 1942. Coléoptères nouveaux du Cameroun. *L'Echange, Revue linnéenne* 58: 1–3.

Pic, M. 1946. Coléoptères scaphidiides nouveaux de la Mission de l'Omo. *Revue française d'entomologie* 13: 82–84.

Pic, M. 1947a. Coléoptères du globe. *L'Echange, Revue linnéenne* 63: 9–12.

Pic, M. 1947b. [without tittle] *Diversités entomologiques* 1: 1–16.

Pic, M. 1947c. [without tittle] *Diversités entomologiques* 2: 1–16.

Pic, M. 1948a. Coléoptères du globe. *L'Echange, Revue linnéenne* 64: 9–12.

Pic, M. 1948b. Nouveaux Scaphidiidae d'Afrique (Col.). *Bulletin de la Société entomologique de France* 53: 71–72.

Pic, M. 1951a. Coléoptères du globe. *L'Echange, Revue linnéenne* 67: 5–8.

Pic, M. 1951b. Nouveaux coléoptères de l'ouest africain. *Trab. XX Conferência Internacion dos Africanistas Ocidentais em Bissau* 1947: 203–210.

Pic, M. 1951c. Coléoptères rares ou nouveaux d'Afrique. *Bulletin de l'Institut français d'Afrique Noire* 13: 1099–1102.

Pic, M. 1953. Coléoptères nouveaux de Madagascar. *Mémoires de l'Institut scientifique de Madagascar*, Série E, 3: 253–278.

Pic, M. 1954a. Coléoptères du globe. *L'Echange, Revue linnéenne* 70: 9–12.

Pic, M. 1954b. Nouveaux Scaphidiidae du Congo Belge (Coleoptera Clavicornia). *Revue de Zoologie et de Botanique africaines* 50: 33–39.

Pic, M. 1954c. Coléoptères nouveaux de Chine. *Bulletin de la Société entomologique de Mulhouse* (1954): 53–59.

Pic, M. 1955. Contribution à l'étude de la faune entomologique du Ruanda-Urundi (Mission P. Basilewsky 1953) VII. Coleoptera Scaphidiidae. *Annales du Musée Royal du Congo belge, Zoologie* 36: 49–54.

Pic, M. 1956a. Nouveaux coléoptères exotiques. *Bulletin de la Société entomologique de France* 60 [1955]: 173–175.

Pic, M. 1956b. Nouveaux coléoptères de diverses familles. *Annales historico-naturales Musei nationalis Hungarici (N.S.)* 7: 71–92.

Plaisier, F. 1986. Zur Präsenz von *Scaphium immaculatum* (Olivier 1790) auf den Ostfriesischen Inseln (Coleoptera: Scaphidiidae). *Drosera* 86: 75–78.

Poggi, R. 1983. Note di caccia. V. Reperti di specie italiane rare o poco note (Coleoptera). *Bollettino della Società entomologica italiana* 115: 156–160.

Poole, R.W. & P. Gentili 1996. Volume 1: Coleoptera and Strepsiptera. *Nomina Insecta Nearctica: A check list of the insects of North America.* Entomological Information Service, Rockville, Maryland, 827 pp.

Porta, A. 1926. *Fauna Coleopterorum Italica. Vol. II. – Staphylinoidea. Staphylinidae, Pselaphidae, Clavigeridae, Scydmaenidae, Silphidae, Liodidae, Clambidae, Leptinidae, Platypsyllidae, Corylophidae, Sphaeriidae, Trichopterygidae, Hydroscaphidae, Scaphidiidae, histeridae. (Con figure nel testo).* Tipografico Piacentino, Piacenze, 405 pp.

Pototskaya, V.A. 1964. Semeistvo Scaphidiidae – tchelnovidki. Pp. 226–227. In: M.S. Ghilarov (ed.). *Opredelitel obyvayushtchikh v potchve litchinok nasekomykh.* Nauka, Moskva, 919 pp.

Prokofiev, A.M. 2013. Contribution to the knowledge of the scaphidiine genus *Scaphisoma* Leach of the Bu Gia Map National Park, Vietnam. *Calodema* 245: 1–6.

Ragusa, E. 1892. Coleopteri nuovi o poco conosciuti della Sicilia. *Naturalista siciliano* 11: 253–256.

Rassi, P., S. Karjalainen, T. Clayhills, E. Helve, E. Hyvärinen, E. Laurinharju, S. Malmberg, I. Mannerkoski, P. Martikainen, J. Mattila, J. Muona, M. Pentinsaari, I. Rutanen, J. Salokannel, K. Siitonen & H. Silfverberg 2015. Kovakuoriaisten maakuntaluettelo 2015 [Provincial list of Finnish Coleoptera 2015]. *Sahlbergia* 21 Supplement 1: 1–164.

Redtenbacher, L. 1847. Heft 1, pp. 1–160. In: *Fauna Austriaca. Die Käfer. Nach der analytischen Methode bearbeitet.* [1847–1849]. Carl Gerold, Wien, xxvii + 883 pp, 2 pls.

Redtenbacher, L. 1868. *Reise der Österreichischen Fregatte Novara um die Erde in den Jahren 1857, 1858, 1859 unter den Befehlen des Commodore B. von Wüllerstorf-Urbair. Zoologischer Theil. Zweiter Band: Coleopteren.* [1867]. Wien. iv + 249 pp, 5 pls.

Redtenbacher, L. 1872. *Fauna Austriaca. Die Käfer. Nach der analytischen Methode bearbeitet. Dritte, gänzlich umgearbeitete und bedeutend vermehrte Auflage. Erste Band.* [1874]. Carl Gerold's Sohn, Wien, 571 + viii pp.

Rehfous, M. 1955. Contribution à l'étude des insectes des champignons. *Mitteilungen der Schweizerischen entomologischen Gesellschaft* 28: 1–106.

Reiche, L. 1864. Espèces nouvelles de coléoptères d'Algérie. *Annales de la Société entomologique de France* (4)4: 233–246.

Reitter, E. 1877. Beiträge zur Käferfauna von Japan. (Drittes Stück). *Deutsche entomologische Zeitschrift* 21: 369–383.

Reitter, E. 1879. Note VII. Descriptions of three new species of Coleoptera collected during the recent scientific Sumatra-expedition. *Notes from the Leyden Museum* 2: 41–49.

Reitter, E. 1880a. Die Gattungen und Arten der Coleopteren-Familie: Scaphidiidae meiner Sammlung. *Verhandlungen des Naturforschenden Vereins in Brünn* 18 [1879]: 35–49.

Reitter, E. 1880b. *Scaphidium nigromaculatum* Reitter, n. sp. *Entomologisches Nachrichtenblatt* 2: 170.

Reitter, E. 1880c. Coleopterologische Ergebnisse einer Reise nach Croatien, Dalmatien und der Herzegovina im Jahre 1879. Unter Mitwirkung von E. Eppelsheim und L. Miller. *Verhandlungen der Kaiserlich-Königlichen Zoologisch-Botanischen Gesellschaft in Wien* 30: 201–228.

Reitter, E. 1880d. Bestimmungs-Tabellen der europäischen Coleopteren III. Enthaltend die Familien: Scaphidiidae, Lathridiidae und Dermestidae. *Verhandlungen der Kaiserlich-Königlichen Zoologisch-Botanischen Gesellschaft in Wien* 30: 41–94.

Reitter, E. 1881. Einige neue Coleopteren. *Mitteilungen des Münchener entomologischen Vereins* 5: 139–141.

Reitter, E. 1883. Neue Coleopteren aus Russland und Bemerkungen über bekannte Arten. *Revue mensuelle d'entomologie* 1: 40–44.

Reitter, E. 1884. Sechs neue Coleopteren aus Italien, gesammelt von Herrn Agostino Dodero. *Annali del Museo civico di Storia naturali di Genova* (2) 1: 369–372.

Reitter, E. 1885a. Coleopterologische Notizen. X. *Wiener entomologische Zeitung* 4: 81–83.

Reitter, E. 1885b. Lieferung 2, pp. 199–362. In: *Naturgeschichte der Insecten Deutschlands begonnen von Dr. W.F. Erichson, fortgesetzt von Prof. Dr. H. Schaum, Dr. G. Kraatz, H. v. Kiesenwetter, Jul. Weise und Edm. Reitter. Erste Abtheilung. Coleoptera. Dritter Band. Zweite Abtheilung*. Nicolaische Verlags-Buchhandlung, Berlin, iv + 362 pp.

Reitter, E. 1886. *Bestimmungs-Tabellen der europäischen Coleopteren. III. Heft. Enthaltend die Familien. Scaphidiidae, Lathridiidae und Dermestidae. II. Aufl.* Edmund Reitter, Mödling, 75 pp.

Reitter, E. 1887. Neue Coleopteren aus Europa, den angrenzenden Ländern und Sibirien, mit Bemerkungen über bekannte Arten. Vierter Theil. *Deutsche entomologische Zeitschrift* 31: 497–528.

Reitter, E. 1889. Note II. Neue Coleopteren aus dem Leydener Museum. *Notes from the Leyden Museum* 11: 3–9.

Reitter, E. 1891. Neue Coleopteren aus Europa, der angrenzenden Ländern und Sibirien, mit Bemerkungen über bekannten Arten. Zwölfter Teil. *Deutsche entomologische Zeitschrift* 35: 17–36.

Reitter, E. 1898. Analytische Uebersicht der *Scaphosoma*-Arten aus der palaearktischen Fauna. *Entomologische Nachrichten* 24: 314–315.

Reitter, E. 1899. Elfter Beitrag zur Coleopteren-Fauna von Europa und angrenzenden Ländern. *Wiener entomologische Zeitung* 18: 155–164.

Reitter, E. 1908. Verzeichnis der von Dr. F. Eichelbaum im Jahre 1903 in Deutsch-Ostafrika gesammelten Scaphidiiden (Col.). *Wiener entomologische Zeitung* 27: 31–35.

Reitter, E. 1909. *Fauna Germanica. Die Käfer des Deutschen Reiches. Nach der analytischen Methode bearbeitet. II. Band.* K.G. Lutz, Stuttgart, 392 pp, pls 41–80.

Reitter, E. 1913. Coleopterologische Novitäten der palaearktischen Fauna. *Coleopterologische Rundschau* 2: 121–125.

Reitter, E. 1915. Zwei neue Käferarten. *Entomologische Blätter* 11: 42–43.

Rogé, J. 2000. Au sujet de *Scaphisoma balkanicum* [sic] Tamanini, 1954 (Coleoptera, Scaphidiidae) 24e note sur les coléoptères du sud-ouest de la France. *L'Entomologiste* 56: 159–160.

Roosileht, U. 2015: Estonian additions to Silfverberg's «Enumeratio renovata Coleopterorum Fennoscandiae, Daniae et Baltiae» Coleoptera Catalog. *Sahlbergia* 21(2): 6–39.

Rossi, P. 1792. *Mantissa Insectorum, exhibens species nuper in Etrusca collectas, adjectis faunae Etruscae illustrationibus ac emendationibus. Tom I.* Polloni, Pisis, 148 pp.

Roubal, J. 1930. *Katalog Coleopter (brouku) Slovenska a Podkarpatska na zakladě bionomickém a zoogeografickém a spolu systematicky doplněk Ganglbauerovych "Die Käfer Mitteleuropas" a Reitterovy "Fauna Germanica". 2. Díl I.* Učená Společnost Safaříkova v Bratislave, Praha, 527 pp.

Rougemont, G. de. 1996. Scaphidiids of Hong Kong. *Porcubine* 15: 12.

Rüschkamp, F. 1929. Beiträge zur Kenntnis der Fauna Südbayerns. (3). *Entomologische Blätter* 25: 35–43.

Ruter, G. 1977. Additiv au "Catalogue des insectes coléoptères de la fôret de Fontainebleau" du Gruardet (premier partie). *L'Entomologiste* 28: 29–39.

Saalas, U. 1917. Die Fichtenkäfer Finnlands. Studien über die Entwicklungsstadien, Lebensweise und geographische Verbreitung der an *Picea excelsa* Link. lebenden Coleopteren nebst einer Larvenbestimmungstabelle. I. Allgemeiner Teil und spezieller Teil 1. *Annales Academiae scientiarum fennicae*, Ser. A, 7: xx + 547 pp., 9 pls.

Sahlberg, J. 1889. Enumeratio coleopterorum clavicornium Fenniae II. Pselaphidae et Clavigeridae. *Acta Societatis pro Fauna et Flora fennica* 6: 13–152.

Samin, N., H. Zhou & S. Imani 2011. A contribution to the Oxyteline group of rove beetles (Coleoptera: Staphylinoidea: Staphylinidae) from Iran. *Entomolofauna Zeitschrift für Entomologie* 32: 277–284.

Say, T. 183. Description of coleopterous insects collected in the late expedition to the Rocky mountains, performed by order of Mr. Calhoun, Secretary of War, under the command of Major Long. *Journal of the Academy of Natural Sciences Philadelphia* 3: 139–216.

Say, T. 1825. Descriptions of new species of coleopterous insects inhabiting the United States. *Journal of the Academy of Natural Sciences of Philadelphia* 5: 160–204.

Schawaller, W. 1974. Bemerkenwerte Funde aus dem Mainzer Raum. *Entomologische Blätter* 70: 60.

Schawaller, W. 1990. Käfer aus Sibirien (Umgebung Novosibirsk) (Insecta: Coleoptera). *Beiträge zur Entomologie* 40: 231–245.

Scheerpeltz, O. & K. Höfler 1948. *Käfer und Pilze*. Verlag für Jugend und Volk, Wien, 351 pp, 9 pls.

Schigel, D.S. 2011. Polypore-beetle associations in Finland. *Annales Zoologici Fennici* 48: 319–348.

Schigel, D.S., T. Niemelä, A. Similä, J. Kinnunen & O. Manninen 2004. Polypores and associated beetles of the North Karelian Biospherer Reserve, eastern Finland. *Karstenia* 44: 35–56.

Schillhammer, H. 1996. Bemerkenswerte Käferfunde aus Österreich (IV). *Koleopterologische Rundschau* 65 [1995]: 229–232.

Schmidl, J., H. Bussler & H. Fuchs 2005. 22. Bericht der Arbeitsgemeinschaft Bayerischer Koleopterologen (Coleoptera). *Nachrichtenblatt der Bayerischen Entomologen* 54: 21–29.

Schülke, M. 2013. Ergänzungen und Berichtigungen zur Staphylinidenfauna von Berlin und Brandenburg (Coleoptera: Staphylinidae) II. *Märkische Entomologische Nachrichten* 15: 123–174.

Schülke, M. & A. Smetana 2015. Staphylinidae. In: I. Löbl & D. Löbl (eds). *Catalogue of Palaearctic Coleoptera. Hydrophiloidea – Staphylinoidea. Revised and updated edition. Volume 1.* Leiden, Boston, Brill, xxv + 900 pp.

Scopoli, J.A. 1763. *Entomologia Carniolica exhibens insecta Carnioliae indigena et distributa in ordines, genera, species, varietates. Methodo Linnaeana.* Ioannis Thomae Trattner, Vindobonae, [30] + 418 + [1] pp.

Scott, H. 1908. Fam. Scaphidiidae. In: H. Sharp & H. Scott. *Fauna Hawaiiensis or the Zoology of the Sandwich* (*Hawaiian Isles: Being results of the explorations institutes by the Joint Committee appointed by the Royal society of London promoting natural knowledge and the British Association for the Advancement of Science and carried on with the assistance of those Bodies and of the Trustees of the Bernice Bishop Museum at Honolulu. Volume III. Part V*). Coleoptera III: 367–579, pls xiii–xvi.

Scott, H. 1922. The Percy Sladen Trust Expedition to the Indian Ocean in 1905, under the leadership of Mr. J. Stanley Gardiner, M.A. Vol. VII. no. IV- Coleoptera: Scydmaenidae, Scaphidiidae, Phalacridae, Cucujidae (Supplement), Lathridiidae, Mycetophagidae (including Propalticus), Bostrychidae, Lyctidae. *Transactions of the Linnean Society of London, 2nd Ser. Zoology* 18(1): 195–260, pls 19–22.

Seidlitz, G.C.M. von. 1888a. Lieferung 3. In: *Fauna Baltica. Die Kaefer (Coleoptera) der Deutschen Ostseeprovinzen Russlands. Zweite neu bearbeitete Auflage. Mit 1 Tafel.* [1887–1891] Hartungsche Verlagsdruckerei, Königsberg,10 +lvi + 192 + 818 pp, 1 pl.

Seidlitz, G.C.M. von. 1888b. Lieferung 3–4. In: *Fauna Transsylvanica. Die Kaefer (Coleoptera) Siebenburgens. Mit 1 Tafel* [1887–1891.] Hartungsche Verlagsdruckerei, Königsberg, lvi + 192 + 918 pp, 1 pl.

Semenov, V.B. 2017. K poznaniyu zhukov-staphilinid (Coleoptera, Staphylinidae) Mordovkogo gosudarstvennogo prirodnogo zapovednika. *Trudy Mordovskogo Gosudarstvennogo Prirodnogo Zapovednika imeni P.G. Smidovida* 18: 190–205.

Sheng, Chun & Fu-Kang Gu, 2009. Two new species of the genus *Episcaphium* Lewis (Coleoptera, Staphylinidae, Scaphidiinae). *Zootaxa* 2325: 35–38.

Shirôzu, T. & K. Morimoto 1963. A contribution towards the knowledge of the genus *Scaphidium* Olivier of Japan (Coleoptera, Scaphidiidae). *Sieboldia* 3(1): 55–90.

Sikes, D.S. 2004. *The beetle fauna of Rhode Island. An annotated checklist. Volume 3 of The Biota of Rhode Island.* The Rhode Island Natural History Survey, 296 pp.

Silfverberg, H. 2004. Enumeration nova coleopterorum Fennoscandiae, Daniae et Baltiae. *Sahlbergia* 9: 1–111.

Singh, U A. & K. Sharma 2016. Pest of mushroom. *Advances in crop science and technology* 4(2): 213. Doi: 10.4172/2329-8863.1000213.

Solodovnikov, I.A. 2016a. Novye i redkie vidy zhestkokrylykh dlya belorusskogo poozerya i respukliki Belarus [New and rare species of Coleoptera (Coleoptera) for the Byelorussian Poozerye and the Republic of Belarus.] Pp. 127–128. In: *Ecological culture and environmental protection: II Dorofeev Readings: Materials of the international scientific and practical conference, Vitebsk, November 29–30, 2016* / Vitebsk. State. Un-t; redact-collective: I.M. Prishchep (editor in chief) [and others]. Vitebsk: VGU named after P.M. Masherov, 222 pp.

Solodovnikov, I.A. 2016b. New and rare species of beetles (Coleoptera) in Belarus Lake Lands (Belarusian Poozeriya) and in the Republic of Belarus. Part 6. *Vestnik VGU* [Newletters of Vitebst State University] 4(94): 53–67.

Solsky, S.M. 1871. Coléoptères de la Sibérie orientale. *Horae Societatis entomologicae Rossicae* 7[1870]: 334–406.

Solsky, S.M. 1874. Zhestkokrylye (Coleoptera). Tetrad 1. In: A.P. Shedtchenko *Puteshestvie v Turkestan. Tom II. Zoogeografitcheskiya isledovaniya. Chast V, Otdel shestoy.* S. Peterburg and Moskva: Imperatorskoe obshchestvo lyubiteley estestvoznaniya, antropologii I etnografii, iv + 222 + [1] pp.

Stephens, J.F. 1829. *A Systematic Catalogue of British Insects: Being an attempt to arrange all the hitherto discovered indigenous insects in accordance with their natural affinities. containing also the references to every English writer on Entomology, and to the principal foreign authors. with all the published British genera to the present time. Insecta Mandibulata. Ordo I. Coleoptera.* Baldwin and Cradock, London, xxxiv + 416 pp.

Stephens, J.F. 1830. *Illustrations of British entomology; or, a synopsis of indigenous insects: containing their generic and specific distinctions; with an account of their metamorphoses, times of appearance, localities, food, and economy, as far as practicable. Embellished with coloured figures of the rarer and more interesting species. Mandibulata. Vol. III.* Baldwin and Cradock, pls. XVI–XIX, 1–374 + [6] pp.

Stierlin, W.G. 1900. *Fauna coleopterorum helvetica. Die Käfer der Schweiz nach der analytischen Methode bearbeitet. I. Theil.* Bolli & Böcherer, Schaffhausen, xii + 667 pp.

Štourač, P. & K. Rébl 2009. Faunistik records from the Chech Republic -277. *Klapalekiana* 45: 121–122.

Strand, A. 1969. Koleopterologiske bidrag XIV. *Norsk entomologisk Tidsskrift* 16: 17–22.

Strand, A. 1975. Koleopterologiske bidrag XVI. *Norsk entomologisk Tidsskrift* 22: 9–14.

Süda, I. 2016. Metsamardikate (Coleoptera) uued liigid Eestis. 2. New woodland beetle species (Coleoptera) in Estonian fauna. 2. *Forestry Studies* 64: 51–69.

Tamanini, L. 1954. Valote tassonomico degli organi genitali nel genere *Scaphosoma* e descrizioni di una nuova specie. *Bulletino della Società entomologica italiana* 84: 85–89.

Tamanini, L. 1955. Richerche zoologiche sul Massiccio del Pollino (Lucania-Calabria). XIV. Coleoptera. 4. Catopidae, Liodidae, Scaphidiidae, Silphidae. *Annuario dell'Institotu e Museo di Zoologia della Università di Napoli* 7(11): 1–19.

Tamanini, L. 1969a. Gli Scaphidiidae del Museo Civico di Storia Naturale di Verona e descrizione di una nuova specie. *Memorie del Museo civico di Storia Naturale di Verona* 16[1968]: 483–489.

Tamanini, L. 1969b. Gli Scaphidiidae del Museo Civico di Storia Naturale di Milano, con appunti sui caratteri specifici e descrizione di una nuova specie (Coleoptera). *Atti della Società italiana di Scienze Naturali e del Museo Civico di Storia Naturale* 109: 351–379.

Tamanini, L. 1969c. Le due tribu' Scaphidiini e Scaphisomini vanno considerate a rango di famiglie a se' stanti (Coleoptera). *Memorie della Società entomologica italiana* 48: 129–137.

Tamanini, L. 1970. Gli scafididi italiani (Coleoptera: Scaphidiidae e Scaphosomidae). *Memorie della Società entomologica italiana* 49: 5–26.

Tang, Liang & Li, Li-Zhen 2009. Three new species of *Ascaphium* Lewis, 1893 from China (Coleoptera, Staphylinidae, Scaphidiinae). *The Pan-Pacific Entomologist* 85: 91–98.

Tang, Liang & Li, Li-Zhen 2010a. On *Scaphidium* grande-complex (Coleoptera, Staphylinidae, Scaphidiinae). *ZooKeys* 43: 65–78.

Tang, Liang & Li, Li-Zhen 2010b. A new species of the genus *Scaphidium* Olivier from China (Coleoptera: Staphylinidae: Scaphidiinae). *Journal of the Kansas entomological Society* 83: 318–321.

Tang, Liang & Li-Zhen Li 2012. Two new species of the genus *Scaphidium* from China (Coleoptera: Staphylinidae: Scaphidiinae). *Acta entomologica Musei nationalis Pragae* 52: 185–192.

Tang, Liang & Li-Zhen Li 2013. More data on Chinese fauna of *Scaphidium* Olivier with description of a new species (Coleoptera: Staphylinidae: Scaphidiinae). Pp. 173–181. In: M.-Y. Lin & C.-C. Chen (eds). *In memory of Mr. Wehnsin Lin*. Formosa Ecological Company, Taiwan.

Tang, Liang, Li-Zhen Li & Wen-Jia He 2014. The genus *Scaphidium* in East China (Coleoptera, Staphylinidae, Scaphidiinae). *ZooKeys* 403: 47–96 [DOI: 10.3897/zookeys.403.7220].

Tang, Liang, Yue-Ye Tu & Li-Zhen Li 2016a. Notes on the genus *Episcaphium* Lewis (Coleoptera, Staphylinidae, Scaphidiinae) with description of a new species from China. *ZooKeys* 595: 49–55 [DOI: 10.3897/zookeys.595.8784].

Tang, Liang, Yue-Ye Tu & Li-Zhen Li 2016b. Notes on *Scaphidium grande*-complex with description of a new species from China (Coleoptera, Staphylinidae, Scaphidiinae). *Zootaxa* 4132(2): 279–282 [DOI: 10.11646/zootaxa.4132.2.9].

Telnov, D. 2004. *Compendium of Latvian Coleoptera. Volume 1. Check-List of Latvian Coleoptera. Second edition*. Entomological Society of Latvia, Riga, 114 pp.

Thayer, M.K. 2005. Staphylinidae. Pp. 296–344. In: R.G. Beutel & R.A.B. Leschen (eds). *Handbook of Zoology. Insecta. Coleoptera. Beetles. Volume 1. Morphology and systematicas (Archostemata, Adephaga, Myxophaga, Polyphagas partim)*. Walter de Gruyter, Berlin, New York. xi + 567 pp.

Theofilova, T.M. 2017. Ground beetles (Coleoptera: Carabidae) and some other invertebrates from the Managed Nature Reserves "Dolna Topchiya" and "Balabana" (Lower valleey of the river Tundzha, Bulgaria). *Ecologia Balcanica* 9: 63–77.

Thomson, C.G. 1862. *Skandinaviens coleoptera, synoptiskt bearbetade. Tom IV*. Lundbergska Boktryckeriet, Lund, 269 pp.

Torrella Allegue, L.P. 2013. Aportación a la biología y corología de *Scaphidium quadrimaculatum* Olivier, 1790 (Coleoptera, Staphylinidae, Scaphidiinae) en la Península Ibèrica. *Arquivos Entomolóxicos* 9: 41–50.

Tronquet, M. 2006. *Cataloque iconographique des coléoptères des Pyrenées-Orientales. Volume 1 Staphylinidae*. Supplément au Tome XV de la Revue de l'Association Roussillonnaise d'Entomologie, Perpignan, 127 pp, 78 pls.

Tsinkevich, V.A. 2004. Zhestkokrylye (Coleoptera) obivateli plodovykh tel bazidialnykh gribov (Basidiomycetes) zapada lesnoy zony Russkoy ravniny (Belarus). *Byulleten' Moskovskogo obshchestva ispytatelei prirody. Otdel Biologicheskii* 109(4): 17–25.

Tsurikov, M.N. 2009. *Zhuki lipetskoy oblasti*. Voronezh State University, Voronzh, 332 pp.

Tu, Yue-Ye & L. Tang. 2017. Supplement to the knowledge of the genus *Scaphidium* Olivier of East China (Coleoptera, Staphylinidae, Scaphidiinae). *Zootaxa* 4268(4): 593–596 [DOI.org/10.11646/zootaxa. 4268.4.11.]

Van Meer, C. 1999. Données entomologiques sur une très vieille fôret de feuillus: la fôret de Sare. *Bulletin de la Société linnéenne de Bordeaux* 27: 1–17.

Vinogradov, E. Yu., L.V. Egorov & V.B. Semenov 2010. Materialy k poznaniya stafilinid (Insecta, Coleoptera, Staphylinidae) Chuvachii. Soobshchenie 2. *Nauchnye Trudy gosudarstvennego prirodnogo zapovednika "Prisursky"* 25: 8–18.

Viñolas, A., J. Muñoz-Baset, J. Bentanachs & G. Masó 2014. Catálogo de los coleópteros del Parque Natural del Cadí-Moixeró, Cataluña, Península Ibérica. *Coleopterological Monographs* 5: 1–155.

Vinson, J. 1943. The Scaphidiidae of Mauritius. *Mauritius Institut Bulletin* 2: 177–209.

Vitale, F. 1929. Fauna coleotterologica sicula Scaphiidae e Histeridae. *Atti dell'Accademia peloritana* (1929): 109–146.

Vlasov, D.V. & N.B. Nikitsky 2017. Fauna zhukov-chelnovidok (Coleoptera, Staphylinidae, Scaphidiinae) Yaroslavskoy oblasti s ukazanyami novykh i maloizvestnykh dlya regiona vidov zhestkokrylykh iz nekotorykh semeistv. *Byulleten' Moskovskogo obshchestva ispytatelei prirody. Otdel Biologicheskii* 122(3): 3–11.

Vollenhoven van Snellen, S.C. 1865. Beschrijving van eenige nieuwe soorten van Curculioniden, uit het geslacht *Apoderus* Oliv. *Nederlandsch Tijdschrift voor de Dierkunde* 2: 158–157.

Vorst, O. 2010. Staphylinidae. Pp. 66–99. In: O. Vorst (ed.). Catalogus van Nederlandse kevers. (Coleoptera) Catalogue of the Coleoptera of Netherlands. *Monografiën van de Nederlandse Entomologische Vereniging* 11: 1–317.

Walter, T. 1990. Käfer des Ruggeller Rietes. Pp. 279–313. In: *Naturmonographie Ruggeller Riet. Naturkundliche Forschung im Fürstertum Liechtenstein. Band 12*. Vaduz, 443 pp.

Webster, R.P., J.D. Sweeney & I. DeMerchant 2012. New Staphylinidae (Coleoptera) records with new collection data from New Brunswick, Canada: Scaphidiinae, Piestinae, Osorinae, and Oxytelinae, *ZooKeys* 186: 239–262.

Weiss, H.B. & E. West 1920. Fungous insects and their hosts. *Proceedings of the Biological Society of Washington* 33: 1–20.

Weiss, H.B. & E. West 1922. Notes on fungous insects. *The Canadian Entomologist* 54: 198–199.

West, A. 1942. *Fortegnelser over Danmarks biller deres udbredelse I Danmark forekomststeder og – tiger biologi m.* P. Haase & sons, København, x+ 664 pp.

Westwood, J.O. 1838. *Synopsis of the genera of British insects.* Pp. 1–48. In: An Introduction to the modern classification of insects. Founded on the natural habits and corresponding organization of the different families. Vol. 2. Longman, Orme, Brown, Green and Longmans, London [1838–1840], xi + 587 + 158 pp.

Weyenbergh, H. Jr. 1869. Sur les insectes fossiles du calcaire lithographique de la Bavière, qui se trouvent au Musée Teyler. *Archives du Musée Teyleer* 2: 247–294, pls 34–37.

Wheeler, Q.D. 1991. Fungus-Coleoptera association of México: Analysis of biodiversity. Pp. 13–44. In: J.L. Navarrete-Heredia & Quiroz-Rocha (eds). *I Simposio nacional sobre la interaccion insecto-hongo*. Veracruz, México.

Winkler, A. 1925. *Catalogus Coleopterorum regionis palaearcticae*. I. Pars 3, pp. 241–368, Albert Winkler, Wien.

Wojas, T. 2016. New data on the distribution of rare rove-beetles (Coleoptera: Staphylinidae) in Southern Poland. *Entomological News* 35: 137–146.

Zachariassen, K.E. 1973. *Scaphidium quadrimaculatum* Oliv. (Col. Scaphidiidae) new to Noway. *Norsk entomologisk Tidskrift* 20: 335.

Zamotajlov, A.S. & N.B. Nikitsky 2010. *Zhestkokrylye nasekomye (Insecta, Coleoptera) respubliki Adygea (annotyrovanyy katalog vidov). Coleopterous insects (Insecta, Coleoptera) of Adyghea (annotated catalog of species)*. Adyghei State University Publishers, Maykop, 404 pp.

Zayas de, F. 1988. *Entomofauna Cubana. Orden Coleoptera. Separata description du nuevas especies*. Editorial Cientifico-Técnica, La Habana, 212 pp.

Zimmermann, C. 1869. Synonymical notes on the Coleoptera of the United States, with descriptions of new species, from the MSS. of the late Dr. C. Zimmermann. *Transactions of the American Entomological Society* 2: 243–259.

Index of Family-group and Genus-group names

*Extinct taxa. Synonyms are in italics

Index of Species-group names

*Extinct taxa. Synonyms are in italics

Printed in the United States
By Bookmasters